CO-CCC-296

T.M

Evolutionary Mechanisms in Sex Determination

Editor

Stephen S. Wachtel, Ph.D.
Professor
Obstetrics and Gynecology,
and Physiology and Biophysics
University of Tennessee, Memphis
Memphis, Tennessee

CRC Press, Inc.
Boca Raton, Florida

Randall Library UNC-W

Library of Congress Cataloging-in-Publication Data

Evolutionary mechanisms in sex determination.

Based on the International Conference on Evolutionary
Mechanisms in Sex Determination, held in Memphis,
May 4-8, 1987.
Includes bibliographies and index.
1. Sex determination, Genetic — Congresses.
2. Evolution — Congresses. 3. Developmental genetics —
congresses. I. Wachtel, Stephen S. II. International
Conference on Evolutionary Mechanisms in Sex Determina-
tion (1987 : Memphis, Tenn.) [DNLM: 1. Sex Chromosomes
— congresses. 2. Sex Determination — congresses.
QH 600.5 E93 1987]
QH600.5.E96 1989   591.1′5   88-5011
ISBN 0-8493-4830-7

This book represents information obtained from authentic and highly regarded sources. Reprinted material is quoted with permission, and sources are indicated. A wide variety of references are listed. Every reasonable effort has been made to give reliable data and information, but the author and the publisher cannot assume responsibility for the validity of all materials or for the consequences of their use.

All rights reserved. This book, or any parts thereof, may not be reproduced in any form without written consent from the publisher.

Direct all inquiries to CRC Press, Inc., 2000 Corporate Blvd., N.W., Boca Raton, Florida, 33431.

© 1989 by CRC Press, Inc.

0-8493-4830-7

Library of Congress Card Number 88-5011
Printed in the United States

QH
600.5
.S96
1989

PREFACE

The chapters of this book are based for the most part on lectures given at our International Conference on Developmental Biology, "Evolutionary Mechanisms in Sex Determination", held May 4 to 8, 1987 in Memphis, Tenn. This is the fourth in a series of conferences on developmental biology sponsored by our laboratory. Although most of the chapters are concerned in the broad sense with the evolutionary biology of sex determination, a few deal with related topics. These are included here as representative of progress in the general study of developmental and molecular genetics.

Recent years have witnessed a virtual explosion of technical advances in biology and medicine, and this has enabled corresponding advances in the study of sex determination. For example, high resolution optics and improved banding techniques have allowed detection of sex chromosome heteromorphism among the "primitive" vertebrates; T cell hybridomas and monoclonal antibodies in conjunction with flow cytometry, radioimmunoassays, and ELISAs have enabled sensitive evaluation of AMH, 12E7, H-Y (Sxs), and other molecules of interest in connection with development of the sex phenotype; and recombinant DNA methods have enabled mapping of the relevant genes, especially in the human, and development of specific DNA probes for further research.

On the practical side, molecular techniques have been applied to the study of Duchenne muscular dystrophy and retinoblastoma; Y-specific DNA probes have been used to trace the etiology of enigmatic conditions such as XX male syndrome and XY gonadal dysgenesis; and in the bovine, similar probes have been applied to the selective implantation of male and female during embryo transfer. With respect to the methodology of recombinant DNA, it seems fair to assume that the complete sequences of the human sex chromosomes will be available within the next several years.

The present volume was organized to provide an introduction to the biology of sex determination and its methodology, a review of recent progress in the field, and an historical perspective for students who come across the volume in the future. The volume comprises six sections: I. Evolution of Sex Determination, including chapters on the development of sex chromosomes in Amphibia and of sex-determining mechanisms in Reptilia; II. The Sex Chromosomes, including discussions of the narrowing search for the male-determining locus; III. Male Antigens, containing reviews on the phylogenetically conserved H-Y antigen and a chapter on the cloning of a candidate H-Y gene; IV. Sex Reversal, including chapters on Sxr and tda-1 of the mouse; V. Biology of Development, including chapters on anti-Müllerian hormone and the development of the early sex phenotype; and VI. Molecular Biology in Medicine and Agriculture, with a chapter on the cloning of the X-linked genes in Duchenne muscular dystrophy, discussions of clinical sex-reversed syndromes, and a review of techniques for the control of sex ratio.

We thank Laura Dillon for help in the preparation of the manuscript and CRC Press for suggesting that it be written. Costs for the conference were defrayed by grants from the March of Dimes Birth Defects Foundation (4-197), the University Physicians Foundation, University of Tennessee, Memphis; and Spring Creek Ranch, Collierville, Tenn. We are grateful to the Chairman of Spring Creek Ranch, David Meyer, M.D., for his expressions of confidence and enthusiasm during the preparation of this volume.

Supported in part by NIH grant AI-23479.

Stephen S. Wachtel

THE EDITOR

Stephen Wachtel is Professor of Obstetrics and Gynecology and Director of the Reproductive Immunogenetics Laboratory at the University of Tennessee, Memphis. He received his AB degree in Biology from Kenyon College in 1959 and entered flight training in the U.S. Air Force in 1960. From 1962 to 1967, he served as an instructor pilot in the Military Assistance Program and Jet Qualification School. He left active service in 1967 and began graduate study at the University of Pennsylvania. After receiving his Ph.D. in Biology in 1971, he joined the staff of the Sloan-Kettering Institute and later held appointments at Sloan-Kettering Institute and Cornell Univeristy Medical College. He joined the faculty of the University of Tennessee in 1985 and now holds joint appointments in the Departments of Obstetrics and Gynecology and Physiology and Biophysics. From 1985 to 1988, he was Director of the Center for Reproductive Biology at Spring Creek Ranch near Memphis; he is currently a consultant of that institution. His research has centered about the immunology and genetics of sex-specific antigens and development of the sex phenotype. He has more than 140 publications including a book, *H-Y Antigen and the Biology of Sex Determination*, published by Grune & Stratton, New York and an edited volume, *Errors of Sex Differentiation*, published as a regular issue of *Human Genetics* by Springer-Verlag, Berlin. Dr. Wachtel was the recipient of an NIH Career Development Award 1976 to 1981. He is a member of the American Association of Immunologists, the International Transplantation Society, the American Genetics Association, the American Society for the Immunology of Reproduction, the American Fertility Society, and the International Embryo Transfer Society.

CONTRIBUTORS

Nacer Abbas
Predoctoral Fellow
Department of Human Immunogenetics
INSERM U276
Institut Pasteur
Paris, France

Keith J. Betteridge, Ph.D.
Professor
Department of Biomedical Sciences
Ontario Veterinary College
University of Guelph
Guelph, Ontario, Canada

Colin E. Bishop, Ph.D.
Director of Research
Department of Human Immunogenetics
INSERM U276
Institut Pasteur
Paris, France

Stan R. Blecher, M.D.
Professor and Director
School of Human Biology
University of Guelph
Guelph, Ontario, Canada

Mark P. Bradley, Ph.D.
Research Fellow
Department of Surgery
University of Otago Medical School
Dunedin, New Zealand

James J. Bull, Ph.D.
Associate Professor
Department of Zoology
University of Texas
Austin, Texas

Paul S. Burgoyne, Ph.D.
Senior Research Scientist
Mammalian Development Unit
Medical Research Council
London, England

Myriam Casanova
Assistant Professor
Department of Human Immunogenetics
INSERM U276
Institut Pasteur
Paris, France

Thomas W. Cline, Ph.D.
Professor
Department of Biology
Princeton University
Princeton, New Jersey

Corinne Cotinot
Assistant Professor
Department of Animal Physiology
Jouy-en-Josas, France

John Anthony Crolla, M.Phil.
Senior Research Technician
Pediatric Research Unit
Guy's Hospital Medical School
London, England

Larry A. Donoso, M.D., Ph.D.
Director
Retina Research Laboratory
Research Department
Wills Eye Hospital
Philadelphia, Pennsylvania

Susan Drucker, M.D.
Pediatric Endocrine Fellow
Department of Pediatrics
The New York Hospital-
  Cornell Medical College
New York, New York

Cecilia Ebensperger, Ph.D.
Institute for Human Genetics and
  Anthropology
University of Freiburg
Freiburg, West Germany

Jörg T. Epplen, M.D.
Leader, Junior Research Unit
Max-Planck-Institut für Psychiatrie
Martinsried, West Germany

Marc Fellous, M.D., Ph.D.
Professor
Department of Human Immunogenetics
INSERM U276
Institut Pasteur
Paris, France

Peter N. Goodfellow, D.Phil.
Department of Human Molecular
  Genetics
Imperial Cancer Research Fund
London, England

Paul J. Goodfellow, Ph.D.
Department of Human Molecular
  Genetics
Imperial Cancer Research Fund
London, England

Jean-Louis Guenet, Ph.D.
Unite de Gentique des Mammiferes
Institut Pasteur
Paris, France

Daniel Guerrier, M.A.
Predoctoral Fellow
Endocrinologie du Développement
Institut National de la Santé et de la
  Recherche Médicale
Paris, France

Thomas Haaf, M.D.
Postdoctoral Fellow
Department of Human Genetics
University of Würzburg
Würzburg, West Germany

Barbara F. Heslop, M.D.
Professor
Department of Surgery
University of Otago Medical School
Dunedin, New Zealand

Vikram L. Jaswaney, Ph.D.
Research Immunologist I
Department of Microbiology and
  Immunology
UCLA School of Medicine
Los Angeles, California

Kenneth W. Jones, Ph.D.
Reader in Genetics
Department of Genetics
Edinburgh University
Edinburgh, Scotland

Nathalie Josso, M.D.
Director of Research
Endocrinologie du Développement
Institut National de la Santé et de la
  Recherche Médicale
Paris, France

Axel Kahn, Ph.D.
Director
Unité de Recherches en Génétique et
  Pathologie Moléculaires
Institut National de la Santé et de la
  Recherche Médicale
Paris, France

Marijo G. Kent, Ph.D.
Veterinary Reproduction Specialist
Department of Equine Research
Center for Reproductive Biology
Collierville, Tennessee

Marek Kirszenbaum, Ph.D.
Assistant Professor
CEA
Jouy-en-Josas, France

Gloria C. Koo, Ph.D.
Senior Research Fellow
Department of Biochemistry and
  Molecular Biology
Merck, Sharp & Dohme Research
  Laboratory
Rahway, New Jersey

Louis M. Kunkel, Ph.D.
Associate Professor
Department of Pediatrics
Division of Genetics
Children's Hospital
Harvard Medical School
Boston, Massachusetts

Yun-Fai Lau, Ph.D.
Associate Investigator
Howard Hughes Medical Institute
University of California
San Francisco, California

Anne McLaren, D.Phil.
Director
Mammalian Development Unit
Medical Research Council
London, England

Jean Luc Michot, Ph.D.
Department of Human Immunogenetics
Institut Pasteur
Paris, France

Anthony P. Monaco, Ph.D.
Predoctoral Fellow
Department of Pediatrics
Division of Genetics
Children's Hospital
Harvard Medical School
Boston, Massachusetts

Ulrich Müller, M.D., Ph.D.
Assistant Professor
Department of Pediatrics
Division of Genetics
Children's Hospital
Harvard Medical School
Boston, Massachusetts

Claude M. Nagamine, Ph.D.
Associate
Department of Medicine and Physiology
Howard Hughes Medical Institute
San Francisco, California

Dean Nakamura, Ph.D.
Molecular Diagnostics Laboratory
Department of Obstetrics and Gynecology
University of Tennessee, Memphis
Memphis, Tennessee

Maria I. New, M.D.
Professor and Chairman
Department of Pediatrics
The New York Hospital-
  Cornell Medical Center
New York, New York

Susumu Ohno, Ph.D.
Department of Theroretical Biology
Beckman Research Institute of the City of
  Hope
Duarte, California

Jean-Yves Picard, Ph.D.
Director of Research
Endocrinologie du Développement
Institut National de la Santé et de la
  Recherche Médicale
Paris, France

Paul E. Polani, M.D.
Geneticist
Pediatric Research Unit
Guy's Hospital Medical School
London, England

Christa Roberts, Ph.D.
Department of Human Immunogentics
INSERM U276
Institut Pasteur
Paris, France

Michael Schmid, Ph.D.
Associate Professor
Department of Human Genetics
University of Würzburg
Würzburg, West Germany

Joe Leigh Simpson, M.D.
Professor and Chairman
Department of Obstetrics and Gynecology
University of Tennessee
Memphis, Tennessee

Phyllis W. Speiser, M.D.
Assistant Professor
Department of Pediatrics
The New York Hospital-
 Cornell Medical College
New York, New York

Gwendolyn Wachtel
Laboratory Supervisor
Department of Obstetrics and Gynecology
University of Tennessee, Memphis
Memphis, Tennessee

Stephen S. Wachtel, Ph.D.
Professor
Departments of Obstetrics and
 Gynecology, and Physiology and
 Biophysics
University of Tennessee, Memphis
Memphis, Tennessee

Andreas Weith, Ph.D.
Institute for Biology
Medical University of Lübek
Lübeck, West Germany

Ulf H. Wiberg, Ph.D.
Research Fellow
Institute for Human Genetics and
 Anthropology
University of Freiburg
Freiburg, West Germany

Laura J. Wilkinson, M.Sc.
Predoctoral Fellow
School of Human Biology
University of Guelph
Guelph, Ontario, Canada

Heinz Winking, Ph.D.
Institute for Biology
Medical University of Lübeck
Lübeck, West Germany

Ulrich Wolf, Ph.D.
Professor
Institute for Human Genetics and
 Anthropology
University of Freiburg
Freiburg, West Germany

TABLE OF CONTENTS

PART VI: MOLECULAR BIOLOGY IN MEDICINE AND AGRICULTURE

Part I: Evolution of Sex Determination

Chapter 1

EVOLUTIONARY ASPECTS OF VERTEBRATE SEX DETERMINATION

Dean Nakamura

TABLE OF CONTENTS

I. INTRODUCTION

Genes of fundamental significance tend to be conservative. This could occur by either of two mechanisms: convergent evolution or persistence of archetypical gene sequences. Because sexuality is basic to the survival of vertebrates, the question arises whether sex-determining genes are conservative among the vertebrate species. In the following paragraphs, we shall discuss briefly certain aspects of vertebrate sex determination that lend support to the notion of a common sex-determining mechanism. This review is not meant to be exhaustive and will highlight phenomena and trends of special interest. It will conclude with a discussion of some preliminary data concerning our initial efforts to understand the evolution of the sex chromosomes.

II. COMMON MECHANISMS IN VERTEBRATE SEX DETERMINATION

A. The Embryonic Gonad Is Bipotential

It was evident early in this century that the vertebrate embryonic gonad is bipotential. According to Witschi[1,2] two regions can be distinguished in the indifferent gonad. One is fated to the development of large discrete germinal elements and the other is fated to the development of small motile germinal elements within cysts or a cyst continuum, the tubule. The superficial region of the gonadal anlage, the cortex, differentiates into ovary; the interior region, the medulla, differentiates into testis. In order to obtain ovary or testis, there is a need for proliferation of one region and regression of the other.

The decision whether to develop as ovary or testis typifies decisions in embryonic development. Ontogeny can be viewed as a continual specialization in cells that arise from the indifferent zygote, and in their progeny. In other words, the commitment of a cell to a differentiative pathway can be viewed as a series of generally irrevocable genetic decisions according to which particular subsets of the genome are activated and others inactivated. Each of those decisions might be no more than a binary choice, as in the gonad: granulosa cell vs. Sertoli cell; ovum vs. spermatozoon. Thus, the orderly formation of tissues and organs implies an orderly pathway of differentiative steps, which in turn implies an orderly expression of genes governing those steps.

B. Developmental Organizers Act as Hormones During Embryogenesis

How do cells become organized during histogenesis and organogenesis? Hans Spemann[3] championed the idea of organizers, diffusible substances that influence the activities of neighboring cells during development. The organizers could operate in alternative ways. The first is induction or recruitment of cells to follow a common developmental pathway. For example, organogenesis requires interaction between germ layers. In salamanders, the oral ectoderm is induced to form teeth by the underlying endoderm. In frogs, the corresponding ectoderm forms horny jaws. When Spemann transplanted incipient flank ectoderm of a frog embryo to the oral region of a salamander embryo, the transplanted tissue developed as horny jaws. A factor capable of activating "ectodermal oral structure" genes was released by cells of the salamander endoderm; and frog ectodermal cells were competent to receive that signal and organize specialized oral tissue.

The other way in which organizers influence embryonic development is by repression of developmental potencies. For example, when frog embryos were cultured with pieces of adult frog heart, the embryos did not develop a normal heart; and when the leg-forming region of a salmander embryo was transplanted to a site immediately adjacent to the leg-forming region of another intact salamander embryo, only a single leg developed at that side.

Table 1
FACTORS THAT SEX
REVERSE THE
VERTEBRATE OVARY

Crowding	Food resources
Temperature	Photoperiod
Social interactions	Sex steroids
Gonad implants	Pharmacologics
Ultraviolet light	Blastophthoria
Water potential	Parabiosis
pH	Carbon dioxide

C. Gonadal Development Is Labile

Could phenomena similar to those described above typify the development of organs such as the gonad? We have said that during development of ovary or testis, one field proliferates and the other regresses. But during development of a testis, the germinal epithelium, the cortical derivative, may persist while Sertoli cells and seminiferous tubules differentiate. We know this from the occurrence of XY hermaphrodites, in humans for example,[4] from the presence of the cortical lobe known as Bidder's organ in males of the Bufonidae;[5,6] and from experiments that prolong retention of the germinal epithelium through the administration of hormones, as in the birds (reviewed in Reference 7).

The antagonistic relationship of ovary and testis is evident in the toads. If the testes are removed, the Bidder's organs develop into functional ovaries,[8,9] whereas the testes grow larger than those of the intact male when the Bidder's organs are removed.[10] In most birds there is a laterality of gonadal development;[11-15] the left gonad shows more vigorous development than the right gonad. The left testis can be induced to retain the germinal epithelium by the administration of estrogens in the developing rooster chick. The result is an ovotestis. But the right testis is more difficult to feminize because it lacks a significant germinal epithelium. By use of higher dosages of a water-soluble estrogen at a much earlier stage of development, Wolff and Wolff[16] could rescue the right germinal epithelium. Those experiments show that cortical regression can be postponed; the result is an ovotestis or ovary in a genetic male.

Similarly, regression of the medulla in the incipient ovary is not immediate. This is borne out by the occurrence of XX true hermaphrodites, in humans for example,[4] by cases in which the medulla of the ovary is developed to a remarkable degree as in the mole,[17] desman,[18] and horse;[19] and by cases in which the medulla is permitted to develop into a testis as in the domestic fowl in which extirpation of the functional left ovary permits the right residual gonad to develop as a testis.[20,21] Again, this is attributed to the scanty development of the right ovary, and in particular, the germinal epithelium.[11] Thus, it is possible to obtain by natural and artificial means an ovotestis or testis in a genetic female.

D. Environment Can Influence Sex Determination

The question arises whether the potential for testicular development is lost in the ovary or persists in the medullary rudiments of the hilar region. In fact, transformation from ovary to testis generally involves a common series of events. Among the agents known to cause masculinization of the ovary (Table 1) are hormones, for example testosterone in *Carassius auratus;*[22] drugs, as in the ranids: cyproterone,[23] thiourea,[24] or methimazole;[25] temperature, as in *Rana sylvatica;*[26] and parabiosis or transplantation, as in *Xenopus laevis*[27] and *Oryzias latipes*.[28-30] The first step is regression of the cortex. As the cortex involutes, the medulla begins to proliferate and develops into a testis. The conversion can occur during or after embryonic development and generally requires prolonged exposure to the sex-transforming agent for complete transformation. It would appear, then, that compensatory development of the testis occurs only after a considerable portion of the ovary is rendered nonfunctional. It follows that the longer the gonad remains indifferent, the more susceptible it is to outside transforming influences.

E. Gonadal Development Occurs in Segments

There is an apparent "segmentation" to gonadal development. In *Xenopus laevis,* a female heterogametic species, the indifferent gonad consists of 15 segments or gonomeres.[31] As in vertebrate development in general, the gonad develops in a cranial-to-caudal sequence. Thus, differentiation is first evident in the most craniad gonomere, and gonadogenesis is completed 7 days later with the differentiation of the most caudad gonomere. It is possible to transform genetic males into functional females in this species by exposing the tadpoles to estrogen during gonadogenesis. Exposure prior to the period of sex differentiation is ineffectual. When male larvae are given a 2-day administration of hormone during the 7-day period of gonadogenesis, sections of the gonad are feminized, corresponding to the segments undergoing differentiation during the hormonal treatment. Thus, the sex-reversing effects of the 2-day treatment shifts caudally with age.[32]

A similar observation is implied in the study of Whitten et al.[33] In their analysis of hermaphroditism in the mouse, ovotesticular gonads were predominantly testicular; the ovarian portions were found preferentially at the poles, particularly the cranial pole.[4] It is not clear whether this was a case of retarded testicular development or whether ovarian development had in fact ensued. From the histologic perspective, in general, the decision on embryonic gonadal sex is a negative one; if there are no tubules, the gonad is considered an ovary. Furthermore, presence of meiotic germ cells is not fully characteristic of the fetal ovary because ectopic germ cells in fetal male mice undergo meiosis at the same stage that oocytes do.[34]

F. The Heterogametic Gonad Is the Dominant Gonad

The experiments with sinistral ovariectomy of the domestic fowl likewise reveal a relationship between the ZW ovary and ZZ testis that may reflect their antagonistic origins. Full development of the right residual gonad occurs only in the absence of the functional left ovary. So it would appear that the left ovary suppresses the development of the right gonad. If the extirpated left ovary is cleared of yolk-bearing follicles, dried in acetone to extract steroids, rehydrated, and then homogenized, the clarified supernatant prevents the masculinization of the right gonad when given parenterally to other recently sinistrally ovariectomized birds.[35] This would seem to indicate that the left ovary releases a "nonsteroidal hormone" that controls the developmental potential of the right gonad.

As another example, Haffen and colleagues, using proteolytic enzymes, were able to separate the germinal epithelium from the medullary region in the left gonad of embryonic ducks.[36] In that way, the cortex and medulla of a gonad could be cultured separately in vitro. For example, the germinal epithelium of the heterogametic ovary developed as an ovarian cortex. Those explants were nearly indistinguishable from whole ovaries sustained in vitro. Thus, the female germinal epithelium alone could develop to form structures that are characteristic histologically of the entire ovary. When the germinal epithelium of the homogametic testis was cultured alone, the result was a structure not unlike the medulla of a normal testis. Germinal epithelium taken from the testes of younger male embryos survived in culture but did not differentiate.

In a later study, Haffen[37] asked whether the medulla and cortex could influence the development in each other. To answer that question she associated the germinal epithelium of indifferent duck gonads with the medullary regions from the ovaries and testes of embryos in which the gonads had already differentiated. In all cases, the medulla maintained its characteristic appearance. But development of the germinal epithelium varied with the source of the medullary tissue. Whereas female germinal epithelium developed into ovarian cortex irrespective of the origin of the medulla, development of male germinal epithelium was modified by the medullary tissue with which it came into contact. The male germinal epithelium developed testicular cords when associated with testicular medulla, and gave rise to cortical tissue when associated with ovarian medulla. If the germinal epithelium was placed in contact with mesonephric tissue, the epithelium survived but did not differentiate.

This trend is evident among the amphibians also. In the 1920s and 1930s, those studying sex determination sometimes joined animals in parabiosis in order to evaluate the influence of one gonad on another.[1] For example it was possible to join a heterogametic male frog side-by-side with a homogametic female frog, with a second female attached head-to-tail to the same male. In such cases the testis had a decided but graded influence on the ovary. In the side-by-side female, the ovary nearest the male was predominantly testicular whereas the farther gonad was an ovotestis. The gonads of the head-to-tail female were unaffected.

It may be noteworthy that hormone-induced sex reversal is not always successful. For example in *Xenopus laevis,* estrogens effect complete sex reversal of the homogametic testis. The result is a functional ovary in a genetic male. In contrast, androgens are ineffective in masculinizing the heterogametic ovary.[2] Masculinization of the ovary is possible following transplantation of a testis into a developing female tadpole.[27] Similarly, estrogens sex reverse the testis of the homogametic rooster (although the result is transient) and androgens are ineffectual in masculinizing the hen ovary.[38]

Development of the homogametic gonad accordingly can be viewed as a constitutive pathway or, to borrow a term from computer jargon, a "default pathway". On the other hand, development of the heterogametic gonad requires activation of a new circuit and thus, diversion from the default circuit. It may be that the many factors that reverse sex are in fact serving as inducers or repressors of genes that comprise a pathway. Indeed the genes responsible for development of ovary or testis are present in both sexes and more than a single gene is required. But it is the initial commitment to one pathway or the other –– no more than a binary decision — that normally determines sex.

G. Transition from Male to Female Heterogamety Requires a Single Mutation

Among the fishes there is an example of a species changing from one mode of heterogamety to the other. The southern platy, *Xiphophorus maculatus,* of Central America is a male heterogametic species, but a strong female-determining mutant is invading the population.[39] As a result, mutant fish that carry the normally male-determining Y chromosome develop as functional females. The sex-reversing mutation occurs on an X chromosome; the modified X is called the W chromosome. Thus, there are two types of males, XY and YY, and three types of females, XX, WX, and WY. The W and X are found in nearly all of the drainages sampled, and there are a few drainages where the frequency of the X is approaching zero. On the other hand, the W has yet to be identified in the western edge of the range of these fishes. It has been suggested that the W chromosome confers a selective advantage to the female, namely drab coloration.[40]

A similar situation is found in the Scandinavian wood lemming, *Myopus schisticolor.* An X chromosome rearrangement may be involved in the atypical sex determination found in this species. Lemmings carrying the mutant X and a Y chromosome develop as functional females.[41] However, degeneration of the mammalian Y chromosome precludes development of YY embryos (the reason for this will become more evident in Section III of this chapter). Thus, it is not likely that this species will undergo a change from male to female heterogamety.

H. Downstream Autosomal Genes Can Influence Sex Development

A few autosomal genes that can affect sex development have been identified in *X. maculatus.*[40] The autosomal genes cannot function autonomously but require interaction with

Table 2
ATYPICAL SEX DETERMINATION IN *Xiphophorus maculatus*[40]

		Offspring class							
		$X^{Sp}X^{Sp}$	$X^{Sp}Y^{Ar}$		$X^{Sp}Y^{Sr}$	$X^{Sp}Y^{Mr}$	$Y^{Sr}Y^{Ar}$	$Y^{Ar}Y^{Ar}$	
Line	Mating type	♀♀	♀♀	♂♂	♂♂	♂♂	♂♂	♀♀	♂♂
1	$JpX^{Sp}X^{Sp}$ ♀♀ × $Y^{Ar}Y^{Mr}$ ♂♂		8[a]	7		13			
2	$JpX^{Sp}X^{Sp}$ ♀♀ × $X^{Sp}Y^{Mr}$ ♂♂	117				123			
3	$X^{Sp}Y^{Ar}$ ♀♀ × $JpX^{Sp}Y^{Sr}$ ♂♂	581	216	385	590		613		
4	$JpX^{Sp}X^{Sp}$ ♀♀ × $X^{Sp}Y^{Ar}$ ♂♂	254	4	245					
5	$X^{Sp}Y^{Ar}$ ♀♀ × $X^{Sp}Y^{Ar}$ ♂♂	75	65	84				2	101

[a]   Number of fish in class.

specific mutant sex chromosomes. Those genes may control steps downstream from the initial commitment to one sex, or may be regulatory elements. Consider a male, the offspring of a gravid female collected in Belize, with two sex-linked pigment markers — red anal fin (Ar) and red mouth (Mr). That male was crossed with an XX female of a reference stock, Jamapa, in which both Xs were marked with the pigment allele-spotted side (Sp). If that male were YY, he would produce all-male progeny. If instead, that male were XY, his progeny would consist of equal numbers of males and females with each sex inheriting a different pigment pattern. Neither expectation was realized (Table 2). The 13 offspring that inherited the Y^{Mr} chromosome developed as males, and when those males were backcrossed with Jamapa females, a 1:1 ratio of male to female occurred (Table 2, line 2). Evidently, that chromosome functioned normally. On the other hand, half the offspring that inherited the Ar-marked sex chromosome developed as females. The mutant phenotypes were reproduced when those fish were crossed with Jamapa males and females (Table 2, lines 3 and 4) and when the F_1 fish were intercrossed (Table 2, line 5). The simplest explanation is that the original male carried two Y chromosomes and that the Y marked with Ar is atypical. Because not all X^{Sp}/Y^{Ar} fish developed as females, it was necessary to invoke an autosomal involvement in the atypical sex determination with the Y^{Ar} chromosome. The autosomal gene shows variable penetrance. (If two autosomal genes were involved, not more than 25% of the XY^{Ar} fish would be female; in those crosses 36% of those fish were female.)[40]

There are indications that autosomal genes may have a subsidiary role in mammalian sex development. When the Y chromosome of the semispecies *Mus musculus domesticus* (obtained from Val Poschiavo in Switzerland) is placed onto the genome of the C57BL/6 (B6) laboratory mouse, XY mice rarely develop as normal males after the first backcross generation. Instead, they develop as hermaphrodites or, more rarely, as females.[42] More than one autosomal (or pseudoautosomal) gene are involved because all XY mice of the F_1 generation are normal males and the segregation ratios in succeeding backcross generations deviate from the expectation of single- or even two-gene models.[43]

I. Sex-Specific H-Y Antigen Is Conserved in Evolution

H-Y antigen, identified by antibodies from male-grafted female mice, has been detected in the heterogametic sex in more than 80 species, representing each of the major vertebrate classes.[44] Widespread occurrence of H-Y and its association with the heterogametic sex led to the proposition that the molecule is the Y chromosome-determined inducer of the heterogametic gonad — the testis in XY species such as mouse and human, and the ovary in ZW species such as domestic fowl and clawed frog (*Xenopus laevis*).

There are indications that H-Y serology might be useful as an adjunct to cytogenetic and molecular methods in tracing the evolution of heterogamety among the vertebrates. In *Emys orbicularis*, a female heterogametic species in which sex determination can be influenced by temperature, H-Y is detected in male and female in the extragonadal tissues. With respect to the gonad, H-Y is found only in the ZW ovary. Studies of this kind might be informative in *Alligator mississipiensis,* another species in which temperature influences sex determination, but one in which heterogamety has not been assigned because the sex chromosomes are homomorphic.[44]

In the apomorphic species, expression of H-Y is in general closely associated with the development of the heterogametic type gonad. In newly hatched females of the domestic fowl, for example, H-Y is found in the left functional ovary, and to a lesser extent in the right residual gonad.[45] When ZZ embryos of the domestic fowl are exposed to estradiol, the gonads are feminized, and this is associated with appearance in the gonad of H-Y.[46]

Recent studies by Lau et al.[47] suggest that the structural gene for H-Y is on chromosome 6 of the human — the site of the major histocompatibility locus. It would follow that the Y chromosome provides a regulatory function in the synthesis of the mammalian H-Y molecule.

J. Summary

Current evolutional theory favors a monophyletic origin of the vertebrates. Given the conservation of housekeeping gene sequences among the vertebrates, it would not seem unreasonable to suppose occurrence of an archetypical genetic sex-determining mechanism. Apparently sexuality is an advantage; and a simple way to accomplish sexuality is to have two sexes. Furthermore, among extant species, it is more the exception than the rule for the sex ratio to be other than one. And a parsimonious mechanism that would assure an even sex ratio from generation to generation is a single gene system with a dominant allele. That sex is, in fact, inherited in simple Mendelian fashion was discovered at the turn of the century.

The vertebrate gonad begins as a bipotential anlage. Because of this, errors in sex development can occur regularly. Whether sex reversal occurs by natural or artificial circumstance, the pathway of transformation is a common one. There is a relationship between ovary and testis that may reflect the way in which the sex-determining genes are regulated and this is somehow related to whether the female or male is the heterogametic sex. With respect to those genes, there is a need for a single genetic decision to trigger ovarian or testicular pathways of development. The similarities of sex development among the diverse vertebrate forms seem to favor a common mechanism rather than convergent evolutionary sex-determining schemes.

III. EVOLUTION OF THE SEX CHROMOSOMES

A. Sex Chromosome Development Parallels Vertebrate Development

There is a trend for sex chromosomes to be distinguishable cytologically in derived species. For example, mammals and birds generally have distinct Y and W chromosomes. On the other hand, the Y and W chromosomes of some fish and amphibians can be identified only by progeny testing of sex-reversed animals or by monitoring the inheritance of sex and sex-linked genes. The sex chromosomes of the lower vertebrates can be viewed as no more than an autosomal pair that carries the sex-determining locus. Thus, sex chromosome evolution appears to parallel vertebrate evolution in general. Coincident with the development of heteromorphic sex chromosomes is the apparent degeneracy of the Y and W chromosomes to the point where their only function is to carry the sex-determining locus. It has been suggested that the evolution of differentiated sex chromosomes is an efficient means to assure the orderly segregation of the opposing sex-determining genes at meiosis. Nevertheless, development of the heteromorphic W and Y constrains the development of the Z and X not

Table 3
SEX CHROMOSOMAL REPEAT
SEQUENCES OBSERVABLE ON
ETHIDIUM BROMIDE-STAINED
AGAROSE GELS

Species	Chromosomal location
Bos taurus	X
B. indicus	X
Felis domesticus	X
Mus domesticus	ND[a]
Gallus domesticus	W
Alligator mississipiensis	ND
Lepidochelys kempi	ND
Rana pipiens	X?[b]

[a] ND — sex-specific repeat sequences were not detected.
[b] Bands found in female samples but not discerned in male samples.

only with respect to the opposing sex-determining locus but also to the proper expression of other genes located on those chromosomes.

B. Molecular Studies Provide New Insights into Sex Chromosome Evolution

As the W and Y evolve, we might anticipate sequence divergence between the W and Z and between the X and Y. Our laboratory has used several sex chromosomal probes in a search for markers of the sex chromosomes in species of interest to us. *Rana pipiens*, for example, can be treated with sex hormones to produce sex-reversed XX males. Moreover, our colleague Christina Richards has found that gynogenetic frogs (XX males having only maternal genes) may occasionally develop as functional males.[48] Because those gynogenetic males were not randomly distributed among egg clutches, it is likely that there is a genetic component in those cases of atypical sex determination. The alligator, *Alligator mississipiensis*, and Kemp's ridley sea turtle, *Lepidochelys kempi*, are species in which temperature can influence the development of sex. Our results, presented below, are based solely on gel electrophoresis and filter hybridizations with DNA from normal animals of each species. Several male-female pairs were used in order to distinguish autosomal polymorphism.

1. Repeat Sequences Are Found on the Sex Chromosomes

In the human, DNA from a normal male can be distinguished from that of a normal female on ethidium bromide-stained agarose gels following disgestion with appropriate restriction enzymes because of a concentration of repeat sequences on the Y. According to Ray-Chaudhuri et al.,[49] heterochromatization may be an initial step in the development of heteromorphic sex chromosomes. They were able to identify the W on the basis of asynchronous replication in snakes in which the sex chromosomes were morphologically indistinct. In the absence of heteromorphism, this technique could provide a relatively rapid means of identifying chromosomal sex. This rationale has been applied in other species with limited success. The results are presented in Table 3. The simplest conclusion is that there are chromosome-specific repeat sequences in some species — here, in species with differentiated sex chromosomes. We were surprised to find an apparent X-chromosomal repeat in the bovines because Kurnit et al.[50] reported that the bovine X and Y lack quantities of heterochromatin. We confirmed the presence of a W-chromosomal Xho I repeat in the domestic fowl as previously reported by Tone et al.[51] However, repeats were not detected in the

Table 4
CHROMOSOMAL LOCATION OF
HPRT AND PGK1[a]

Species	Chromosomal location
Homo sapiens	X
Mus domesticus	X
Bos taurus	X
B. indicus	X
Felis domesticus	X
Gallus domesticus	A[b]
Alligator mississipiensis	X?[c]
Lepidochelys kempi	X?[c]
Rana pipiens	X

[a]   HPRT — hypoxanthine phosphoribosyl transferase, PGK1 — phosphoglycerate kinase-1.
[b]   A — autosomal or pseudoautosomal.
[c]   Bands in female samples were twice as intense as those of male samples.

alligator and turtle, the two species of special interest. In *Rana pipiens,* a ladder of bands was found in female DNA but was not apparent in male DNA. The pattern was observed among four pairs of animals; that will be investigated further.

2. HPRT and PGK1 May Be Sex Linked in a Frog Species

Because of the apparent conservation of X-chromosomal sequences among the mammals,[52] it is possible to acertain sex by observing a dosage difference between genomic DNA samples from male and female. Thus, probes for PGK1 (phosphoglycerate kinase-1) and HPRT (hypoxanthine phosphoribosyl transferase) would be expected to reveal a single dose in the DNA of male mammals and a double dose in the DNA of female mammals (Table 4). We have extended the survey to nonmammalian vertebrates. A dosage difference was detected in some of those species. For example, we found a clear-cut dosage effect in *Rana pipiens,* a male heterogametic species; the female sample contained bands twice as intense as those in the male sample. DNA samples from XX male and XY female frogs and from related species are currently under examination to further evaluate those results. In the alligator and turtle, we observed dosage effects that indicate male heterogamety. We are increasing our sample size in order to validate those observations.

3. Human Y Probes Detect Conserved Sex-Linked Sequences

Many human Y-chromosomal probes recognize homologous sequences on the X or autosomes in the human genome.[53] In fact, some probes reveal X or autosomal linkage in primates indicating a transposition of sequences to the Y during human evolution.[54] We have used several Y-chromosomal probes obtained from human genomic libraries on DNA from other species in order to determine the conservation and distribution of those sequences. For example, with probe pDP34,[54] we found male-female differences in hybridizing sequences in mouse and bovine (Table 5). But pDP34 sequences are present in nonmammalian species as well, and the distribution of homologous sequences varies with the species examined.

4. GATA-GACA Repeats are Valuable Sex Chromosomal Markers

We have used a probe that recognizes the GATA-GACA repeats which characterize the Bkm (banded krait minor) satellite DNA. Those sequences are generally abundant in the genome of vertebrates and quite often are present on the sex chromosomes (Table 6). The presence of sex-specific GATA-GACA sequences in the alligator is noteworthy because

Table 5 CHROMOSOMAL LOCATION OF pDP34-LIKE SEQUENCES	
Species	**Chromosomal location**
Mus domesticus	X, Y
Bos taurus	X
B. indicus	X
Gallus domesticus	Z, W
Lepidochelys kempi	A[a]
Rana pipiens	X, Y

[a]   A — autosomal or pseudoautosomal.

Table 6 CHROMOSOMAL LOCATION OF Bkm-LIKE SEQUENCES	
Species	**Chromosomal location**
Gallus domesticus	W
Alligator mississipiensis	Y
Lepidochelys kempi	Y
Rana pipiens	X, Y

temperature can influence sex determination, and preliminary cytogenetic studies failed to reveal heteromorphic sex chromosomes in that species. Thus, we may now have a means of identifying the genetic sex of alligators during the course of studies aimed at clarification of the way in which temperature can influence sex. The sex-linked marker in the turtle is of particular importance to conservationists. Currently Kemp's ridley, an endangered species, breeds on a single beach in Mexico. As soon as a clutch is laid, the eggs are retrieved and placed into sand-filled foam chests as a conservation practice. Because sex determination in this species is also influenced by temperature, there is a fear that the artificial incubator may be skewing the sex ratio in favor of males. This is another species that lacks heteromorphic sex chromosomes and lacks sex dimorphism when the hatchlings are released.

C. Summary

Cytogenetic data have allowed us to propose that the evolutional development of the sex chromosomes parallels the development of the vertebrates in general. The Ophidia are a prime example. Among the constrictors — the group of snakes considered the most primitive — heteromorphic sex chromosomes are rare, whereas among the highly derived group consisting of the vipers, heteromorphic sex chromosomes are common. The colubrids represent an evolutionary transition group between the constrictors and vipers, and in this group we find many examples of what are thought to be developing heteromorphic sex chromosomes.

Application of molecular techniques to the study of the sex chromosomes offers new insights into the development of those chromosomes. Homologies between the X and Y are becoming apparent as are the relationships of the sex chromosomes between species. It seems reasonable to assume that molecular study of the sex chromosomes will lead ultimately to the identification and characterization of the sex-determining genes themselves.

ACKNOWLEDGMENT

Supported in part by NIH Grant AI 23479.

REFERENCES

1. **Witschi, E.,** Sex deviations, inversions, and parabiosis, in *Sex and Internal Secretions,* Allen, E., Ed., Williams & Wilkins, Baltimore, 1932, 160.
2. **Witschi, E.,** The inductor theory of sex differentiation, *J. Fac. Sci. Hokkaido Univ. Ser. B,* 13, 428, 1957.
3. **Spemann, H.,** *Embryonic Development and Induction,* Yale University Press, New Haven, Conn., 1938.
4. **van Niekerk, W. A. and Retief, A. E.,** The gonads of human true hermaphrodites, *Hum. Genet.,* 58, 117, 1981.

5. **Dubois, R.,** On the distribution of bidder's organ in bufonids, *Zool. Med.,* 28, 275, 1947.

6. **Zaccanti, F. and Tognato, G.,** Effects of different doses of diethylstilbestrol dipropionate on the bidder's organ of intact or orchiectomized adult males of *Bufo bufo* (L), *Monogr. Zool. Ital.,* 10, 105, 1976.

7. **Burns, R. K.,** Role of hormones in the differentiation of sex, in *Sex and Internal Secretions,* Vol. 1, Young, W. C., Ed., Williams & Wilkins, Baltimore, 1961, 76.

8. **Ponse, K.,** L'evolution de l'organe de Bidder et la sexualite chez le Crapaud, *Rev. Suisse Zool.,* 34, 217, 1927.

9. **Witschi, E.,** Studies in sex differentiation and sex determination in amphibians, VI. The nature of Bidder's organ in the toad, *Am. J. Anat.,* 52, 461, 1933.

10. **Jorgensen, C. B. and Billeter, E.,** Growth differentiation, and function of the testes in the toad *Bufo bufo bufo* (L), with special reference to regulatory capacities: effects of unilateral castration, hypophysectomy, and excision of bidder's organs, *J. Exp. Zool.,* 221, 225, 1982.

11. **Brode, M. D.,** The significance of the asymmetry of the ovaries of the fowl, *J. Morphol. Physiol.,* 46, 1, 1928.

12. **Burke, W. H.,** Testicular asymmetry in the turkey, *Poult. Sci.,* 52, 1652, 1973.

13. **Didier, E. and Fargeix, N.,** Aspects quantitatifs du peuplement des gonades par les cellules germinales chez l'embryon de Caille (*Coturnix coturnix japonica*), *J. Embryol. Exp. Morphol.,* 35, 637, 1976.

14. **Dubois, R. and Cuminge, D.,** Sur les aspects statistiques de l'asymetrie primaire de repartition des cellules germinales primordiales chez l'embryon de Poulet, *C. R. Acad. Sci.,* 286, 1613, 1978.

15. **Witschi, E.,** Origin of asymmetry in the reproductive system of birds, *Am. J. Anat.,* 56, 119, 1935.

16. **Wolff, Et. and Wolff, Em.,** Sur l'induction experimentale de l'ovaire droit chez l'embryon d'oiseau, *C. R. Acad. Sci.,* 226, 1140, 1948.

17. **Godet, R.,** Recherches d'anatomie et d'embryologie normale et experimental sur l'appareil genitale de la taupe, *Bull. Biol. Fr. Belg.,* 83, 25, 1949.

18. **Peyre, A.,** Note sur la structure histologique de l'ovaire du Desman des Pyrenees (*Galamys pyrenaicus G*), *Bull Soc. Zool. Fr.,* 77, 441, 1952.

19. **Cole, H. H., Hart, G. H., Lyons, W. R., and Catchpole, H. R.,** The development and hormonal content of fetal horse gonads, *Anat. Rec.,* 56, 275, 1933.

20. **Domm, L. V.,** New experiments on ovariotomy and the problem of sex inversion in the fowl, *J. Exp. Zool.,* 48, 31, 1927.

21. **Goodale, H. D.,** Gonadectomy in relation to the secondary sexual characters of some domestic birds, *Carnegie Inst. Washington Publ.,* p. 243, 1916.

22. **Yamamoto, T. and Kajishima, T.,** Sex hormone induction of sex reversal in the goldfish and evidence for male heterogamety, *J. Exp. Zool.,* 168, 215, 1968.

23. **Chieffi, G., Iela, L., and Rastogi, R. K.,** Effect of cyproterone, cyproterone acetate and ICI 46,474 on gonadal sex differentiation in *Rana esculenta, Gen. Comp. Endocrinol.,* 22, 532, 1974.

24. **Iwasawa, H.,** Effects of thiourea on the gonadal development of *Rana temporaria ornativentris* larvae, *Zool. Mag. (Tokyo),* 68, 251, 1959.

25. **Hsu, C. Y., Huang, H.-C., Chang, C.-H., and Liang, H.-M.,** Independence of ovarian masculinization and hypothyroidism in frog tadpoles after methimazole treatment, *J. Exp. Zool.,* 189, 235, 1974.

26. **Witschi, E.,** Studies on sex differentiation and sex determination in amphibians. II. Sex reversal in female tadpoles of *Rana sylvatica* following the application of high temperature, *J. Exp. Zool.,* 52, 267, 1929.

27. **Mikamo, K. and Witschi, E.,** Functional sex-reversal in genetic females of *Xenopus laevis,* induced by implanted testes, *Genetics,* 48, 1411, 1963.

28. **Satoh, N.,** Sex differentiation of the gonad of fry transplanted into the anterior chamber of the adult eye in the teleost, *Oryzias latipes, J. Embryol. Exp. Morphol.,* 30, 345, 1973.

29. **Satoh, N.,** An ultrastructural study of sex differentiation in the teleost, *Oryzias latipes, J. Embryol. Exp. Morphol.,* 32, 195, 1974.

30. **Satoh, N. and Egami, N.,** Preliminary report on sex differentiation in germ cells of normal and transplanted gonads in the fish, Oryzias latipes, in *Genetics and Mutagenesis of Fish,* Schroder, J. H., Ed., Springer-Verlag, Berlin, 1973, 29.

31. **Chang, C. Y. and Witschi, E.,** Genic control and hormonal reversal of sex differentiation in *Xenopus, Proc. Soc. Exp. Biol. Med.,* 93, 140, 1956.

32. **Witschi, E.,** Biochemistry of sex differentiation in vertebrate embryos, in *The Biochemistry of Animal Development,* Vol. 2, Weber, R., Ed., Academic Press, New York, 1967, 193.

33. **Whitten, W. K., Beamer, W. G., and Byskov, A. G.,** The morphology of fetal gonads of spontaneous mouse hermaphrodites, *J. Embryol. Exp. Morphol.,* 52, 63, 1979.

34. **Upadhyay, S. and Zamboni, L.,** Ectopic germ cells: natural model for the study of germ cell sexual differentiation, *Proc. Natl. Acad. Sci. U.S.A.,* 79, 6584, 1982.

35. **Taber, E., Knight, J. S., Ayers, C., and Fishburne, J. I.,** Some of the factors controlling growth and differentiation of the right gonad in female domestic fowl, *Gen. Comp. Endocrinol.,* 4, 343, 1964.

36. **Haffen, K.,** Sexual differentiation and intersexuality in vitro, in *Organ Culture,* Thomas, J. A., Ed., Academic Press, New York, 1970, 121.

37. **Haffen, K.,** La culture in vitro de l'epithelium germinatif isole des gonades males et femelles de l'embryon de Canard, *J. Embryol. Exp. Morphol.,* 8, 414, 1960.

38. **Willier, B. H.,** The modification of sex development in the chick embryo by male and female sex hormones, *Physiol. Zool.,* 10, 101, 1937.

39. **Kallman, K. D.,** Genetics and geography of sex determination in the poeciliid fish, Xiphophorus maculatus, *Zoologica (N.Y.),* 50, 151, 1965.

40. **Kallman, K. D.,** A new look at sex determination in poeciliid fishes, in *Evolutionary Genetics of Fishes,* Turner, B. J., Ed., Plenum Press, New York, 1984, 95.

41. **Schempp, W., Wiberg, U., and Fredga, K.,** Correlation between sexual phenotype and X-chromosome inactivation pattern in the X*XY wood lemming, *Cytogenet. Cell Genet.,* 39, 30, 1985.

42. **Eicher, E. M., Washburn, L. L., Whitney, J. B., and Morrow, K. E.,** *Mus poschiavinus* Y chromosome in the C57BL/6J murine genome causes sex reversal, *Science,* 217, 535, 1982.

43. **Eicher, E. M. and Washburn, L. L.,** Inherited sex reversal in mice: identification of a new primary sex-determining gene, *J. Exp. Zool.,* 228, 297, 1983.

44. **Nakamura, D., Wachtel, S. S., Lance, V., and Beçak, W.,** On the evolution of sex determination, *Proc. R. Soc. London, Ser. B,* 232, 159, 1987.

45. **Nakamura, D. and Wachtel, S. S.,** Vertebrate sex determination: an immunologic perspective, in *Biology of Fertilization,* Vol. 1, Metz, C. B. and Monroy, A., Eds., Academic Press, Orlando, Fla., 1985, 95.

46. **Müller, U., Zenzes, M. T., Wolf, U., Engel, W., and Weniger, J.-P.,** Appearance of H-W (H-Y) antigen in the gonads of oestradiol sex-reversed male chicken embryos, *Nature (London),* 280, 142, 1979.

47. **Lau, Y.-F., Chan, K., Kan, Y. W., and Goldberg, E.,** Isolation of a male-specific and conserved gene using an anti-H-Y antibody, *Am. J. Hum. Genet.,* Suppl. 39, A142, 1987.

48. **Richards, C.,** personal communication, 1986.

49. **Ray-Chaudhuri, S. P., Singh, L., and Sharma, T.,** Evolution of sex-chromosomes and formation of W-chromatin in snakes, *Chromosoma,* 33, 239, 1971.

50. **Kurnit, D. M., Shafit, B. R., and Maio, J. J.,** Multiple satellite deoxyribonucleic acids in the calf and their relation to the sex chromosomes, *J. Mol. Biol.,* 81, 273, 1973.

51. **Tone, M., Sakaki, Y., Hashiguchi, T., and Mizuno, S.,** Genus specificity and extensive methylation of the W chromosome-specific repetitive DNA sequences from the domestic fowl, *Gallus gallus domesticus, Chromosoma,* 89, 228, 1984.

52. **Ohno, S.,** Evolution of sex chromosomes in mammals, *Annu. Rev. Genet.,* 3, 495, 1969.

53. Human gene mapping 8, 8th international workshop on human gene mapping, *Cytogenet. Cell Genet.,* 40 (1-4), 6, 1985.

54. **Page, D. C., Harper, M. E., Love, J., and Botstein, D.,** Occurrence of a transposition from the X-chromosome long arm to the Y-chromosome short arm during human evolution, *Nature (London),* 311, 119, 1984.

Chapter 2

FROM GATA-GACA REPEATS TO DICTYOSTELIUM CELL ADHESION PROTEIN TO C-CAM, N-CAM, AND GONAD-ORGANIZING PROTEINS

Susumu Ohno

TABLE OF CONTENTS

I. INTRODUCTION

In 1975, Wachtel et al.[1] proposed that the apparent evolutionary conservation of the H-Y plasma membrane antigen in the heterogametic sex indicated its involvement in primary sex determination. In recent years, however, the sex specificity of this antigen has come to be questioned, as have the reliability of serological tests for H-Y, and the identity of this antigen with the classical male antigen of Eichwald and Silmser[2] recognized by major histocompatibility complex (MHC)-restricted T cells. Thus, it seems appropriate to start this paper with the reminder that our laboratory identified the male-specific plasma membrane antigen of subunit molecular weight 18,000 as H-Y not by serological means, but via the specific receptor residing on the plasma membrane of bovine fetal ovarian cells.

II. IN THE ABSENCE OF CLASS I MHC ANTIGENS, A NUMBER OF CELL SURFACE PROTEINS INVOLVED IN DEVELOPMENT CANNOT REMAIN ON THE PLASMA MEMBRANE

MHC antigens were originally discovered as the major cause of graft rejection. Inasmuch as nature has not practiced organ grafts, and MHC-incompatible fetuses are readily tolerated by mammalian mothers, the *raison d'etre* of MHC antigens is clearly not graft rejection. The true role of MHC antigens was finally clarified in 1974, with the discovery of MHC restriction in T cell recognition of alien antigens.[3]

On the plasma membrane of infected target cells, *cytotoxic* T cells do not recognize a viral antigen per se, but rather a complex formed between the viral antigen and a self class I MHC antigen. It follows that viruses are inviting their own destruction by immune prosecution in choosing to present their proteins on the host cell membrane in association with host class I MHC antigens. Yet unless viruses are masochistic by nature, they must gain considerable benefits from association between their proteins and class I MHC antigens of the host, because such benefits must necessarily outweigh the disadvantage of provoking T cell responses from the host. We reasoned that by association with host class I MHC antigens, viral proteins must be displacing host plasma membrane proteins that are involved in organogenesis, cell differentiation, and cell growth. Thus, it was proposed that those host proteins are normally in association with class I MHC antigens.[4]

The corollary of the above is that those proteins will not be found on the plasma membrane of mutant cell lines that fail to express class I MHC antigens. In a particular human male Burkitt lymphoma cell line called Daudi, one of the two β-2-microglobulin loci has been deleted, while the other has been silenced by a mutation. In the absence of this dimeric partner, class I MHC antigens (HLA-A, -B, and -C in the human) are not expressed on the plasma membrane of Daudi.[5] Thus, we reasoned that the male-specific but ubiquitously expressed H-Y antigen should not be maintained on the plasma membrane in this cell line. Accordingly, the exhausted culture medium of the Daudi human male Burkitt lymphoma became the source of our H-Y antigen.[6] More recently, it has been shown that the insulin receptor is not expressed by mutant cell lines that have lost class I MHC antigen; restoration of class I MHC antigen expression by transfection results in the expression of insulin receptors, and thereby, restoration of the insulin response.[7] It would thus appear that the proposed association between plasma membrane proteins involved in organogenesis, cell differentiation, and cell growth, and class I MHC antigens is a general phenomenon.

III. SPECIFIC RECEPTOR FOR TESTIS-ORGANIZING H-Y ANTIGEN RESIDING ON BOVINE FETAL OVARIAN CELLS

Complete sex reversals in both directions are readily attainable in some of the gonochoristic species of teleost fish.[8] Even among avian species, complete sex reversal in one direction is readily achieved. Removal of the left functional ovary from the ZW female chicken starts the compensatory growth of the right residual gonad in the direction of a testis in which complete spermatogenesis and spermiogenesis may take place.[9] Such sex-reversed ZW cocks may even be fertile if the right Wolffian duct persists during the embryonic development of the precursor female chick. In mammals, however, only secondary sexual development, which depends on the presence or absence of androgenic steroids, can be reversed. The reversal of gonadal sex is not known. Yet, among even-toed ungulate mammals, the fetal ovaries of XX females may undergo testicular transformation under a specific circumstance.

The fact that heifers born as the twin of a bull are invariably sterile and somewhat masculinized must have been known even to prehistoric breeders of cattle. This masculin-ization involving transformation of the ovary is due to vascular anastomosis between het-erosexual dizygotic twins. In this condition, the ovary develops normally until day 50 to 60 of gestation, whereupon its growth stops; the ovary may then be transformed into a small testis at about day 90.[10] Such transformed testes are devoid of germ cells. Nevertheless they synthesize testosterone, albeit too late to cause the complete development of the Wolffian ducts.[11] Accordingly, masculinization of affected heifers (known as "freemartins") is more evident in the gonads than it is in the external genitalia.

Interestingly, in humans, marmosets, and horses, the ovary of XX females is not affected at all under the same conditions, i.e., in cases of vascular anastomosis between heterosexual dizygotic twins. Thus, we reasoned that there must be two conditions specific to cattle and other members of the family Bovidae: (1) after the completion of testicular organogenesis, but not before, XY males disseminate daily a considerable amount of testis-organizing protein or proteins into the blood circulation; and (2) even after the initiation of ovarian differen-tiation, XX gonadal cells in the ovary must continue to express specific receptors for testis-organizing protein. We accordingly decided to utilize this specific receptor in bovine fetal ovarian cells for the identification of H-Y antigen contained in the exausted culture medium of Daudi Burkitt lymphoma cells.

IV. IDENTIFICATION OF THE EXTRAMEMBRANE PORTION OF H-Y ANTIGEN AS A CYSTEINE-CONTAINING HYDROPHOBIC SUBUNIT OF MOLECULAR WEIGHT 18,000

As shown in Figure 1, when exhausted Daudi culture medium containing ^{3}H-lysine-labeled proteins was incubated with a large number of bovine fetal ovarian cells (mostly follicular cells in suspension), only one peak with the subunit molecular weight of 18,000 was spe-cifically removed from the culture medium. Once this subunit had been completely removed, there was no further uptake of any other subunit by fresh ovarian cells upon further incubation (Figure 1, top). Moreover, the plasma membrane fraction of a smaller number of fetal ovarian cells, previously exposed to a large excess of labeled Daudi proteins, exclusively yielded a single subunit peak identical to that removed from the culture medium (Figure 1, bottom).[12] It was thus concluded that the extramembrane portion of the testis-organizing H-Y antigen excreted by Daudi human male Burkitt lymphoma cells has a subunit molecular weight of 18,000. This subunit remained a monomer only in the presence of a reducing agent (β-mercaptoethanol), thus revealing the presence of active cysteine residues. Not only was this subunit about as hydrophobic as dissociated immunoglobulin light-chains, but our preliminary amino acid composition analysis revealed the relative abundance of Leu, Val, Ile, Ser, Thr, and Pro.

It should be noted here, however, that since we utilized the abnormal absence of class I MHC antigen for identification, the identified subunit might have been modified in a some-what unusual way. For example, the subunit might have been endowed with N-glycosylation sites near its N-terminal that were not glycosylated in Daudi cells. The fact that this subunit

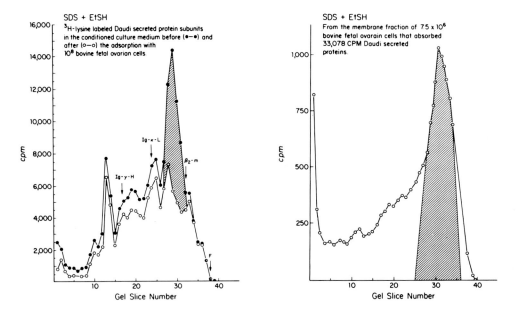

FIGURE 1.   Demonstration of specific uptake, by bovine fetal ovarian cells of male-specific plasma membrane antigen secreted by Daudi β-2-microglobulin (−), HLA (−) human male Burkitt lymphoma cells. Top: SDS gel electrophoresis profile after prolonged treatment with 5% β-mercaptoethanol (EtSH) of Daudi secreted protein subunits labeled with [³H]-lysine before (–●–●–) and after (–○–○–) exposure to 10^8 bovine fetal ovarian cells per 5 mℓ. CPM were adjusted per 5 mℓ. Exponentially growing Daudi cells (100 to 140 × 10^6) were placed in 60 to 100 mℓ RPMI 1640 culture medium free of protein, as well as cold lysine containing 6 Ci/mℓ [³H]-lysine (specific activity: 2.1 Ci/mmol). The cells were maintained for 16 hr. Treated cells were allowed to recover for 20 hr in 10% Ig-free fetal calf serum. Twice-washed cells were then allowed to secrete their proteins into protein free but cold lysine containing RPMI 1640 for 16 hr. This was the Daudi-conditioned culture medium containing [³H]-lysine labeled secreted proteins. Bovine fetal ovarian cells for the specific uptake study were derived from fetuses of CR length, 16 to 60 cm; 10^8 freely suspended fetal ovarian cells were incubated with 5 mℓ labeled Daudi-conditioned medium, noted above, for the first 5 min on ice and for the next 30 min at 37°C.[12] F marks the migration front indicated by pyronin Y dye. Positions of three molecular weight markers are also indicated. Bottom: SDS gel electrophoresis profile of the membrane fraction from 7.5 × 10^6 bovine fetal ovarian cells that absorbed 33,078 TCA precipitable CPM from 5 mℓ Daudi-conditioned culture medium noted above. Comparison of top and bottom graphs leaves little doubt that the fetal ovarian cell is endowed with specific plasma membrane receptors that bind the 18,000-dalton subunit secreted by Daudi human male Burkitt lymphoma cells.

was found in the culture medium, indicates that it lacked the transmembrane domain. In the absence of associating class I MHC antigens, the normally protected protease cleavage site might have become exposed, thus resulting in dissemination of the extramembrane portion. Of course it is possible that the identified subunit was not disseminated by live Daudi cells; perhaps being unable to be expressed on the plasma membrane, the subunits accumulated within the cytoplasm and were released into the culture medium upon cell death. Nevertheless, the fact is that in organ culture, these subunits were able to induce tubular differentiation in bovine XX embryonic gonads; treated gonads came to resemble the miniature-testis-like freemartin gonad.[12]

V. THE TESTIS-ORGANIZING PLASMA MEMBRANE PROTEIN MUST BE ONE OF THE CAM PROTEINS

Slime molds such as *Dictyostelium discoideum* can be viewed as straddling the border between unicellular and multicellular eukaryotes. In a rich nutrient environment, they live as unicellular amoeboid entities, but when the environment becomes unfavorable, the cells aggregate to form a stalk and fruiting body somewhat resembling a miniature mushroom.

The plasma membrane protein, that by homologous recognition causes cell aggregation and the subsequent formation of the stalk and fruiting body, is known as CsA CAM (contact site cell-adhesion molecule). The amino acid sequence of this 494-residue-long protein has recently been deduced.[13] Its amino terminal region is endowed with N-glycosylation sites, whereas its carboxyl terminal either is anchored to the plasma membrane or remains in the cytoplasm with the transmembrane domain preceding. This protein is rich in Leu, Ile, Val, Thr, Pro, and Ser, with Thr-X-Thr-X dipetidic repeats recurring throughout. There are two pairs of cysteine residues. The first pair occurs near the amino terminal occupying the 38th and 51st positions, whereas the second pair occurs just above the presumptive transmembrane domain occupying the 386th and 397th positions.

On the other hand neuronal-CAM (N-CAM) protein, which mediates neuronal organization of the chicken, is at least 964 residues long, although the N-terminal residue has not been ascertained.[14] Below the transmembrane domain represented by the 586th to 602nd residues, there exists a huge 362-residue-long intracytoplasmic domain. It is in this domain that one finds the greatest homology between chicken N-CAM and the above-noted *Dictyostelium* CsA CAM, for this region is replete with Thr-X-Thr-X dipeptidic repeats. The intracytoplasmic domain of N-CAM, however, may be of little functional significance, for a bulk of it, represented by the 687th to 932nd residues, may be lost by a differential splicing during the processing of messenger RNA. In contrast, its amino terminal portion, which protrudes outside the plasma membrane, is occupied by four successive β-2-microglobulin-like domains. The presence of these domains in N-CAM indicates that the oriented aggregation of neurons that leads to organogenesis of the chick central nervous system occurs through homologous recognition via these domains — a case of like attracting like.

The adaptive immune system possessed uniquely by vertebrates is often spoken of as the most sophisticated form of cell-cell interaction. Therefore, it seems fitting that the β-2-microglobulin-like domain first arose in one of the CAM proteins of the ancient lineage traditionally engaged in cell-cell interactions leading to organogenesis. It should be recalled that nearly all of the components of the adaptive immune system are comprised of this β-2-microglobulin-like domain.

The currently popular concept of evolution of new proteins via domain exchanges should be kept in mind, however, for it could be argued that these four successive β-2 microglobulin-like domains did not arise anew *in situ,* but rather were borrowed from a component of the adaptive immune system through domain exchange. Yet this is highly unlikely because the β-2-microglobulin-like domains of N-CAM appear to be more primitive than any of those of the adaptive immune system. For one thing, the distance between two invariant cysteines is about 50 residues, as already noted, and this is distinctly shorter than the distance between two invariant cysteines in the β-2-microglobulin-like domains of the adaptive immune system. Domains of the adaptive immune system not involved in antigen-binding, such as the constant regions of immunoglobulin and T cell receptors and class I and II MHC antigens, already have an average distance of 60 residues separating two invariant cysteines, whereas that distance is increased to 65 to 78 residues in the antigen-binding *variable* regions of immunoglobulin and T cell receptors.

The tendency for progressive increase is also evident in the distance separating the first cysteine from the nearly invariant tryptophane. This distance is from 9 to 13 residues in four β-2-microglobulin-like domains of chicken N-CAM, whereas in the variable regions of immunoglobulin and T cell receptors, the distance is from 12 to 15 residues. Indeed, four β-2-microglobulin-like domains of chicken N-CAM have apparently evolved *in situ* for Thr-X-Thr-X dipeptidic repeats are also prominent in these regions. For example, the first cysteine of the third domain is a part of Thr-Leu-Thr-Cys repeats, whereas the invariant tryptophane of the second domain is a part of the Thr-Met-Thr-Trp-Thr-Lys repeats. In view of the above, it would appear that the testis-organizing plasma membrane protein of subunit

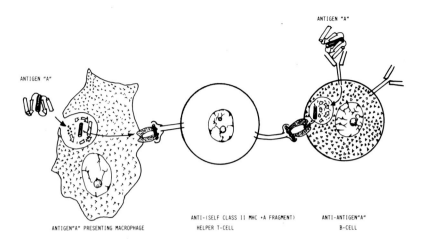

FIGURE 2.   Schematic demonstration of antigen-specific help offered by *helper* T cells to antibody-producing B cells. Left: macrophage presenting a specific black fragment of phagocytized antigen A to the outside world together with self class II antigen. Center: T cells equipped with the receptor directed against this complex. Right: B cell presenting membrane-bound anti-antigen A IgM antibody encounters antigen A. The antigen-antibody complex is taken inside the cell and the same black fragment of antigen A is presented to the outside world by B cells together with class II MHC antigen. This complex is recognized by T cells for antigen-specific help.

molecular weight of 18,000 that we identified as H-Y antigen must be a member of the CAM family, and therefore, replete with Thr-X-Thr-X dipeptidic repeats. It would not be surprising if it consisted of two successive β-2-microglobulin-like domains of the primitive type represented in N-CAM.

VI. WHAT *HELPER* T CELLS SEE VS. WHAT B CELL ANTIBODIES SEE IN CONNECTION WITH THE POOR IMMUNE RESPONSE TO H-Y ANTIGEN

From the very discovery of H-Y antigen by Eichwald and Silmser,[2] it was clear that the anti-H-Y response is mounted only by female mice of H-2[b] haplotype. Female mice of other MHC haplotypes were H-Y nonresponders. Genes in the MHC complex that determine responder or nonresponder status against specific antigens were known originally as immune response genes, but they are now known to encode class II MHC antigens. Whereas receptors of *cytotoxic* T cells recognize the complex formed between alien antigens and ubiquitously expressed self-class I MHC antigens on the plasma membrane of target cells, receptors of *helper* T cells, recognize the complex formed between alien antigens and self-class II MHC antigens on the plasma membranes of antigen-presenting macrophages. Antibodies produced by B cells, on the other hand, directly recognize the various epitopes of alien antigens in total indifference to self-class II MHC antigens. Although it was known that class II MHC antigens are expressed not only by macrophages but also by B cells, it nevertheless remained a long-standing puzzle as to how *helper* T cells equipped with such class II-restricted receptor could offer antigen-specific help to specific clones of B cells producing antibodies directed against various epitopes of the same antigen.

This puzzle has recently been solved, as schematically illustrated in Figure 2. At the left, a macrophage phagocytizes antigen A, and once inside the cell, antigen A is digested by cathepsin-like protease. Of various fragments thus produced, the one with amphipathic α-helical configuration (''A'', black cylinder) is preferentially chosen and presented to the outside world together with self-class II MHC antigen. Meanwhile, a T cell that happens to be endowed with the receptor for this complex (Figure 2, center) recognizes the complex on the plasma membrane of the antigen-presenting macrophage, and, stimulated by this recognition, begins clonal expansion. Elsewhere (Figure 2, right) there is a B cell presenting membrane-bound IgM directed against one of the epitopes of antigen A. When this B cell encounters antigen A, an antigen-antibody complex is formed on the plasma membrane. The formed complex is taken into the B cell cytoplasm by pinocytosis. Once inside the cytoplasm, the same cathepsin-like protease begins the work on the antigen-antibody complex. The black cylinder of amphipathic α-helical complex derived from antigen A is again chosen and presented to the outside world by B cells. Since this is the same complex that T cells in the center of Figure 2 have seen on the plasma membrane of the antigen-presenting macrophage, clonally expanded *helper* T cells are now in a position to offer antigen-specific help to the B cell at the right of the figure in order to prompt its clonal expansion and antibody secretion.

In view of the above, the generally poor immune response to H-Y antigen is readily understandable: CAM proteins in general appear to be and β-2-microglobulin-like domains definitely are essentially β-sheet forming proteins. Thus, the amphipathic α-helical configuration has to be secondarily assumed by certain digested pieces. While this apparently happens in the case of MHC antigens and immunoglobulins (they are good immunogens), this may not happen in the case of H-Y; only a particular class II MHC antigen of a certain MHC haplotype is able to pick up a fragment which is a poor imitation of amphipathic α-helix. Without antigen-specific help from T cells, B cells can hardly produce sufficient amounts of anti-H-Y antibody.

VII. SUMMARY

We have defined testis-organizing H-Y antigen not by serological means, but via the specific receptor residing on bovine fetal ovarian cells. The antigen is a hydrophobic disulfide bridge-forming protein of subunit molecular weight 18,000.

Two specific circumstances were utilized for this identification. (1) In the absence of class I MHC antigens, some of the other plasma membrane antigens cannot be expressed on the plasma membrane. Thus, the exhausted culture medium of Daudi human male Burkitt lymphoma cells became the source of our antigen. (2) Only in cattle and other members of the family, Bovidae, vascular anastomoses between heterosexual dizygotic twins lead to the transformation of the fetal ovary into a minute testis relatively late in fetal development. This suggested secretion by the bull fetus of a testis-organizing protein into the blood circulation, and also the persistence of a specific receptor for that protein in bovine fetal ovarian cells. The protein so identified is obviously a member of the CAM (cell-adhesion molecule) family of proteins, engaged in organogenesis via oriented cell-cell aggregation. The reason for generally poor immune responses against H-Y antigen has also been given.

REFERENCES

1. **Wachtel, S. S., Ohno, S., Koo, G. C., and Boyse, E. A.,** Possible role of H-Y antigen in primary sex determination, *Nature (London),* 257, 235, 1975.
2. **Eichwald, E. J. and Silmser, C. R.,** Untitled communication, *Transplant. Bull.,* 2, 148, 1955.
3. **Zinkernagel, R. M. and Doherty, P. D.,** Immunological surveillance against altered self components by sensitized T-lymphocytes in lymphocytic choriomeningitis, *Nature (London),* 251, 547, 1974.
4. **Ohno, S.,** The original function of MHC antigens as the general plasma membrane anchorage site of organogenesis-directing proteins, *Immunol. Rev.,* 33, 59, 1977.

5. **Klein, G., Terasaki, P., Billing, R., Honig, R., Jondal, M., Rosen, A., Zeuthen, J., and Clements, G.,** Somatic cell hybrids between human lymphoma lines. III. Surface markers, *Int. J. Cancer,* 19, 66, 1977.

6. **Beutler, B., Nagai, Y., Ohno, S., Klein, G., and Shapiro, I.,** The HLA dependent expression of testis-organizing H-Y antigen by human male cells, *Cell,* 13, 509, 1978.

7. **Due, C., Simonsen, M., and Olsson, L.,** The major histocompatibility complex class I heavy chain as a structural subunit of the human cell membrane insulin receptor: implications for the range of biological functions of histocompatibility antigens, *Proc. Natl. Acad. Sci. U.S.A.,* 83, 6607, 1986.

8. **Yamamoto, T.,** Progeny of artificially induced sex-reversals of male genotype (XY) in the medaka, *Oryzias latipes,* with special reference to the YY-male, *Genetics,* 40, 406, 1955.

9. **Miller, R. A.,** Spermatogenesis in a sex-reversed female and in normal males of the domestic fowl, *Gallus domesticus, Anat. Rec.,* 70, 155, 1938.

10. **Jost, A., Perchellet, J. P., Prepin, J., and Vigier, B.,** The prenatal development of bovine freemartins, in *Intersexuality in the Animal Kingdom,* Reinboth, R., Ed., Springer-Verlag, Berlin, 1975, 392.

11. **Short, R. V., Smith, J., Mann, T., Evans, E. P., Hallet, J., Fryer, A., and Hamerton, J. L.,** Cytogenetics and endocrine studies of a freemartin heifer and its bull co-twin, *Cytogenetics,* 8, 369, 1969.

12. **Nagai, Y., Ciccarese, S., and Ohno, S.,** The identification of human H-Y antigen and testicular transformation induced by its interaction with the receptor site of bovine fetal ovarian cells, *Differentiation,* 13, 155, 1979.

13. **Noegel, A., Gerisch, G., Stadler, J., and Westphal, M.,** Complete sequence and transcript regulation of a cell adhesion protein from aggregating *Dictyostelium* cells, *EMBO J.,* 5, 1473, 1986.

14. **Hemperly, J. J., Murray, B. A., Edelman, G. M., and Cunningham, B. A.,** Sequence of a cDNA clone encoding the polysialic acid-rich and cytoplasmic domains of the neural cell adhesion molecule N-CAM, *Proc. Natl. Acad. Sci. U.S.A.,* 83, 3037, 1986.

Chapter 3

EXPLORING THE ROLE OF THE GENE, *SEX-LETHAL*, *(Sxl)*, IN THE GENETIC PROGRAMMING OF *DROSOPHILA* SEXUAL DIMORPHISM

Thomas W. Cline

TABLE OF CONTENTS

I. A "TEXTBOOK PROBLEM"

In the study of *Drosophila* development, the question of how sex is determined has been pursued longer than any other. The rapid pace of progress in this area over the past few years can be gauged from a comparison between the work reviewed in this chapter and a summary statement that was added to a popular genetics text when that text was last revised in 1984:

> The genetic analysis of sex determination and dosage compensation in *Drosophila* is currently an area of active research. A large number of mutants affecting the development of sexual phenotype and/or the sex-specific activity of genes on the X-chromosome have been identified. An understanding of the interrelated actions of these genes has not yet been attained. Surprisingly, a more coherent picture currently exists of how sex determination and dosage compensation are achieved during mammalian development.[1]

This text revision mirrored the state of the field that was presented in a comprehensive review of *Drosophila* sex determination published in 1980.[2] Pieces of the sex-determination puzzle were just beginning to fall into place when that review was being written. The view of the genetic programming of fruit fly sexual dimorphism presented in Baker and Belote's review[3] of the same subject published just 3 years later is far more encouraging, particularly with respect to the classical "sex determination genes" that regulate sexual differentiation. Progress since then has continued apace.

The discussion of sex determination presented in the pages that follow centers on the X-linked gene, *Sex-lethal (Sxl)*. Analysis of this gene perhaps more than any other has led to a highly integrated view of *Drosophila* sex determination and dosage compensation, since it is these very processes that Sxl^+ integrates in vivo. It has helped the field move beyond an outline (still incomplete) of the regulatory gene hierarchy that leads to sexual dimorphism — the question of which gene controls which — toward an understanding of fundamental qualitative differences in the nature of the regulatory roles of these genes in this process.

A difficulty with the experimental basis for this new understanding is that it relies not only on sophisticated tools of *Drosophila* developmental genetics forged over nearly a century of work, but also on a large number of highly specialized genetic tools custom designed for the analysis of one particular aspect of this organism's development. As a consequence, it has become progressively more difficult to communicate even to the specialist the experimental logic, details, and criticism on which rapid progress in this area depends. In some respects, the situation is coming to resemble that of phage lambda genetics. On the other hand, the phage lambda experience should give hope to the nonspecialists that their patience eventually will be rewarded by the elucidation of elegant biological principles that are accessible and generally applicable. Even before that point, however, it is hoped that a perspective on *Drosophila* may stimulate those working on sex determination in other organisms to think about their own experimental systems in new and productive ways.

II. THE HISTORICAL CONTEXT: AWAY FROM COHERENCE AND BACK AGAIN

The picture of *Drosophila* sex determination was quite clear in 1921 just after Bridges showed that *Drosophila* sex was determined by the ratio of the number of X chromosomes to the number of sets of autosomes,[4] a parameter known as the X/A balance. Male development follows when the X/A balance is 0.5, female when it is 1.0. Bridges showed that the sexually ambiguous intermediate value of 0.67 leads to a sexually intermediate phenotype. From this he proposed that sexual development was a consequence of a balance between genes on the autosomes that "tend towards the production of male characters" and genes on the X that promote female development. Soon systematic attempts were made to discover the identity of the "sex factors" that "promoted" male vs. female development.[5] At about the same time, Sturtevant[6] discovered the first of a number of mutations that behaved in a simple Mendelian fashion and profoundly affected sexual phenotype. For decades, however, it remained unclear how these Mendelian "sex genes" were related to the X/A balance.

For understanding the primary events of sex determination in *Drosophila*, it is essential to distinguish clearly among (1) genetic elements, whose relative dose actually represents the X/A balance; (2) genetic elements, whose products function to transduce that balance signal; and (3) regulatory genes that respond to the signal, either primarily or secondarily, to elicit sexual differentiation. Since a clear appreciation of the distinction between regulatory and structural genes was only developed in the 1960s through the study of microorganisms, it is understandable that none of the classical studies of sex determination were carried out within such a conceptual framework. Classically trained geneticists thought of the gene in terms of the enzyme that it specified, and of development as a complex physiological process involving a network of biochemical pathways. The distinction between gene products controlling other genes and gene products controlling metabolic substrates is nontrivial: a change in the view one takes can completely reverse how one interprets a series of epistatic interactions among loss-of-function mutations.

Historically, mutations affecting sex determination were commonly lumped into a single vague category variously referred to as "sex factors", "sex determiners", or "sex genes".[7] Goldschmidt did not follow this practice, but he went to the other extreme of dismissing most such mutations as irrelevant "modifiers". His attitude also reflected the fact that enhancers and suppressors, so important to the progress of microbial genetics, had not been particularly useful in the study of *Drosophila* development:

Among innumerable examples (all the inhibitors, suppressors, enhancers in genetic literature) are the accurately studied dominance modifiers for vestigial wings in *Drosophila*. One of them . . . is the mutant locus purple, so called for its effect on eye color. Shall we now say that purple is one of the wing-determining loci, or is it not rather true that the action of purple collides at some point with one of the processes or reaction chains which take place in wing determination? This collision may mean a common substrate or precursor substance or condition for the kinetics of the reaction. Exactly the same situation is known from the realm of sex determination . . . Thus it is not surprising that in *Drosophila* species a number of mutant loci which alone produce intersexuality have been found; further, that combinations of ordinary mutants exist which make the carriers intersexual . . . Whatever their specific features — and these are expected, since they probably act through different coordinates of the developmental system — their interpretation is always the same in principle.[8]

Missing from these early attempts to understand sex determination was an appreciation of perhaps the most important point of all: the relationship between sex determination and X chromosome dosage compensation. The latter is an essential process, the disruption of which has organismal and even cell-lethal consequences.[9,10] It is ironic that the first mutation in the gene that was eventually shown to link sex determination to dosage compensation was reported in 1960 by Müller himself,[11] the individual who discovered dosage compensation. Twenty years later, in the 1980 review on sex determination mentioned above, this gene was still not viewed as relevant to either process. The problem was that organismal and cell lethality arising from the genetic link between these two processes masked the sex-transforming effects of perturbations in the primary process of sex determination.[12,13] As a consequence, key genes operating in the primary process of sex determination were first identified not by their effects on sexual phenotype, but rather by their differential effects on male and female viability.

Müller discovered dosage compensation by comparing the phenotypes of leaky mutant alleles as a function of gene dose and X/A balance.[14] As a consequence of dosage compensation, diploid males with one copy of an X-linked gene achieve the same level of the gene's function as diploid females with two copies of the same gene. Both copies of the gene in females are expressed, but the level of expression per copy in females is half that in males.[9,10]

The fact that the classical sex-transforming mutations had no effect on dosage compensation may have delayed discovery of the link between sex determination and dosage compensation. The conclusion that followed from this fact, namely that dosage compensation is not simply an aspect of sexual physiology per se but is instead an independent process, may have deflected attention away from the idea of a link and of the possible relevance of sex-specific lethal mutations. Apparently not considered was the point that if the extant "sex-determination" mutants *had* affected dosage compensation, it is unlikely they would have been identified as sex-determination genes in the first place. Obscuring the situation even further was the fact that the two most attractive theories for how a level of X chromosome gene expression appropriate to the number of X chromosomes might be achieved required no sex-specific — or even X/A balance specific — elements.[15] Two reports in 1979 dealing with sex-specific lethal mutations in the *Sxl* gene provided the first indication that sex determination and the vital process of dosage complication were linked to the functioning of a single gene with their control only diverging downstream of *Sxl*.[12,13] The results of these two studies suggested that *all* cells of the organism (at least all cells that must dosage compensate) must "know their sex", not just those in tissues that appear overtly different between males and females.

III. COPING WITH LETHAL SEX CHANGES

From an experimental standpoint, there are advantages and disadvantages to the vital link between sex determination and dosage compensation. One can take advantage of viability effects to devise extremely powerful positive selection methods for the isolation of new mutant alleles of sex-determination genes and for their characterization at the genetic fine-structure level.[16] In the study of pleiotropic genes that have one role in the primary process of sex determination and another in some other vital process that is not sex specific, one can often focus on the sex-specific aspect of a gene's function separately from the other functions by following sex ratios alone;[17,18] in effect, the less affected sex serves as the control for aspects of the phenotype that are not related to sex determination.

On the other hand, viability effects are the crudest measure of upsets in development. In contrast, effects on sexual phenotype can be a rich source of information (see Section IV). In order to be able to see effects on sexual differentiation of perturbations that affect both processes, some means must be used to overcome viability effects arising as a consequence of accompanying dosage compensation upsets. It is possible to devise approaches that exploit the fact that dosage compensation is a very special class of "cell vital" process. Its disruption leads not to the loss of essential gene products, but rather to an imbalance of gene products. The ability of cells and whole organisms to tolerate such an imbalance can be expected to depend on its magnitude and duration, on the extent to which cells suffering from an imbalance must compete in the same developmental compartment with cells that are not, and on the tissue type in which the imbalance occurs.[19]

The first approach exploited genetic mosaics. XX/XO sexual mosaics produced by early loss of X chromosomes from XX zygotes allowed individuals to survive otherwise lethal effects because the dosage compensation upsets occurred in only a fraction of the cells of the animals.[12] Genetic mosaics produced by radiation-induced mitotic recombination survived sex-specific lethal effects not just because only a small fraction of the animals' cells were affected, but also because even those cells were affected for only a part of their developmental history.[13,20] These two kinds of mosaics first led to the discovery of *Sxl*'s role in sex determination.

A second approach used to cope with lethality arising from dosage compensation upsets was to study mutations in a situation in which one could anticipate that the *magnitude* of the genetic imbalance caused by perturbations in genes controlling dosage compensation would be less than in the standard diploid situation, and thus, would be tolerated better. Triploid intersexes (XX AAA) met this condition and were used to demonstrate directly that the maternal-effect lethal gene, daughterless (da), is involved in progeny sex determination.[21]

The first and second approaches mentioned above are limited in the number of experimental individuals that can be generated conveniently. The next two approaches are affected far less by this problem.

The third approach reduces upsets that arise in mutant diploid females from overexpression of X-linked genes by eliminating the activity of some of the genes responsible for that overexpression, i.e., genes downstream of those that control the primary sex-determination process. These are the autosomal male-specific lethals (msl's).[22] They are required in males for the elevation of X-linked gene expression that is appropriate for cells with an X/A balance of 0.5, but they are not involved in sex determination per se.[23] Soon after the isolation of these msl's, it was observed that they could interact with female-lethal mutations in other genes to masculinize females.[24,25] Subsequently, it was demonstrated that masculinization can result from the partial rescue of diplo-X cells or, in some cases, entire individuals that fail to express Sxl[+] properly:[26] female-specific lethal effects of upsets in the primary sex-determination process appear to stem from expression of such genes in an inappropriate (X/A = 1) situation.

A puzzling aspect to these apparent sex-transforming interactions was why the known msl's only partially suppressed viability effects of female-specific lethals in most situations. There were a variety of reasons to favor the explanation that incomplete rescue of diplo-X cells was a reflection of the fact that the known male-specific ''dosage-compensation genes'' did not control as many aspects of dosage compensation (whether as many genes, or in as many tissues, or at as many points in development) as the female-specific genes directing them from upstream in the regulatory hierarchy.[26] The validity of this interpretation has been confirmed recently by Gergen who demonstrated that genes (Sxl and da) upstream of the known msl's are responsible for dosage compensation of the blastoderm-stage-specific segmentation gene, runt, but the msl's are not.[27]

The msl's are not useful in the reciprocal situation — for the rescue of haplo-X (male) cells that express their dosage-compensation machinery in the female mode. An alternative strategy in this situation may be to interfere with the growth of neighboring haplo-X cells that are compensating properly, thereby reducing the competitive disadvantage suffered by the haplo-X cells, the dosage compensation of which is faulty.[28]

A fourth approach that has been used to overcome the lethal effects of perturbations in the primary sex-determination process utilizes mutations in Sxl that affect its sex determination and dosage-compensation functions differentially. For example, the mutant allele $Sxl^{fm7,M1}$ eliminates the feminizing activity of Sxl, but leaves the gene with sufficient female dosage-compensation function to support survival to the adult stage (survival requires two doses of the mutant allele alone, or one dose in combination with the msl's).[26] Because the mutant allele lacks feminizing activity, any feminization that is observed in a particular experimental situation must arise from the expression of some other Sxl allele that is present — but the diplo-X individual will survive regardless. A combination of the third and fourth approaches led to the discovery of the positive autoregulatory aspect to Sxl[+] function.[26]

In view of the efforts that had been required to reveal what appeared to be cryptic masculinizing effects of female-lethal mutations in genes that act in primary sex-determination steps, one aspect of a recent report claiming the identification of a zygotic component of the X/A signal seemed unexpected.[29] A masculinizing interaction between Sxl[−] and a deletion of this putative X/A element was observed in females without the need for any additional mutations to rescue masculinized cells from the upsets in dosage compensation one might have anticipated. However, subsequent findings have cast doubt on the proposition that this region does in fact contain such an X/A signal element;[30,31] the sex transformation observed instead seems more likely to arise from interference with *Sxl* function at a somewhat later point in development (see Section V.B).[28]

IV. WHERE SHOULD ONE LOOK FOR ABNORMAL SEX?

Arguments have been advanced for why the sexually dimorphic region of the first leg that gives rise to the distinctive male "sex comb" is optimal for assaying genetic perturbations in sexual phenotype.[32] In this region, the phenotypic difference between the sexes involves the number, orientation, and morphology of bristles, each shaft of which represents the differentiated product of a single cell. Thus, sexual phenotype in the foreleg can be assessed unambiguously at the level of individual cells. It is important to remember that here, as in all fruit fly epidermal structures, sexual differentiation is a cell-autonomous process: events taking place within one cell that affect its sexual phenotype do not influence the sexual phenotype of adjacent cells.[33]

The particular significance of the foreleg's cellular sexual dimorphism is that it allows one to distinguish easily between two qualitatively different types of intersexual phenotypes, and this difference can indicate which step of sexual development is perturbed in a particular experimental situation: (1) "true" or "cellular" intersexes in which individual cells display an abnormal sexual phenotype intermediate between normal male and female; and (2) "mosaic" intersexes which are sexually intermediate at the level of the whole individual, but at the level of individual cells are composed of a mixture of phenotypically normal male and normal female structures. Information on the timing of developmental events can be deduced from the degree of interspersion of male and female or of normal and abnormal cells.[26]

V. A COHERENT VIEW OF THE *DROSOPHILA* SEX DETERMINATION REGULATORY GENE HIERARCHY

Figure 1 summarizes our current view of the placement of *Sxl* in the regulatory gene hierarchy that controls the development of sexual dimorphism. *Sxl* can be thought of as a master feminizing gene: when its functions are expressed in a cell, that cell will follow the female pathway of development, regardless of its X/A balance. When those functions are not expressed, the cell will develop male by default.

This conclusion follows from the reciprocal phenotype of two opposite sex-specific lethal classes of *Sxl* mutations.[34] Loss-of-function mutations behave as female-specific lethals because they transform genotypic females into phenotypic males.[26] Males are unaffected by such mutations; they are viable and fertile even if they lack *Sxl* DNA.[35] The reciprocal phenotype is exhibited by gain-of-function (constitutive) mutations[12,35] which transform genetic males (X/A = 0.5) into phenotypic females, generally killing them in the process. Genotypic females homozgyous for a male-lethal gain-of-function mutant allele develop normally. The gain-of-function mutation identifies *Sxl* as a key "switch gene," the product of which not only is necessary for dictating a particular developmental alternative, but also is sufficient (everything else being equal).

It is useful to consider three regulatory levels in what is loosely termed "sex determination": sexual pathway initiation, sexual pathway maintenance, and sexual pathway expression.[32,35] The first level, sexual pathway initiation, includes the process by which cells assess their X/A balance during the embryonic period and become stably committed to the male or female pathway of development. In some cells this occurs long before any overt sexual differentiation. It seems proper to restrict the term "primary sex determination" in its strictest sense in *Drosophila* to this level of the overall process. The second level, sexual pathway maintenance, includes the process by which cells maintain their sexually determined state during the period of cell division prior to metamorphosis, independent of the X/A signal

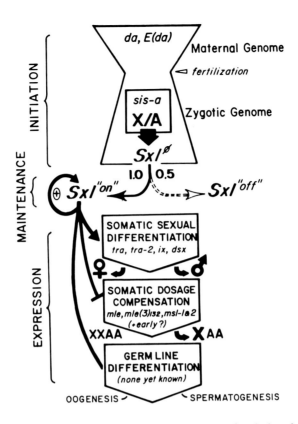

FIGURE 1. A general scheme to account for the functioning of
regulatory genes known to operate in the control of *Drosophila*
sexual dimorphism. The scheme accommodates differences among
these genes with respect to the specific aspects of differentiation
that are controlled and the cellular level of that control: initiation,
maintenance, or expression of the sexual pathway commitment.
The schematic of the initiation step is designed to emphasize distinctions among genetic elements of the X/A signal itself, the primary genetic target of the signal, and genes that are not part of the
signal per se, but instead function to transduce that signal to affect
the primary target's activity state. This model is oversimplified
insofar as it does not reflect possible differences in the control of
Sxl between germline and soma.[43,44] All genes shown except $E(da)$[57]
are discussed in the text. Details of the functional relationships
among genes in the "somatic sexual differentiation" box are presented in References 51 and 58.

that initiated that state. The third level, sexual pathway expression, includes the processes
by which cells differentiate according to the sexual pathway commitment that they have
made and maintained. Such differentiation may be as subtle as a slight difference in cell
growth rate, or as overt as the production of sex-specific proteins such as the vitellogenins,
or organs such as the male accessory glands. The differentiation may be terminal or not; it
may be reversible or not. In its broadest sense, sexual pathway expression can include dosage
compensation as well as sexual differentiation per se.

These three processes are temporally and genetically distinguishable. Moreover, they are
all under the control of a single gene, *Sxl*. The actions of this gene in coordinating all three
levels of the sex-determination process and in functioning from the earliest to the latest
stages of development are some of the features that distinguish it from all other elements
known to operate in the sex-determination gene hierarchy.[35,36] Mutations in *Sxl* that affect

these functions differentially are especially useful for generating a holistic understanding of sexual development in *Drosophila*.

A. Sexual Pathway Initiation

There is now a variety of evidence (not yet reported fully in the literature) that cells do in fact assess their X/A balance only during the embryonic period, many cell divisions prior to sexual differentiation. Two groups have reported experiments designed to change the X/A balance from 1 to 0.5 at various times during development with results that are consistent with, but do not unambiguously demonstrate, a heritable sexual pathway commitment.[3,37] Ironically, this approach, which would seem rather straightforward, requires the most indirect arguments to support the hypothesis, as a consequence of lethality accompanying the X/A balance change. Experiments of this type have suffered from the lack of an unequivocal measure of change in X chromosome number, and/or lack of clone recovery in regions where sexual phenotype can be assessed with certainty.

We used alternative approaches that were not complicated by cell lethality to establish that cells become committed early to a particular sexual pathway and can maintain that commitment over a large number of cell divisions. The more indirect evidence involved a comparison of phenotypes between triploid intersexes (X/A = 0.67) and gynandromorphs[26] — animals that are mosaics of XX and XO cells as a consequence of X chromosome loss from XX zygotes soon after fertilization. The more direct evidence for a heritable sexual pathway commitment involved a test of the sexual potential of cells in mosaic intersexes during development. A preliminary report of this approach has appeared[32] and subsequent results and a variety of important controls not yet published have confirmed that the precursor cells of the adult foreleg become committed to the male or female pathway at least by the time they resume growth during the early larval period (growth of these cells is suspended after the first half of embryogenesis), and can maintain that commitment over as many as 13 cell divisions.[28]

Several features should be characteristic of the genetic elements involved in the sexual pathway initiation step. The elements should act either as positive or negative regulators of Sxl^+. They should act on Sxl^+ only early in development if their function is restricted to the pathway initiation process. They should have a dramatic effect on sexual phenotype of triploid intersexes (XX AAA), animals whose X/A balance is ambiguous. Lastly, intersexual phenotypes generated by mutations in these elements should be of the mosaic, rather than the cellular type.

The genes *da* and sisterless-a (*sis-a*) appear to meet these criteria.[17,18,21,34,38] These criteria are also met by a partial-loss-of-function mutation in *Sxl* itself: Sxl^{f9} appears to be defective only in this early process of sexual pathway initiation.[26,35,38] Sxl^{fLS} may represent the complementary class of allele defective in all functions except this early one.[35,38,39]

Once the roles of some elements in the sexual pathway initiation step are established, they can be used as tools to fit new mutations into this picture of sex determination. In effect, each attempt to fit a new genetic element into the growing network of observations in a consistent and coherent fashion constitutes an extremely detailed test of the model. Apparent departures from expectations identify those aspects of the model which deserve special attention and are the observations most likely to lead to a refinement of the overall picture.

For understanding the X/A signal and how it acts in sexual pathway initiation, it seems useful to distinguish two classes of genetic elements: (1) "signal elements", the relative dose of which represents the X/A balance per se; and (2) "signal transduction elements", the products of which are required to transduce this dose difference so that it affects the probability of a cell stably activating the feminizing functions of Sxl^+. An operational distinction between these two classes of elements relies on analysis of the relationship between gene dose and phenotype. For example, components of the numerator of the X/A balance signal can be expected not only to interfere with female (XX AA) development when decreased in dose, but also to interfere with male development when increased in dose. From classical studies, one anticipates that signal elements may be part of a polygenic, additive system;[40] hence, an increase in the dose of one element may be compensated for by a corresponding decrease in another. Since it is the X/A balance of the zygote that determines sex, X/A signal elements should be strictly zygotic in their action, not maternally acting. These distinctions place the maternally acting gene, *da*, into the "signal transduction" class,[18] and the gene *sis-a* into the "signal element" class,[28,32] as shown in Figure 1.

B. Sexual Pathway Maintenance and Positive Autoregulation

A distinction between perturbation of sexual pathway initiation steps and disruption of steps in sexual pathway maintenance — the cells' "sexual memory" — can be made on the basis of developmental timing. Effects on initiation steps will affect the probability of cells choosing the male or female pathway, but once the choice is made, that pathway should be maintained in a heritable fashion throughout development. In contrast, perturbation of sexual pathway maintenance steps should break down that heritable restriction. Incomplete disruption of sexual pathway maintenance steps should generate mosaic intersex phenotypes *unless* the breakdown happens to occur immediately prior to sexual differentiation, in which case perdurance (persistence of gene product in the absence of further synthesis) could result in an intermediate level of feminizing product that would generate an individual (or cells) with a cellular intersex phenotype. Cellular intersexes are produced when functional Sxl^+ alleles are removed from foreleg precursor cells by somatic recombination at the end of the larval stage immediately prior to foreleg bristle differentiation,[28] or when flies mutant for temperature-sensitive alleles of *Sxl* or *tra-2* are shifted to the restrictive temperature at about the same point in development.[28,35,41] In contrast to disruptions of sexual pathway initiation, disruption of sexual pathway maintenance in females could be expected to allow some masculinized diplo-X cells to survive as a consequence of their having had vital wild type Sxl^+ dosage compensation function during a significant fraction of their developmental lifetime.

The partial-loss-of-function allele, Sxl^{fPb}, has been shown to cause a breakdown in somatic sexual pathway maintenance.[35] The possibility that the zygotic function of an element in the 4F5-11 region of the X may also be involved in sexual pathway maintenance is currently being tested.[28,30] The results of this test will bear on the validity of the proposal that upsets in the primary sex-determination process — sexual pathway initiation — are not expected to produce a visible sex transformation in the absence of additional mutations to ameliorate effects on dosage compensation.

How the feminizing mode of Sxl^+ expression is maintained independent of the initiating X/A balance signal is an intriguing question. Some time ago it was shown that the *Sxl* gene appeared able to act in *trans* to elicit the feminizing functions of an Sxl^+ allele that would not otherwise be expressed.[26] That demonstration was in females in which the X/A signal was necessarily correct for Sxl^+ expression. If this unusual *positive* autoregulatory function of *Sxl* acts to maintain the sexual pathway commitment, it should be possible to demonstrate such transactivation activity even when the X/A signal itself is inappropriate for normal Sxl^+ activation; moreover, it should be possible to demonstrate that transactivation can occur relatively late in development, long after the sexual pathway initiation steps normally would be complete. These predictions are fulfilled by a complex mutant allele, $Sxl^{M1,fPa-ra}$ [35] the feminizing functions (including autoregulation) of which depend on temperature, rather than on the X/A balance.[28]

C. Sexual Pathway Expression in Soma and Germline

Expression of the sexual pathway commitment has been shown to occur at various points in development, many quite late, depending on when sexual differentiation takes place in a particular tissue.[41] Intersexes generated by disruption of these steps are of the cellular, rather than the mosaic class. Sexual pathway expression results from the action of Sxl^+ on at least the four "classical" sex determination genes downstream: *tra, tra-2, ix,* and *dsx.* The assignment of *Sxl* to a position upstream of these genes is based in part on the epistatic relationships among pairs of mutations. For example the feminizing effect of the gain-of-function allele, Sxl^{M1}, is blocked by the masculinizing action of a loss-of-function *tra* mutation.[12] Conversely, the transformation of females to cellular intersexes by loss-of-function mutations in *dsx* is epistatic to the masculinizing effect of $Sxl^{fm7,M1}$ [28] an allele which has lost the normal feminizing activity of Sxl^+.[26] Moreover, the cellular intersex phenotypes that can be generated by low-level expression of *Sxl*'s feminizing functions in females by Sxl^{2593} [26] and in males by $Sxl^{M1,fPa-ra}$ [35] and the extreme sensitivity of these intersexual phenotypes to the dose of tra^+ and $tra-2^+$,[28] are fully consistent with our understanding of the relationship between Sxl^+ and the four downstream sex-differentiation regulatory genes shown in Figure 1.

Among the set of genes known to control sexual pathway expression, *Sxl* is unique as the only one that functions in the germ line in a fashion even remotely consistent with its somatic functions:[42-44] It is required for the development of female but not male germ cells. On the other hand, little is known regarding the regulation of *Sxl* or its specific developmental role in this tissue type. Practically nothing is known with respect to germline dosage compensation.[45] It has been shown that diplo-X germ cells that lack Sxl^+ are *not* equivalent to haplo-X germ cells;[44] moreover, alleles of *Sxl* that disrupt its germline but not its somatic functions do not display an obviously male germline phenotype even when the soma is completely masculinized by the mutation *tra.*[46] Thus, it is possible that *Sxl* in the germline functions at a later step in the sex-determination process — perhaps only in a maintenance or an expression step — and that other genes (perhaps *ovo?*[31]) have a more primary regulatory role in this tissue at an earlier point. Mutant alleles of Sxl^+ that affect only germline development may be useful in identifying these other genes.[46-48]

VI. MOLECULAR ANALYSIS OF *Sxl*: WHAT MUST THE X/A SIGNAL DO?

In order to understand the mechanism by which the X/A signal acts we must know what exactly it does, at the molecular level, to its target, *Sxl.* Although genetic analysis indicates that Sxl^+ is functionally on when X/A = 1 and off when X/A = 0.5, molecular analysis shows that after the first few hours following fertilization, the distinction between "on" and "off" is not the distinction between the presence or absence of transcription per se, but seems instead to be a difference in the pattern of splicing of the primary transcripts.[35,49,50] Thus, sexual pathway initiation appears to involve the eventual establishment of one of two alternative stable models of Sxl^+ transcript processing: (female) a pattern of three female-specific transcripts that are synthesized from the first half of embryogenesis through the adult stage (plus one additional germline-dependent female transcript), or (male) a pattern of three slightly larger male-specific mRNA's that overlap the female-specific transcripts.

Molecular information is also available on the genes such as *tra* and *dsx* which *Sxl* regulates.[51-54] In these cases as well, sex-specific regulation appears to be at the level of transcript processing. Thus, it seems that *Sxl* may control itself (through autoregulation) and the genes downstream of it by the same mechanism: RNA splicing. Consistent with this possibility is the recent discovery of an amino acid sequence homology between the conceptual translation product of an adult female-specific Sxl^+ cDNA and the products of genes encoding proteins of mammalian heterogeneous nuclear ribonucleoprotein particles,[50] proteins implicated in RNA processing.[55]

Although the adult transcript pattern of Sxl^+ is present during most of development, the pattern very early is quite different. After about the second hour of development when the maternally synthesized Sxl^+ transcripts have disappeared, three "embryonic" transcripts of as yet unknown sex specificity appear transiently.[35,49] How the synthesis of these early mRNA's is controlled, and how this leads to the later, stable pattern of mRNA's is obviously of considerable importance with respect to sexual pathway initiation and the mode of action of the X/A signal. Analysis of the effects of intragenic deletions on Sxl functioning suggests that the promoter for at least one product acting early in the sexual pathway initiation step, and the promoter for at least one Sxl germline function, may be different from the promoter for the sexual pathway expression functions.[48] The process by which a small quantitative difference in relative chromosome number (the X/A balance) leads to a striking qualitative difference in gene expression and development is clearly complex. Analysis of the effects of various mutations upstream of Sxl and in Sxl itself on Sxl's transcription pattern should help in understanding this process.

While genetic and developmental analyses certainly facilitate molecular studies, the converse is particularly true for Sxl. For example, inferences from the interaction of the gain-of-function allele, Sxl^{M1}, with other mutations in the sex-determination pathway depend in important ways on our knowledge of the specific molecular nature of the mutation responsible for this allele's constitutive behavior and its effects on the timing and level of feminizing products. The initial autoregulation studies which were based on mutant derivatives of Sxl^{M1} led to the suggestion that transactivation of Sxl^+ alleles might occur only early in development, a result inconsistent with a role of positive feedback in sexual pathway maintenance and inconsistent with subsequent studies (see Section V.B). It may be possible to reconcile these two different sets of experiments in light of molecular studies suggesting that Sxl^{M1} might have a considerably higher level of expression during the embryonic period than even the wild-type allele;[35] hence, if a mutant derivative of Sxl^{M1} is relatively ineffective at transactivation of Sxl^+, one could expect to see such a marginal process taking place only during the early period when the expression of the mutant allele is unusually high. Structural analysis of Sxl^{M1} showing that its behavior is due to the insertion of a transposon (and not just any foreign DNA) into a particular region of the gene also provides a plausible explanation for the observation that this "constitutive feminizing" allele fails to impose the female rate of dosage compensation on the blastoderm-stage-specific gene, $runt$.[27,39]

Molecular analysis of the dosage-compensation genes downstream of Sxl may be extremely helpful in studies of the genetic and developmental link between this process and sex determination. Studies of dosage compensation, particularly those involving triploid intersexes, have been limited by the fact that dosage-compensation levels cannot be measured in single cells. A thorough understanding of Sxl regulation requires that one be able to assess dosage compensation and sex determination in the same cells, and be able to do this in moribund animals. It may be possible to accomplish this with cloned molecular probes if the expression of dosage compensation genes downstream of Sxl, like that of Sxl, tra, and dsx, has an aspect that is qualitatively different between the sexes.

VII. FILLING OUT THE NETWORK

Many of the genes shown in Figure 1 were discovered fortuitously, rather than from systematic searches for components of the sex-determination system. On the other hand, it could be argued that systematic searches for such elements would have been premature until our understanding of those products of good fortune advanced to its present state. In any event, experience with the pleiotropic gene, da, provides a clear illustration of importance of designing systematic screens for regulatory elements so that they not exclude elements with essential functions in more than one vital regulatory process.[56] Screens based on dominant interactions with known elements of the network would seem most likely to meet this requirement, and such screens appear to be available now for elements at all three levels of the sex-determination regulatory network.[28,32,50,56] Even so, considerable effort may be required to determine whether the modifiers identified in such screens have a primary or only a secondary involvement in the processes of interest.

Synergism between molecular and genetic analysis of *Drosophila* sex determination seems likely to lead to an exponential increase in information available on this subject. Indeed gene regulation networks elaborated in model organisms such as yeast, nematodes, and fruit flies may soon resemble the metabolic pathways charts that were once ubiquitous in biological labs. This may arouse in some a nostalgia for the time when the picture of fruit fly sex determination was coherent simply because it had to include only a single basic fact: Bridges' X/A balance.

ACKNOWLEDGMENTS

I thank I. Greenwald for helpful comments on this manuscript. The unpublished work of references 28, 49, and 50 was supported by grants from the U.S. NIH to T.W.C. and to P. Schedl.

REFERENCES

1. **Ayala, F. J. and Kiger, J. A.,** *Modern Genetics,* 2nd ed., Benjamin/Cummings, Menlo Park, Calif., 1984, 564.
2. **Laugé, G.,** Sex determination, in *The Genetics and Biology of Drosophila,* Vol. 2, Ashburner, M. and Wright, T. R. F., Eds., Academic Press, New York, 1980, chap. 29.
3. **Baker, B. S. and Belote, J. M.,** Sex determination and dosage compensation in *Drosophila melanogaster, Annu. Rev. Genet.,* 17, 345, 1983.
4. **Bridges, C. B.,** Triploid intersexes in *Drosophila melanogaster, Science,* 54, 252, 1921.
5. **Dobzhansky, T. and Schultz, J.,** Evidence for multiple sex factors in the X-chromosome of *Drosophila melanogaster, Proc. Natl. Acad. Sci. U.S.A.,* 17, 513, 1931.
6. **Sturtevant, A. H.,** Intersexes in *Drosophila simulans, Science,* 51, 325, 1920.
7. **Gowen, J. W.,** Biologic basis of sex, in *Sex and Internal Secretions,* Vol. 1, 3rd ed., Young, W. C. and Corner, G. W., Eds., Robert E. Krieger, Huntington, N.Y., 1973, chap. 1.
8. **Goldschmidt, R. B.,** *Theoretical Genetics,* University of California Press, Berkeley, 1955, 449.
9. **Lucchesi, J. C. and Manning, G.,** Genetics of dosage compensation in *Drosophila, Adv. Genet.,* 24, 371, 1988.
10. **Jaffe, E. and Laird, C.,** Dosage compensation in *Drosophila, Trends in Genet.,* 2, 316, 1986.
11. **Müller, H. J. and Zimmering, S.,** A sex-linked lethal without evident effect in *Drosophila* males but partially dominant in females, *Genetics,* 45, 1001, 1960.
12. **Cline, T. W.,** A male-specific lethal mutation in *Drosophila* that transforms sex, *Dev. Biol.,* 72, 266, 1979.
13. **Cline, T. W.,** A product of the maternally influenced Sex-lethal gene determines sex in *Drosophila melanogaster, Genetics,* 91 (Suppl), s22, 1979.
14. **Müller, H. J.,** Evidence for the precision of genetic adaptation, *Harvey Lecture Series,* 43, Charles C Thomas, Springfield, Ill., 1950.
15. **Stewart, B. and Merriam, J.,** Dosage compensation, in *The Genetics and Biology of Drosophila,* Vol. 2, Ashburner, M. and Wright, T. R. F., Eds., Academic Press, New York, 1980, chap. 30.
16. **Cline, T. W.,** Positive selection methods for the isolation and finestructure mapping of cis-acting homeotic mutations at the Sex-lethal locus of *D. melanogaster, Genetics,* 97(Suppl), s23, 1981.
17. **Cline, T. W.,** Maternal and zygotic sex-specific gene interactions in *Drosophila melanogaster, Genetics,* 96, 903, 1980.
18. **Cronmiller, C. and Cline, T. W.,** The relationship of relative gene dose to the complex phenotype of the daughterless locus in *Drosophila, Dev. Genet.,* 7, 205, 1986.
19. **Ripoll, P.,** Effect of terminal aneuploidy on epidermal cell viability in *Drosophila melanogaster, Genetics,* 94, 135, 1980.

20. **Sanchez, L. and Nothiger, R.,** Clonal analysis of Sex-lethal, a gene needed for female sexual development in *Drosophila melanogaster, Wilhelm Roux' Arch. Entwicklungsmech Org.,* 191, 211, 1982.

21. **Cline, T. W.,** The interaction between daughterless and Sex-lethal in triploids: a lethal sex-transforming maternal effect linking sex determination and dosage compensation in *Drosophila melanogaster, Dev. Biol.,* 95, 260, 1983.

22. **Belote, J. M. and Lucchesi, J. C.,** Male-specific lethal mutations of *Drosophila melanogaster, Genetics,* 96, 165, 1980.

23. **Belote, J. M.,** Male-specific lethal mutations of *Drosophila melanogaster.* II. Parameters of gene action during male development, *Genetics,* 105, 881, 1983.

24. **Skripsky, T. and Lucchesi, J. C.,** Intersexuality resulting from the interaction of sex-specific lethal mutations in *Drosophila melanogaster, Dev. Biol.,* 94, 153, 1982.

25. **Uenoyama, T., Uchida, S., Fukunaga, A., and Oishi, K.,** Studies on the sex-specific lethals of *Drosophila melanogaster.* V. Sex transformation caused by interactions between a female-specific lethal, Sxl^{f1}, and the male-specific lethals $mle(3)132$, $msl-2^{27}$, and *mle, Genetics,* 102, 233, 1982.

26. **Cline, T. W.,** Autoregulatory functioning of a *Drosophila* gene product that establishes and maintains the sexually determined state, *Genetics,* 107, 231, 1984.

27. **Gergen, J. P.,** Dosage compensation in *Drosophila:* evidence that daughterless and Sex-lethal control X-chromosome activity at the blastoderm stage of embryogenesis, *Genetics,* 117, 477, 1987.

28. **Cline, T. W.,** unpublished data, 1987.

29. **Steinmann-Zwicky, M. and Nöthiger, R.,** A small region on the X chromosome of *Drosophila* regulates a key gene that controls sex determination and dosage compensation, *Cell,* 42, 877, 1985.

30. **Cline, T. W.,** Reevaluation of the functional relationship in *Drosophila* between a small region on the X chromosome (3E8-4F11) and the sex-determination gene, Sex-lethal, *Genetics,* 116 (Suppl.), s12, 1987.

31. **Oliver, B., Perrimon, N., and Mahowald, A. P.,** The *ovo* locus is required for sex specific germ line maintenance in *Drosophila melanogaster, Genetics,* 116 (Suppl.), s15, 1987.

32. **Cline, T. W.,** Primary events in the determination of sex in *Drosophila melanogaster,* in *Origin and Evolution of Sex,* Halvorson, H. O. and Monroy, A., Eds., Alan R. Liss, New York, 1985, 301.

33. **Tokunaga, C.,** Cell lineage and differentiation on the male foreleg of *Drosophila melanogaster, Dev. Biol.,* 4, 489, 1962.

34. **Cline, T. W.,** Two closely linked mutations in *Drosophila melanogaster* that are lethal to opposite sexes and interact with daughterless, *Genetics,* 90, 683, 1978.

35. **Maine, E. M., Salz, H. K., Schedl, P., and Cline, T. W.,** Sex-lethal, a link between sex determination and sexual differentiation in *Drosophila melanogaster, Cold Spring Harbor Symp. Quant. Biol.,* 50, 595, 1985.

36. **Tompkins, L.,** Genetic control of sexual behavior in *Drosophila melanogaster, Trends Genet.,* 2, 14, 1986.

37. **Sanchez, L. and Nöthiger, R.,** Sex determination and dosage compensation in *Drosophila melanogaster:* production of male clones in XX females, *EMBO J.,* 2, 485, 1983.

38. **Cline, T. W.,** A female-specific lethal lesion in an X-linked positive regulator of the *Drosophila* sex determination gene, Sex-lethal, *Genetics,* 113, 641, 1986.

39. **Maine, E. M., Salz, H. K., Cline, T. W., and Schedl, P.,** The Sex-lethal gene of *Drosophila:* DNA alternations associated with sex-specific lethal mutations, *Cell,* 43, 521, 1985.

40. **Dobzhansky, T. and Schultz, J.,** The distribution of sex factors in the X-chromosome *Drosophila melanogaster, J. Genet.,* 28, 233, 1934.

41. **Belote, J. M. and Baker, B. S.,** Sex determination in *Drosophila melanogaster:* analysis of transformer-2, a sex-transforming locus, *Proc. Natl. Acad. Sci. U.S.A.,* 79, 1568, 1982.

42. **Schüpbach, T.,** Autosomal mutations that interfere with sex determination in somatic cells of *Drosophila* have no direct effect on the germline, *Dev. Biol.,* 89, 117, 1982.

43. **Cline, T. W.,** Functioning of the genes daughterless and Sex-lethal in *Drosophila* germ cells, *Genetics,* 104 (Suppl.), s16, 1983.

44. **Schüpbach, T.,** Normal female germ cell differentiation requires the female X-chromosome-autosome ratio and expression of Sex-lethal in *Drosophila melanogaster, Genetics,* 109, 529, 1985.

45. **Bachiller, D. and Sanchez, L.,** Mutations affecting dosage compensation in *Drosophila melanogaster:* effects in the germline, *Dev. Biol.,* 118, 379, 1986.

46. **Kolodny, M. R. and Cline, T. W.,** unpublished data, 1986.

47. **Perrimon, N., Mohler, D., Engstrom, L., and Mahowald, A. P.,** X-linked female-sterile loci in *Drosophila melanogaster, Genetics,* 113, 695, 1986.

48. **Salz, H. K., Cline, T. W., and Schedl, P.,** Functional changes associated with structural alterations induced by mobilization of a P element inserted in the Sex-lethal gene of *Drosophila, Genetics,* 117, 221, 1987.

49. **Salz, H. K., Maine, E. M., Samuels, M. E., Cline, T. W., and Schedl, P.,** unpublished data 1987.

50. **Bell, L. R., Maine, E. M., Schedl, P., and Cline, T. W.,** unpublished data, 1987.
51. **Belote, J. M., McKeown, M. B., Andrew, D. J., Scott, N. T., Wolfner, M. F., and Baker, B. S.,** Control of sexual differentiation in *Drosophila melanogaster, Cold Spring Harbor Symp. Quant. Biol.,* 50, 605, 1985.
52. **Butler, B., Pirrotta, V., Irminger-Finger, I., and Nothiger, R.,** The sex-determining gene *tra* of *Drosophila:* molecular cloning and transformation studies, *EMBO J.,* 5, 3607, 1986.
53. **McKeown, M., Belote, J. M., and Baker, B. S.,** A molecular analysis of transformer, a gene in *Drosophila melanogaster* that controls female sexual differentiation, *Cell,* 48, 489, 1987.
54. **Baker, B. S., Nagoshi, R. N., and Burtis, K. C.,** Molecular genetic aspects of sex determination in *Drosophila, Bioessays,* 6, 66, 1987.
55. **Swanson, M. S., Nakagawa, T. Y., LeVan, K., and Dreyfuss, G.,** Primary structure of human nuclear ribonucleoprotein particle C proteins: conservation of sequence and domain structures in heterogeneous nuclear RNA, mRNA, and pre-rRNA-binding proteins, *Mol. Cell. Biol.,* 7, 1731, 1987.
56. **Cronmiller, C. and Cline, T. W.,** The *Drosophila* sex determination gene daughterless has different functions in the germ line versus the soma, *Cell,* 48, 479, 1987.
57. **Mange, A. P. and Sandler, L.,** A note on the maternal effect mutants daughterless and abnormal oocyte in *Drosophila melanogaster, Genetics,* 73, 73, 1973.
58. **Nöthiger, R. and Steinmann-Zwicky, M.,** Sex determination in *Drosophila, Trends Genet.,* 1, 209, 1985.

Chapter 4

ORIGIN AND EVOLUTION OF SEX CHROMOSOMES IN AMPHIBIA: THE CYTOGENETIC DATA

Michael Schmid and Thomas Haaf

TABLE OF CONTENTS

I. INTRODUCTION

A. General Comments

Twenty years ago the classic monograph of Susumu Ohno,[1] *Sex Chromosomes and Sex-Linked Genes,* was published. The cytogenetic data available at that time led to the generalization that sex chromosomes in fishes, amphibians, and reptiles are still in a very primitive state of morphological differentiation. This meant that in the heterogametic sex, the X and Y or Z and W chromosomes had identical morphology and could not be identified by microscopy. Exceptions to this rule were found in the females of some snake species belonging to the families Colubridae, Crotalidae, Elapidae and Viperidae.[2,3] The contention that in the lower vertebrate classes the sex chromosomes were identical morphologically was supported in 1949 by one of the founders of comparative cytogenetics, Robert Matthey.[4]

The early studies on amphibian chromosomes were performed with the conventional methods of classical cytogenetics (squash preparations and uniformly stained chromosomes). Advances in chromosome technology made it possible to improve considerably the quality of the preparations. Banding techniques allowed a more precise analysis of the fine structure of chromosomes and the identification of sex chromosomes. Chromosomal homologies among different species and chromosome rearrangements could be recognized using multiple banding patterns (Q-, G-, or R-bands). Constitutive heterochromatin could be differentially labeled by C-banding or staining with AT- and GC-base pair specific fluorochromes. Furthermore, methods were developed allowing demonstration of the sequence of DNA replication in the chromosomes (BrdU-banding patterns), and specific staining of active nucleolus organizers or centromeres, etc. Finally, *in situ* nucleic acid hybridization to metaphase chromosomes inaugurated what now is called molecular cytogenetics.

Modern cytogenetic techniques were developed primarily on human chromosomes and later applied with great success to many other mammalian species. In comparison with the enormous number of reports dealing with banded mammalian chromosomes, there are only a few completed studies on amphibians, fishes, and reptiles. However, it has become possible to demonstrate definitively that heteromorphic sex chromosomes also developed during the evolution of amphibian genomes. Although the number of species with cytologically detected sex chromosomes is small when compared with the other vertebrate classes, it can be assumed that further examples will be found.

B. Phylogeny and Classification of the Amphibia

According to the current view of vertebrate evolution, *Ichthyostega* from the late upper Devonian period is one of the first amphibian-like tetrapods. *Ichthyostega* represents a transitional stage between the crossopterygian fishes of the suborder Rhipidistia and the early amphibians. The adaptations necessary for a terrestrial way of life (atmospheric breathing, prevention of dessication, tetrapod locomotion) were accomplished by the Amphibia during the Carboniferous and Permian ages. In these periods, several groups differentiated (Ichthyostegalia, Rhachitomi, Stereospondyli, Embolomeri, Lepospondyli) with the Lepospondyli considered the ancestors of the modern Amphibia.[5,6] Unfortunately, very few fossils exist from the time between the well-documented Paleozoic records and the recent Amphibia. Therefore, it has not been possible to reconstruct the immediate ancestry of the Amphibia. The abundant development of species among the recent amphibians is thought to have begun in the Tertiary period. The Amphibia in existence today are classified into three orders (Apoda, Urodela, and Anura) comprising about 28 families, 320 genera, and 3000 species. The Apoda (caecilians, 160 species) occur exclusively in the tropical zones of America, Africa, and Asia. The Urodela (salamanders and newts, 350 species) are restricted to the temperate and subtropical regions of the Northern Hemisphere. In contrast, the Anura (frogs and toads, 2500 species) are almost ubiquitous.[7]

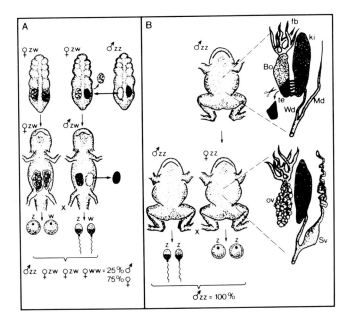

FIGURE 1. (A) Schematic representation of the sex-reversal experiment demonstrating female heterogamety in the Mexican axolotl (*Ambystoma mexicanum*). A detailed explanation is given in the text. The granular tissue in the female embryos represents the primordium of the ovary, and the black tissue represents the primordium of the testis. The cross of a normal female (left) with a sex-reversed female (right) must yield a progeny of 25% males and 75% females if the female sex is heterogametic (ZW). (B) Representation of the sex-reversal experiment in the European toad *Bufo bufo* demonstrating male homogamety. Further explanation in the text. The structure of the gonads in the normal and sex-reversed male is shown in the drawings on the right. Bo: Bidder's organ; fb: fat body; ki: kidney; Md: Müllerian duct; ov: ovary; Sv: seminal vesicle; te: testis; Wd: Wolffian duct. The cross of a normal male (left) with a sex-reversed male (right) yields only male progeny if the male sex is homogametic (ZZ).

II. SEX-DETERMINING MECHANISMS IN AMPHIBIA

A. Homomorphic Sex Chromosomes

As described above, the early pioneering work on amphibian karyotypes failed in the demonstration of differentiated sex chromosomes or yielded contradictory results.[8] Moreover, as no sex-linked genes with their characteristic mode of inheritance were known in the amphibians, other approaches were chosen to reveal the type of sex-determining mechanisms in these organisms. These experiments were time-consuming and difficult, but offer a most appealing and convincing method of proof.

The first successful demonstration of a sex-determining mechanism with noncytogenetic methods was made by Humphrey[9-11] for the salamanders *Ambystoma mexicanum* and *A. tigrinum*. The experiment is presented in Figure 1A. The primordium of the ovary was excised from one side of the body of a genetically female embryo and replaced by the primordium of a testis taken from a genetically male embryo. Since at this stage of embryogenesis (tail-bud stage) the sex could not be recognized, the appropriate combination (male primordium into female embryo) was achieved in approximately a quarter of all transplantations.

During further development of the embryo, the transplanted testis hormonally modified the contralateral ovary into a functional testis. As soon as this condition was achieved, the transplanted testis was removed. In this way, genetically female embryos were induced to develop into male animals (sex-reversed females) with a testis producing sperm cells from genetically female germ cells. When these sex-reversed females were mated with normal females, the ratio of their male and female progeny was 1:3. This result could be obtained only if, as shown in Figure 1A, the normal *Ambystoma* female is the heterogametic sex (ZW) and the male, the homogametic sex (ZZ).

Among the female progeny, two genotypes (ZW and WW), which cannot be differentiated phenotypically, must occur. Indeed, one third of the female animals (WW individuals) when crossed with normal males (ZZ) produced only female progeny (ZW). The fact that the ZW and WW females are phenotypically identical is a strong indication that the sex chromosomes of this species are not only morphologically but also genetically largely homologous; the only genes in which the Z and the W chromosomes ought to differ are the opposing sex-determining factors.[1]

Sex-reversed males also permit demonstration of the type of sex-determining mechanism. Thus, a similar experiment was performed on the toad *Bufo bufo*. The technique is summarized in Figure 1B. In the sexually mature male toad, the Bidder's organ is located in the anterior part of the testis. This tissue is the incompletely involuted cortex of the embryonic gonad; it has been compared to a rudimentary ovary. Furthermore, the Müllerian duct has been conserved. When the testes of an adult toad were removed, the Bidder's organ developed into a functional ovary and the Müllerian ducts enlarged. When a sex-reversed male was mated with a normal male, only male progeny ensued. This could be explained if, as represented in Figure 1B, the male *Bufo* is homogametic (ZZ) and the female, heterogametic (ZW).

In the clawed frog, *Xenopus laevis*, sex reversal was readily induced by treating developing larvae with estradiol. All tadpoles exposed to estradiol developed as functional females whatever their sex chromosome constitution. Half of the female animals (ZZ) when crossed with normal males (ZZ) produced only male progeny. In this way, female heterogamety could be demonstrated for *X. laevis*.[12-14]

The examples so far described have shown the occurrence of a ZW/ZZ sex-determining mechanism. However, there are species in which breeding experiments show a mechanism of the XX/XY type. Thus, analyses of the sex ratio in the progeny of sex-reversed females and parthenogenetically bred animals demonstrated that the males are heterogametic in the following species of Asiatic frogs: *Rana nigromaculata*, *R. japonica*, *R. brevipoda*, *R. rugosa*, *Hyla arborea japonica*, and *Bombina orientalis*.[15]

The cell-membrane-associated H-Y antigen or a cross-reactive antigen is found in the gonad of the heterogametic (XY or ZW) sex in all vertebrates so far studied.[16-18] It can be inferred that this antigen is the product of a gene that has been extremely conserved during vertebrate evolution, dating from a common ancestor of fishes, amphibians, reptiles, birds, and mammals.

Because H-Y antigen is found in the heterogametic (XY or ZW) sex, it can function as a useful tool in evaluating the type of sex-determining mechanism (XX/XY or ZZ/ZW) in those amphibian species in which no heteromorphic sex chromosomes have evolved or in which sex-reversal experiments have not yet been done. In this way, the heterogametic sex was deduced in five species of Amphibia (Tables 1 and 2).

A summary of the species of Urodela and Anura in which the type of sex-determining mechanism has been assigned by sex-reversal and breeding experiments or by H-Y antigen typing, is given in Tables 1 and 2. Although these species constitute a very small sample, it becomes clear that in the primitive and the highly evolved families of the Urodela and Anura XX/XY and ZZ/ZW sex-determining types coexist. Within the same genus, however, the type appears to be uniform. This indicates that during the evolution of Urodela and Anura, the XX/XY − or ZZ/ZW type of sex determination was established after the various ancestors of the recent genera had diverged. For the order Apoda, breeding experiments or H-Y antigen-typing studies have not yet been performed.

Table 1
SEX-DETERMINING MECHANISMS AND SEX-SPECIFIC CHROMOSOMES FOUND IN THE ANURA

Family	Species	Genetic sex[a]	Sex chromosomes[b]	Morphology[c]	Ref.
Pipidae	*Xenopus laevis*	ZZ/ZW			12—14, 16
Discoglossidae	*Bombina orientalis*	XX/XY			15
Ranidae	*Rana brevipoda*	XX/XY			15
	R. clamitans	XX/XY			54
	R. esculenta		XX/XY	X = Y; rbp	19, 20
	R. japonica	XX/XY			15
	R. nigromaculata	XX/XY			15
	R. pipiens	XX/XY			55
	R. ridibunda	XX/XY			59
	R. rugosa	XX/XY			15
	R. temporaria	XX/XY			60
	Pyxicephalus adspersus	ZZ/ZW	ZZ/ZW	Z > W	25—27
	P. delalandii		ZZ/ZW	Z = W; ci; ch	25
Hylidae	*Hyla arborea japonica*	XX/XY			15
	Gastrotheca riobambae		XX/XY	X < Y	37—39
	G. pseustes		XX/XY	X = Y; ch	Present report
Bufonidae	*Bufo bufo*	ZZ/ZW			15, 27, 58
Leptodactylidae	*Eupsophus migueli*		XX/XY	X = Y; ci	42

[a] Determined by analysis of the sex ratio in the progeny of sex-reversed animals, parthenogenetically bred individuals, patterns of inheritance, and expression of isozymes encoded by the sex chromosomes, or H-Y antigen typing.
[b] Determined by cytogenetic studies.
[c] X = Y or Z = W: both sex chromosomes have the same size but differ in their replication banding patterns (rbp), centromeric index (ci), or amount of constitutive heterochromatin (ch). X < Y: X is smaller than Y; Z > W: Z is larger than W.

B. Heteromorphic Sex Chromosomes
1. Initial Stages of Differentiation

An amphibian species in which the XY/XX chromosomes were demonstrated in a very primitive stage of differentiation is the European water frog, *Rana esculenta*.[19] In this species, which belongs to the advanced amphibians, male heterogamety was determined early on by breeding experiments.[20] The precondition for the demonstration of the Y chromosome was a technique based on the incorporation of the thymidine analog, bromodesoxyuridine, into the replicating chromosomal DNA during the synthesis (S) phase of the cells (BrdU-replication banding). With this method, the "homomorphic" chromosome pair no. 4 of *R. esculenta* could be identified as sex chromosomes. All males have an extremely late replicating band in the long arm of the Y (Figure 2a, b, e) which is lacking in the X (Figure 2e). In this species, the sex chromosomes could not be distinguished by other banding techniques (Figure 2d). This is the first example of sex chromosomes in vertebrates in which heteromorphism could be recognized exclusively by a difference in the replication patterns. The actual cause of the asynchrony in the replication of the small band in the Y must be a specific DNA sequence. Since repetitive DNA sequences generally begin their replication in the S phase later than the other regions, it is likely that the late-replicating Y band consists of such repetitive DNA.

Some of the sex chromosomes so far found in Amphibia indicate that heterochromatini zation of the Y or W precedes the actual morphologic differentiation. The European newts of the genus *Triturus* are highly evolved Urodela with 24 meta- to submetacentric chro-

Table 2
SEX-DETERMINING MECHANISMS AND SEX-SPECIFIC CHROMOSOMES
FOUND IN THE URODELA

Family	Species	Genetic sex[a]	Sex chromosomes[b]	Morphology[c]	Ref.
Proteidae	*Necturus alabamensis*		XX/XY	X > Y	35
	N. beyeri		XX/XY	X > Y	35
	N. lewisi		XX/XY	X > Y	35
	N. maculosus		XX/XY	X > Y	34, 35
	N. punctatus		XX/XY	X > Y	35
Ambystomatidae	*Ambystoma mexicanum*	ZZ/ZW			9—11
	A. tigrinum	ZZ/ZW			9—11
Salamandridae	*Pleurodeles waltlii*	ZZ/ZW	ZZ/ZW	Z = W; lbc	47—49, 53
	P. poireti		ZZ/ZW	Z = W; lbc	47—49
	Triturus alpestris	XX/XY	XX/XY	X = Y; ch	21, 23
	T. cristatus		XX/XY	X = Y; ch	22
	T. helveticus		XX/XY	X = Y; mm	21
	T. italicus		XX/XY	X = Y; ch	23
	T. marmoratus		XX/XY	X = Y; ch	22
	T. vulgaris	XX/XY	XX/XY	X = Y; ch	21, 23, 27
Plethodontidae	*Aneides ferreus*		ZZ/ZW	Z = W; ci	43
	Chiropterotriton abscondens		XX/XY	X > Y	28—33
	C. bromeliacea		XX/XY	X > Y	28—33
	C. cuchumatanos		XX/XY	X > Y	28—33
	C. rabbi		XX/XY	X > Y	28—33
	Bolitoglossa subpalmata		XX/XY	X > Y	28—33
	Hydromantes ambrosii		XX/XY	X = Y; ci; ch	24
	H. flavus		XX/XY	X = Y; ci; ch	24
	H. imperialis		XX/XY	X = Y; ci; ch	24
	H. italicus		XX/XY	X = Y; ci; ch	24
	H. sp. nova		XX/XY	X = Y; ci; ch	24
	Oedipina bonitaensis		XX/XY	X > Y	28—33
	O. poelzi		XX/XY	X > Y	28—33
	O. syndactyla		XX/XY	X > Y	28—33
	O. uniformis		XX/XY	X > Y	28—33
	Thorius pennatulus		XX/XY	X > Y	28—33
	T. subitus		XX/XY	X > Y	28—33

[a] Determined by analysis of the sex ratio in the progeny of sex-reversed animals, parthenogenetically bred individuals, patterns of inheritance, and expression of isozymes encoded by the sex chromosomes, or H-Y antigen typing.

[b] Determined by cytogenetic studies.

[c] X = Y or Z = W: both sex chromosomes have the same size but differ in their centrometric index (ci), amount of constitutive heterochromatin (ch), loop patterns in lampbrush chromosomes (lbc), or pairing arrangement in male meiosis (mm). X > Y: X is larger than Y.

mosomes. After conventional staining, the 12 chromosome pairs appear homomorphic. The XY sex chromosomes can only be recognized by specific staining of the constitutive heterochromatin.[21,22] In *Triturus alpestris* and *T. vulgaris* the males have one heteromorphic chromosome pair, of which only one homolog (Y) displays heterochromatic telomeres in the long arms, whereas the telomeres of the other homolog (X) are euchromatic (Figure 3a to c). The same results have been obtained for the heteromorphic long arms of the XY pair no. 2 of *T. italicus*.[23] In the males of *T. cristatus* and *T. marmoratus*, the telomeres of the long arms in the chromosome pairs no. 4 are heteromorphic.[22] In the Y, there is always more heterochromatin than there is in the X (Figure 4). Of special interest is *T. helveticus* in which no heteromorphism could be demonstrated in the male karyotypes (Figure 3d).

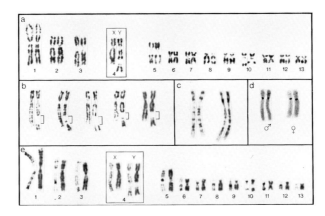

FIGURE 2. (a) Male karyotype of the European water frog *Rana esculenta* showing the replication banding patterns obtained by BrdU treatment of the cells. The XY chromosome pair no. 4 is framed. Note the late replicating band in the long arm of the Y. Due to the incorporation of BrdU, the condensation of this region is distorted, so that the long arm of the Y appears longer than that in the X. (b) Time sequence of replication bands in the XY pair from early (left) to the very late replication stages (right). Note that in all stages the asynchrony of replication between X and Y is restricted to the late-replicating band in the Y (small brackets). (c) Synchronous replication banding patterns in the XX chromosomes from two females. (d) C-banded sex chromosome pairs no. 4 from a male and a female specimen. (e) The 13 chromosome pairs of *R. esculenta* exhibiting early (left) and late (right) replication banding patterns.

However, as described in Section III, in the meiosis of male animals, the long arms of pair no. 5 show a highly decreased frequency of chiasmata formation. It is conceivable that in the Y long arm of *T. helveticus* repetitive DNA sequences are already concentrated, although no heterochromatin was detected.[21]

In the newly discovered species of the South American marsupial frog *Gastrotheca pseustes* from Ecuador, two different forms of the Y chromosome were found.[61] In some male animals, the sex chromosome pairs no. 5 (XY_a) exhibit the same pattern of constitutive heterochromatin as the XX chromosomes of the females (Figure 5a, b). In contrast, all other males from this species show a distinctly heterochromatic telomere in the long arm of the Y_b chromosome (Figure 5c, d). This example clearly shows that several distinct stages of morphological differentiation can exist within a single species of Amphibia.

A slight difference in the morphology of the X and Y was recently observed in the European salamanders *Hydromantes italicus, H. ambrosii, H. imperialis, H. flavus,* and *H. sp. nova.*[24] In the karyotypes of all females both homologs of pair 14 are acrocentric with identical C-banding patterns (Figure 6). In the males, one homolog 14 (X) is acrocentric and indistinguishable from the female chromosomes no. 14, while the other homolog (Y) is submetacentric (Figure 6). The X possesses heterochromatin close to the centromere in the long arm and at the centromere itself, whereas in the Y, the heterochromatin is located at the centromere and within the short arm (Figure 6).

2. Advanced Stages of Differentiation

The first certified highly heteromorphic sex chromosomes in Anura were discovered in the South African bull frog *Pyxicephalus adspersus.*[25-27] All males examined had ZZ sex chromosomes and all females had ZW sex chromosomes. The W chromosome is considerably smaller than the Z chromosome and its short arm is completely heterochromatic (Figure 7a,

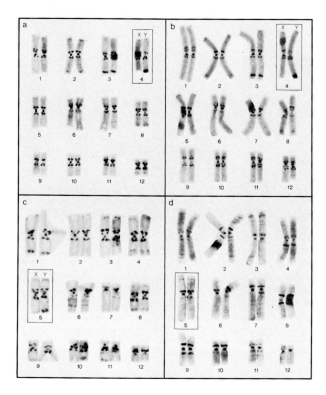

FIGURE 3.   C-banded male karyotypes of the European newts. (a, b) *Triturus alpestris,* (c) *T. vulgaris,* and (d) *T. helveticus* showing the regions of constitutive heterōchromatin. The XY chromosomes are framed. In (a, b) *T. alpestris* and (c) *T. vulgaris,* the heteromorphism between X and Y consists of a heterochromatic region present at the telomeres in the long arm of the Y but absent in the X. In (d) *T. heleveticus,* no sex-specific heteromorphism of the constitutive heterochromatin can be detected in the sex chromosome pair no. 5; however, the long arms of these chromosomes have a distinctly decreased chiasmata frequency in male meiosis (see Figure 11).

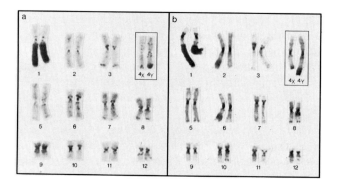

FIGURE 4.   C-banded karyotypes of the European newts (a) *Triturus cristatus carnifex* and (b) *T. marmoratus.* The telomeric heterochromatin in the long arm of the Y is considerably larger than in the X. (From Sims, S. H., *Chromosoma,* 89, 169, 1984. With permission.)

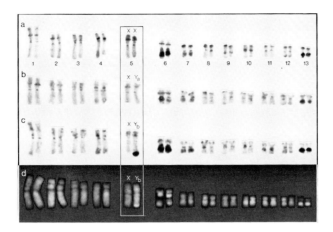

FIGURE 5. Karyotypes of the South American marsupial frog *Gastrotheca pseustes*. The sex chromosome pairs no. 5 are framed. (a to c) C-banding of the constitutive heterochromatin. All females possess homomorphic XX chromosomes. Among the males there are individuals with unconspicuous Y_a chromosomes, and others with telomeric heterochromatin in the long arm of the Y_b chromosome. (d) Quinacrine mustard banding. Note the bright quinacrine fluorescence of the telomeric heterochromatin in the Y_b.

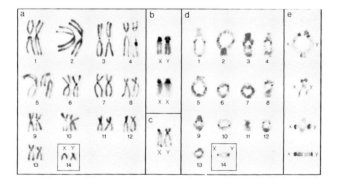

FIGURE 6. (a) C-banded male karyotype of the European salamander *Hydromantes italicus* showing the sites of constitutive heterochromatin. The sex chromosome pair no. 14 is framed. (b) Enlarged XY/XX sex chromosomes of *H. italicus* and (c) *H. flavus*. (d) C-banded bivalents from male meiosis of *H. flavus* and (e) four examples of XY pairing configurations in the diakineses of *H. italicus*. Note that the X and Y show only chiasmata in an interstitial or distal region of their long arms. (From Nardi, I., *Chromosoma*, 94, 377, 1986. With permission.)

b). There is evidence that the same chromosome pair no. 8, which in *P. adspersus* represents the highly heteromorphic ZW pair, is still in an initial stage of morphologic differentiation in the closely related *P. delalandii*. Although the chromosomes no. 8 of *P. delalandii* are still of the same length, they differ from each other by a pericentric inversion and by the amount of heterochromatin.[25]

The American salamanders of the family Plethodontidae have been cytogenetically analyzed extensively by James Kezer, who has studied the species of most of the 23 genera of this family, Unfortunately, most of Kezer's results are still unpublished, although they have been the subject of many bibliographic references.[28-33] Well-differentiated XY sex chro-

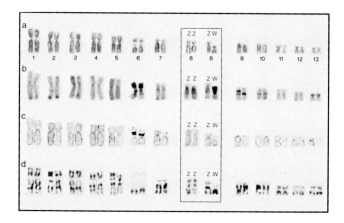

FIGURE 7. Karyotypes of the South African bull frog *Pyxicephalus
adspersus*. The sex chromosome pairs no. 8 are framed. (a) Conventional
orcein staining showing homomorphic ZZ chromosomes in the male and
highly heteromorphic ZW chromosomes in the female. (b) C-banding of
the constitutive heterochromatin. The W chromosome has a completely
heterochromatic short arm. (c) Silver staining showing specific labeling
of the nucleolus organizer regions in the short arms of chromosome pair
no. 6. (d) Chromosomes with replication banding patterns obtained by
BrdU treatment of the cells. Note the synchronous replication patterns in
the homologous autosomes and in the ZZ chromosomes of the male and
the very late replicating short arm of the W in the female.

mosomes characterize the males of several species of the genera *Chiropterotriton, Oedipina,
Thorius,* and *Bolitoglossa* (Table 1). These four genera are spread from Mexico to South
America and are grouped into the subfamily Bolitoglossinae due to common taxonomic
characteristics. The best phylogenetic interpretation is that the common ancestor of the recent
Bolitoglossinae possessed the XX/XY mechanism of sex determination. However, since the
structure of the sex chromosomes of *Thorius* and *Chiropterotriton* was found to be different
from that of *Oedipina* and *Bolitoglossa,* it is also possible that there were independent lines
of XY differentiation within these salamanders.

The five species of the salamander genus *Necturus* have the most highly differentiated
sex chromosome heteromorphism yet discovered in the Urodela.[34,35] These neotenic, per-
ennibranch, and permanently aquatic salamanders with larval morphology belong to the
primitive family Proteidae and inhabit the lakes and streams of eastern North America. The
genomes of the *Necturus* species are among the largest of the vertebrates (165 pg DNA per
diploid cell in *N. maculosus*).[36] All species (*N. alabamensis, N. beyeri, N. lewisi, N.
maculosus,* and *N. punctatus*)[35] have 19 pairs of extremely large chromosomes and highly
differentiated XY sex chromosomes (Figure 8a). The Y chromosomes are about one quarter
the size of the X chromosomes and composed almost completely of constitutive heterochromatin (Figure 8a, b).

Exceptional XY sex chromosomes were found in the South American marsupial frog
Gastrotheca riobambae.[37-39] These frogs, belonging to the highly evolved family Hylidae,
are very specialized such that the back skin of the females is folded to form a pocket into
which the fertilized eggs are placed; here, the embryos spend a large part of their larval
development. The sex chromosomes of *G. riobambae* are highly heteromorphic; the Y is
the largest element in the karyotype and almost completely heterochromatic (Figure 9a, b).
This is one of the very few examples of a vertebrate species having a Y larger than the X.
Another peculiarity in the karyotype of *G. riobambae* is the nucleolus organizer region which
is located on the short arm of the X. Since nucleolus organizers are X linked, there is a

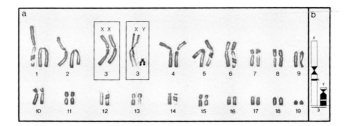

FIGURE 8. (a) Conventionally stained female karyotype of the North American salamander *Necturus maculosus* showing the homomorphic XX sex chromosome pair no. 3. The highly heteromorphic XY sex chromosomes from a C-banded metaphase of a male have been included. Note the almost completely heterochromatic Y. (b) Diagrammatic representation of the C-banded XY chromosomes. The arrowhead designates a secondary constriction. (From Sessions, S. K., *Chromosoma*, 77, 157, 1980. With permission.)

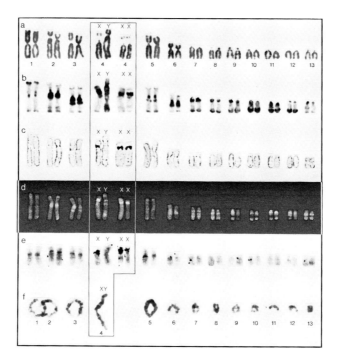

FIGURE 9. Karyotypes of the South American marsupial frog *Gastrotheca riobambae*. The sex chromosome pairs no. 4 are framed. (a) Conventional orcein staining. In the males the sex chromosomes are heteromorphic (XY), whereas in the females they are homomorphic (XX). The Y is the largest chromosome in the karyotype and the nucleolar constriction is located in the short arm of the X. (b) C-banding of the karyotype demonstrating the almost completely heterochromatic Y. (c). Silver staining showing the nucleolus organizer region in the short arm of the X. (d) Mithramycin staining. Note the bright mithramycin fluorescence of the nucleolus organizer regions. (e) Autoradiograph of chromosomes after *in situ* hybridization with ^{3}H-18S + 28S ribosomal DNA. The only site of rDNA hybridization is the nucleolus organizer in the short arm of the X. (f) Orcein-stained bivalents from male meiosis showing end-to-end pairing between the XY chromosomes.

dose ratio of 1:2 in male and female with respect to the number of 18S + 28S ribosomal genes (Figure 9c to e). Sex-specific numbers of nucleolus organizers have been demonstrated only in one other vertebrate, the bat *Carollia perspicillata*.[40,41]

In the Amphibia, as well as in the other vertebrates, heteromorphism of sex chromosomes can be restricted to isolated populations, as shown by the following examples. *Eupsophus roseus* is a highly evolved leptodactylid frog from the temperate forests of southern Chile. No heteromorphic sex chromosomes could be found in this frog. However, in a local population restricted to a small coastal valley (*E. migueli*), differentiated XY sex chromosomes were found in male animals.[42] The Y was a small metacentric element the same size as the telocentric X. The plethodontid salamander genus *Aneides* comprises five species widely distributed in temperate North America. *Aneides ferreus* is found along the west coast from northern California to northern Oregon, with an isolated population on Vancouver Island in British Columbia (Figure 10). Extensive cytogenetic studies showed that in the populations of Vancouver Island and northern California two different modes of ZZ/ZW sex determination operate, whereas in the Oregon population, sex-specific chromosomes were not identified. This variability of ZZ/ZW sex chromosomes is explained in detail in Figure 10.

C. Evolution of Sex Chromosomes

The evolution of heteromorphic sex chromosomes from one originally homomorphic chromosome pair was not the result of a single structural change but developed instead in several subsequent steps.[1] Since evolutionary processes cannot easily be reproduced experimentally, the individual changes taking place in the course of sex chromosome evolution can only be reconstructed by means of comparative studies on different species. In this case, one would assume that several of the initial stages of differentiation of the sex chromosomes have been conserved in some primitive vertebrates Thus, it becomes possible to create evolutionary series in which the increasing structural complexity of the sex chromosomes can be reconstructed. Beçak et al.[2] recognized for the first time an evolutionary series of this kind in the karyotypes of snakes, which possess W chromosomes in three different stages of differentiation. In the snakes of the primitive family Boidae, all chromosome pairs of male and female are still homomorphic. In the specialized family Colubridae, there are many species in which females exhibit heteromorphic ZW sex chromosomes; although the W chromosomes are still of the same length as the Z chromosomes, they differ from the latter by a pericentric inversion. Finally, in the highly evolved families Crotalidae and Viperidae, the W chromosomes are reduced to very small heterochromatic elements similar to the W chromosomes in birds. The hypothesis on the origin of heteromorphic sex chromosomes, derived from these findings, oriented all further work.[1] According to this hypothesis, the primary step in the differentiation of Z and W, or X and Y is the creation of an isolation mechanism that partly or completely prevents free meiotic crossing over between these chromosomes. An inversion, as found in the W chromosome of the colubrid snakes, is thought to be the cause of such an isolation mechanism. The isolation would provide the condition for the differential accumulation of opposing sex-determining genes on the originally homologous chromosomes. The pericentric inversion does not inhibit crossovers, but it does create a situation in which the complete lack of crossing-over is of great selective advantage. If a crossover should occur within the inverted segment, gametes with partially deleted and duplicated chromosomes would be produced. Following the development of the pericentric inversion on the W or Y chromosome, heterochromatinization occurs, whereas the structure of the Z and X chromosomes is largely preserved. Finally the W and Y chromosomes are reduced to small heterochromatic elements by successive deletions.[1]

More recent studies on constitutive heterochromatin and a satellite DNA associated with the W chromosome of snakes indicate that the primary step in the differentiation of the W

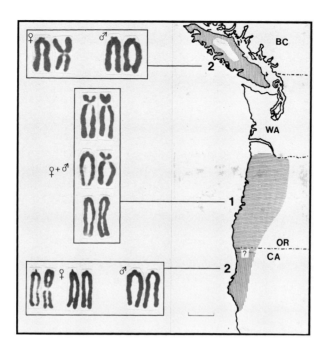

FIGURE 10. Conventionally stained chromosome pairs no. 13 of the North American salamander *Aneides ferreus* and the geographic distribution of its different karyomorphs. The bar represents 100 km. The *A. ferreus* population 1 occurs in western Oregon and northern California. This population is polymorphic for a pericentric inversion in the chromosome pair no. 13 which is either telocentric or submetacentric. The population is in Hardy-Weinberg equilibrium with respect to the three alternative forms of chromosome pair 13. The *A. ferreus* population 2 has a widely disjunct distribution. One group is found on Vancouver Island in British Columbia. In this population, all females are heterozygous for a telocentric and a metacentric chromosome 13 (ZW sex chromosomes), whereas males are homozygous for telocentric chromosomes no. 13 (ZZ sex chromosomes). The other group of the *A. ferreus* population 2 inhabits northern California. In this area, all males are homozygous for the telocentric chromosome no. 13, but the females may be either identical to males or heterozygous for a telocentric and a metacentric chromosome no. 13. It is unknown whether this *A. ferreus* population 2 contacts the *A. ferreus* population 1 in northern California. (From Sessions, S. K., *Chromosoma*, 95, 17, 1987. With permission.)

chromosome was not pericentric inversion, but the development of specific repetitive DNA sequences (heterochromatinization).[44-46] Thus, the *in situ* hybridization technique revealed a specific satellite DNA located on the highly differentiated W chromosomes of the advanced snake families. The same W-associated DNA was also demonstrated in those highly evolved snake species that still possess undifferentiated homomorphic ZW chromosomes. Therefore, the pericentric inversion and the deletions seem to be the results and not the causes of sex chromosome differentiation.[44]

The various, hitherto known sex chromosomes of Amphibia support the assertion that the initial step in the differentiation of sex chromosomes was an accumulation of repetitive DNA in the W and Y chromosomes. Thus, in the primitive Y chromosomes of the genus *Triturus* and of *Gastrotheca pseustes* the only visible difference between X and Y chromosome is a very small heterochromatic band in the Y. Of special importance is the even more primitive Y of *Rana esculenta,* which differs from the X by only a single late-replicating euchromatic band. Singh et al.[44] have proposed that the sex chromosome-associated satellite DNA in the primitive homomorphic W of snakes could bring about asynchrony in the DNA replication pattern of Z and W and thus, reduce the frequency of crossing-over between them, which is the prerequisite of sex chromosome differentiation. Therefore, it should be determined whether such repetitive DNA sequences are enriched in the late-replicating Y band of *R. esculenta*.

In the more advanced Y and W chromosomes of the Amphibia, inversions are already present as shown by examples from the genera *Hydromantes* and *Aneides* or from *Pyxicephalus delalandii* and *Eupsophus migueli*. Finally, most of the highly evolved Y and W chromosomes were reduced to small, almost completely heterochromatic elements (*Necturus, Pyxicephalus adspersus*) very similar to the Y and W chromosomes in mammals and birds. The Y- and W-linked genes, which were once the alleles of the X- and Z-linked genes, were lost in the course of heterochromatinization. Since this process was very slow, a regulatory system compensating for the monosomy of the X- and Z-linked genes in the heterogametic sex could have evolved simultaneously. However, those genes determining the development of the heterogametic sex could not be permitted to be affected by the heterochromatinization. Nothing can as yet be said about the location of these sex-determining genes within the Y and W chromosomes; it is possible that they are situated in the few and small euchromatic regions still preserved in these chromosomes.

It is important to note that in Amphibia, the stage of differentiation of the sex chromosomes is independent of the evolutionary (taxonomic) status of the species. Thus, in the primitive families, as well as the more evolved families, there are sex chromosomes in initial, advanced, and highly evolved stages of morphological differentiation.

III. SEX CHROMOSOMES IN MEIOSIS

The pairing arrangement of chromosomes in prometaphase and metaphase of the first meiotic division can yield important information about the presence of sex chromosomes. This is especially necessary in those species where demonstration of heteromorphic sex chromosomes in mitosis is negative. The more structural differences there are between the X and Y or the Z and W chromosomes, the less is their contact during meiosis. In *Triturus*, it was shown that the small heterochromatic band at the telomeres of the Y chromosomes (described in Section II.B.1) reduces or sometimes even suppresses completely crossing-over between the Y and X along the entire stretch of the long arms up to the centromere.[21,22] Figure 11 a to d presents various examples of these pairing configurations. The homology between the short arms of these XY chromosomes has not been influenced by the differentiation of the long arms. Therefore, the short arms pair in the same way as the autosomes. In *Triturus helveticus,* no sex-specific heteromorphism of the heterochromatin was found (Figure 3d).[21] However, the long arms of the chromosomes no. 5 have a greatly reduced crossover frequency in male meiosis (Figure 11e, f). It can be concluded that these are the XY chromosomes. There already may be a structural change in the long arm of the Y which is not yet detectable cytologically. The reduced crossover frequency between the XY long arms in the *Triturus* species promotes further morphological differentiation of these sex chromosomes.

In male diakinesis of *Hydromantes*, the advanced XY sex chromosomes are usually joined by one chiasma only, which occurs in a terminal or an intercalary region of the long arms (Figure 6). Chiasmata within the heteromorphic short arms or in the pericentromeric regions of the long arms have not been observed.[24]

In *Gastrotheca riobambae*, homology between the highly heteromorphic XY sex chromosomes is almost completely lost. In diakinesis a long, stretched sex bivalent can be found, as in the meiosis of mammals (Figure 9f). This sex bivalent consists of terminally connected

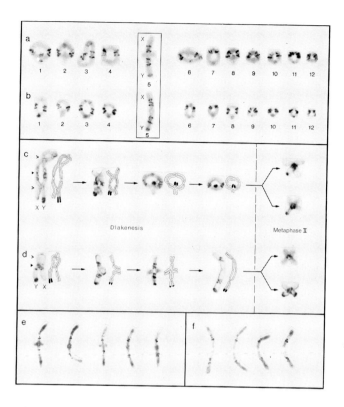

FIGURE 11.   (a, b) C-banded bivalents from two diakineses in the male meiosis of the newt *Triturus vulgaris*. The XY bivalents no. 5 are framed. The homomorphic short arms of X and Y pair and form chiasmata, whereas the heteromorphic long arms always remain unpaired and point in opposite directions. The autosomes nos. 1 to 4 and 6 to 12 form bivalents with chiasmata in both arms; the terminalization of the chiasmata causes the ring-like form of these bivalents. (c, d) Successive stages of C-banded XY bivalents and metaphase II chromosomes no. 4 from male meiosis in *Triturus alpestris*. The positions of the centromeres, chiasmata, and telomeric Y heterochromatin are elucidated in the schematic drawings. White arrowheads indicate chiasmata, black arrowheads point to centromeres. (c) Chiasmata in both short and long XY arms lead after their terminalization to ring-like bivalents and to heteromorphic telomeres in the chromatids of the long arm in metaphase II. (d) Whenever chiasmata occur only in the XY short arms, end-to-end paired bivalents with diametrically opposed long arms ensue; in the metaphase II the telomeres in the chromatids of the long arm have to be homomorphic (either euchromatic or heterochromatic). (e, f) C-banded XY bivalents from the diakinetic stage of male meiosis in *T. helveticus* showing (e) the characteristic end-to-end pairing between the XY short arms with terminalizing chiasmata, and (f) the beginning separation of X and Y in anaphase I.

XY chromosomes. There is no information so far as to whether a sex vesicle is formed during the meiotic prophase of *G. riobambae*.[37,38]

In female meiosis, the lampbrush chromosomes with their manifold and complex structures permit a very accurate morphologic comparison between the homologous chromosomes. In the salamanders *Pleurodeles waltlii* and *P. poireti*, a heteromorphic region was identified within the lampbrush bivalent no. 4 and female heterogamety concluded therefrom.[47-49] In these species, no heteromorphism could be discovered in the mitotic metaphase chromosomes no. 4 by banding techniques.[50,51]

IV. SEX-LINKED GENES IN AMPHIBIA

A. Absence of Dosage Compensation

In amphibian species with recognizable heteromorphic sex chromosomes, there is no indication of a dosage compensation mechanism in the homogametic (XX or ZZ) sex. Typical sex-chromatin bodies have not been found in the XX or ZZ nuclei of somatic tissues in species with sex chromosomes in an initial stage of differentiation, or in species with highly heteromorphic sex chromosomes.[19,21,25,37]

In the somatic prophase stages of mitosis of male *Pyxicephalus adspersus* (ZZ) and female *Gastrotheca riobambae* (XX), no chromosome shows a tendency toward positive hetero-pyknosis as does one of the two X chromosomes in female mammals. BrdU banding allowed the analysis of the time sequence of DNA replication in the sex chromosomes of female *Rana esculenta* (XX) and male *Pyxicephalus adspersus* (ZZ) during S-phase of the cell cycle.[19] Both the X and the Z chromosomes replicated synchronously and there was no sign of delayed replication as in the late replicating inactivated X of female mammals. These cytogenetic data provide strong evidence that in amphibians, as in female birds,[1,52] dosage compensation of sex-linked genes is not accomplished by random inactivation of one of the two X or Z chromosomes in the homogametic sex. As will be shown below, the same holds true for amphibian species with undifferentiated homomorphic sex chromosomes.

B. Gene Mapping on Sex Chromosomes

Various problems exist in the study of genetic linkage in Amphibia. There are only few laboratory stocks of known genotype at multiple loci. Most amphibians have a long generation time; therefore, the crossing of stocks that differ in two or more loci to produce F_1 hybrids and test crossing for linkage are time-consuming. The first known example of sex linkage of an enzyme phenotype was in the salamander *Pleurodeles waltlii*.[53] In this species, both alleles of a sex-linked peptidase gene are expressed in erythrocytes. The peptidase is a dimeric enzyme and a hybrid enzyme can be detected in heterozygotes. Unfortunately, no other genes have been localized on sex chromosomes of the Urodela. In contrast, by extensive electrophoretic studies, Elinson[54] and Wright[55] mapped several genes on the sex chromosomes of frogs of the genus *Rana*. An example of a sex-linked gene, *aconitase-1*, in *Rana clamitans* is illustrated in Figure 12. When males heterozygous for *aconitase-1* were crossed with homozygous females, all the F_1 males were homozygous and all the F_1 females were het-erozygous. When male and female were heterozygous for the same *aconitase-1* alleles, 27% of the F_1 were males homozygous for one allele, 27% were females homozygous for the other allele, and 46% were heterozygous males and females. Finally, in crosses between heterozygous female and homozygous male, the F_1 showed no relationship between aconitase-1 phenotype and sex.[54] These results clearly show that the gene *aconitase-1* is located in the homomorphic sex chromosomes of *R. clamitans* and that the male is the heterogametic (XY) sex.

In *Rana pipiens*, sex linkage was demonstrated for *peptidase-C (Pep-C)* and *superoxide dismutase-1 (SOD-1)*.[55] Figure 13 shows that the F_1 offspring of a heterozygous male for *Pep-C* crossed with a homozygous female are mostly of the two parental types, heterozygous males and homozygous females. Recombinant males and females are rare. This mode of inheritance was not found when *Pep-C* heterozygous females were crossed with homozygous males. The analysis of the F_1 offspring of males heterozygous for *SOD-1* shows a mode of inheritance similar to that for *Pep-C* (Figure 13).[55] These results demonstrate that the male is heterogametic (XY) in *Rana pipiens*. Furthermore, the recombinant frequencies between the sex-determining gene and the *Pep-C* and *SOD-1* loci, as well as between each of the *Pep-C* and *SOD-1* loci, allowed construction of a linkage map. The sex-determining gene is more distant from *Pep-C* (12.1% recombinants) than it is from *SOD-1* (8.6% recombinants).[55]

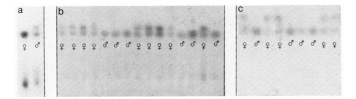

FIGURE 12. Starch gel plates stained for aconitase-1 isozymes in the North American frog *Rana clamitans*. In (a), the patterns of the homozygous female crossed to a heterozygous male are shown. (b) and (c) demonstrate the patterns of 24 of their offsprings. Note that all heterozygous offspring are female, whereas all homozygous offspring are male. This pattern is consistent with the assumption that in the parents the aconitase-1 alleles *c* and *a* were linked to the sex chromosomes as follows: female X^cX^c and male X^aY^c. Gels in (a) and (b) were run in EDTA-borate-Tris buffer, the gel in (c) in Tris-citrate buffer. (From Ellinson, R. P., *Biochem. Genet.*, 21, 435, 1983. With permission.)

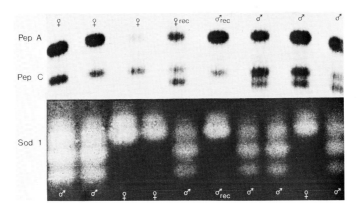

FIGURE 13. Starch gel plates stained for peptidase isozymes (above) and super oxide dismutase-1 isoenzymes (below)in the North American frog *Rana pipiens*. Pep-C shows either a single band (alleles *b/b*) or a double band (*b/c*). The double-band pattern is typical of a monomeric enzyme. The samples are from the offsprings of a homozygous X^bX^b female and a heterozygous X^bY^c male. The female offspring are predominantly of the b/b type, whereas the male offsprings are predominantly of the b/c type. Two rare recombinants are indicated by 'rec'. The Pep-A is invariant in this cross. SOD-1 shows either three-band (*b/c*) or single-band (*b/b*) patterns typical of a dimeric enzyme structure. The samples are from offspring of a homozygous X^bX^b female and a heterozygous X^bY^c male. All of the female offspring are X^bX^b homozygous and all but one of the male offspring are X^bY^b heterozygous. The single male X^bY^b recombinant is indicated by 'rec'. (From Wright, D. A., *Genetics*, 103, 249, 1983. With permission.)

Experiments using *Rana* hybrids have provided evidence for sex linkage of the genes for phosphoglucomutase, fructose-1,6-diphosphatase, alcohol dehydrogenase-2, β-glucosidase,[56] lactate dehydrogenase-B, and mannose phosphate isomerase.[57] Other data indicate that the sex-determining genes may be on different chromosome pairs in related *Rana* species although their karyotypes are very similar.[56]

It appears that there is no common ancestral or conserved sex linkage group in amphibians. Thus, the peptidase gene which is sex linked in *Pleurodeles waltlii* is not homologous with the *Pep-C* gene in *Rana pipiens*. In *R. catesbeiana* the gene coding for lactate dehydrogenase-

B is sex linked, but in *R. clamitans* it is located on an autosome. Also, the fact that an *aconitase* locus is found on the sex chromosomes of the frog *R. clamitans* and the Z chromosomes of birds seems to be coincidental.[54]

ACKNOWLEDGMENTS

Our studies on amphibian chromosomes were funded by the Deutsche Forschungsgemeinschaft (grants Schm 484/2-2, Schm 484/2-3, and 484/4-1) and the Universitätsbund Würzburg, West Germany. The following colleagues kindly provided material for reproduction: Dr. Irma Nardi (Pisa), Dr. Richard P. Elinson (Toronto), Dr. Stanley K. Sessions (Irvine), Dr. Simon H. Sims (Seattle), and Dr. David A. Wright (Houston).

REFERENCES

1. **Ohno, S.**, *Sex Chromosomes and Sex-Linked Genes,* Springer-Verlag, Berlin, 1967.
2. **Beçak, W., Beçak, M. L., Nazareth, H. R. S., and Ohno, S.**, Close karyological kinship between the reptilian suborder Serpentes and the class *Aves, Chromosoma,* 15, 606, 1964.
3. **Ray-Chaudhuri, S. P., Singh, L., and Sharma, T.**, Evolution of sex chromosomes and formation of W chromatin in snakes, *Chromosoma,* 33, 239, 1971.
4. **Matthey, R.**, *Les Chromosomes des Vertébrés,* Rouge, Lausanne, Switzerland, 1949.
5. **Romer, A. S.**, *Vertebrate Paleontology,* University of Chicago Press, Chicago, Ill., 1966.
6. **Schmalhausen, I. I.**, *The Origin of Terrestrial Vertebrates,* Academic Press, New York, 1968.
7. **Ziswiler, V.**, *Die Wirbeltiere 1,* Thieme, Stuttgart, 1976.
8. **Singh, L.**, Present status of sex chromosomes in amphibians, *Nucleus (Calcutta),* 17, 17, 1974.
9. **Humphrey, R. R.**, Sex of the offspring fathered by two *Amblystoma* females experimentally converted into males, *Anat. Rec.,* 82 (Suppl. 77), 469, 1942.
10. **Humphrey, R. R.**, Sex determination in ambystomid salamanders: a study of the progeny of females experimentally converted into males, *Am. J. Anat.,* 76, 33, 1945.
11. **Humphrey, R. R.**, Male homogamety in the Mexican axolotl: a study of the progeny obtained when germ cells of a genetic male are incorporated in a developing ovary, *J. Exp. Zool.,* 134, 91, 1957.
12. **Chang, C. Y. and Witschi, E.**, Breeding of sex-reversed males of *Xenopus laevis* Daudin, *Proc. Soc. Exp. Biol. Med.,* 89, 150, 1955.
13. **Chang, C. Y. and Witschi, E.**, Gene control and hormonal reversal of sex differentiation in *Xenopus, Proc. Soc. Exp. Biol. Med.,* 93, 140, 1956.
14. **Gallien, L.**, Inversion totale du sexe chez *Xenopus laevis* Daud, á la suite d'un traitement gynogéne par le benzoate d'oestradiol administré pendant la vie larvaire, *C. R. Acad. Sci. (Ser. D),* 237, 1565, 1953.
15. **Kawamura, T. and Nishioka, M.**, Aspects of the reproductive biology of Japanese anurans, in *The Reproductive Biology of Amphibians,* Taylor, D. H. and Guttman, S. I., Eds., Plenum Press, New York, 1977, 103.
16. **Wachtel, S. S., Koo, G. C., and Boyse, E. A.**, Evolutionary conservation of H-Y ("male") antigen, *Nature (London),* 254, 270, 1975.
17. **Wachtel, S. S. and Ohno, S.**, The immunogenetics of sexual development, *Prog. Med. Genet.,* 3, 109, 1979.
18. **Wachtel, S. S., Wachtel, G. M., Nakamura, D., and Gilmour, D.**, H-Y antigen in the chicken, *Differentiation,* 23, S107, 1983.
19. **Schempp, W. and Schmid, M.**, Chromosome banding in Amphibia. VI. BrdU-replication patterns in Anura and demonstration of XX/XY sex chromosomes in *Rana esculenta, Chromosoma,* 83, 697, 1981.
20. **Witschi, E.**, Ergebnisse der neueren Arbeiten über die Geschlechtschromosomen bei Amphibien, *Z. Induk. Abstamm. Vererbungsl.,* 31, 287, 1923.
21. **Schmid, M., Olert, J., and Klett, C.**, Chromosome banding in Amphibia. III. Sex chromosomes in *Triturus, Chromosoma,* 71, 29, 1979.
22. **Sims, S. H., Macgregor, H. C., Pellat, P. S., and Horner, H. A.**, Chromosome 1 in crested and marbled newts (*Triturus*): an extraordinary case of heteromorphism and independent chromosome evolution, *Chromosoma,* 89, 169, 1984.

23. **Mancino, G., Ragghianti, M., and Bucci-Innocenti, S.,** Cytotaxonomy and cytogenetics in European newt species, in *The Reproductive Biology of Amphibians,* Taylor, D. H. and Guttman, S. I., Eds., Plenum Press, New York, 1977, 411.

24. **Nardi, I., Andronico, F., De Lucchini, S., and Batistoni, R.,** Cytogenetics of the European plethodontid salamanders of the genus *Hydromantes* (Amphibia, Urodela), *Chromosoma,* 94, 377, 1986.

25. **Schmid, M.,** Chromosome banding in Amphibia. V. Highly differentiated ZW/ZZ sex chromosomes and exceptional genome size in *Pyxicephalus adspersus* (Anura, Ranidae), *Chromosoma,* 80, 69, 1980.

26. **Schmid, M. and Bachmann, K.,** A frog with highly evolved sex chromosomes, *Experientia,* 37, 243, 1981.

27. **Engel, W. and Schmid M.,** H-Y Antigen as a tool for the determination of the heterogametic sex in Amphibia, *Cytogenet. Cell Genet.,* 30, 130, 1981.

28. **Kezer, J. and Macgregor, H. C.,** A fresh look at meiosis and centromeric heterochromatin in the red-backed salamander, *Plethodon c. cinereus* (Green), *Chromosoma,* 33, 146, 1971.

29. **León, P. E. and Kezer, J.,** Localization of 5S RNA genes on chromosomes of plethodontid salamanders, *Chromosoma,* 65, 213, 1978.

30. **Mancino, G.,** Osservazioni cariologiche sull' Urodelo della Sardegna *Euproctus platycephalus:* morfologia dei bivalenti meiotici e dei lampbrush chromosomes, *Rend. Accad. Naz. Lincei,* 39, 540, 1965.

31. **Schmid, M.,** Evolution of sex chromosomes and heterogametic systems in Amphibia, *Differentiation,* 23, S13, 1983.

32. **Morescalchi, A.,** Chromosome evolution in the caudate Amphibia, in *Evolutionary Biology,* Vol. 8, Dobzhansky, T., Hecht, M. K., and Steere, W. C., Eds., Plenum Press, New York, 1975, 233.

33. **Schmid, M.,** Chromosome evolution in Amphibia, in *Cytogenetics of Vertebrates,* Müller, H., Ed., Birkhäuser, Basel, 1980, 4.

34. **Sessions, S. K.,** Evidence for a highly differentiated sex chromosome heteromorphism in the salamander *Necturus maculosus* (Rafinesque), *Chromosoma,* 77, 157, 1980.

35. **Sessions, S. K. and Wiley, J. E.,** Chromosome evolution in salamanders of the genus *Necturus, Brimleyana,* 10, 37, 1985.

36. **Morescalchi, A. and Serra, V.,** DNA renaturation kinetics in some paedogenetic urodeles, *Experientia,* 30, 487, 1974.

37. **Schmid, M., Haaf, T., Geile, B., and Sims, S.,** Chromosome banding in Amphibia. VIII. An unusual XY/XX-sex chromosome system in *Gastrotheca riobambae* (Anura, Hylidae), *Chromosoma,* 88, 69, 1983.

38. **Schmid, M., Haaf, T., Geile, B., and Sims, S.,** Unusual heteromorphic sex chromosomes in a marsupial frog, *Experientia,* 39, 1153, 1983.

39. **Schmid, M., Sims, S. H., Haaf, T., and Macgregor, H. C.,** Chromosome banding in Amphibia. X. 18S and 28S ribosomal RNA genes, nucleolus organizers and nucleoli in *Gastrotheca riobambae, Chromosoma,* 94, 139, 1986.

40. **Goodpasture, C. and Bloom, S. E.,** Visualization of nucleolar organizer regions in mammalian chromosomes using silver staining, *Chromosoma,* 53, 37, 1975.

41. **Hsu, T. C., Spirito, S. E., and Pardue, M. L.,** Distribution of 18 + 28S ribosomal genes in mammalian genomes, *Chromosoma,* 53, 25, 1975.

42. **Iturra, P. and Veloso, A.,** Evidence for heteromorphic sex chromosomes in male amphibians (Anura: Leptodactylidae), *Cytogenet. Cell Genet.,* 31, 108, 1981.

43. **Sessions, S. K. and Kezer, J.,** Cytogenetic evolution in the plethodontid salamander genus *Aneides, Chromosoma,* 95, 17, 1987.

44. **Singh, L., Purdom, I. F., and Jones, K. W.,** Satellite DNA and evolution of sex chromosomes, *Chromosoma,* 59, 43, 1976.

45. **Jones, K. W. and Singh, L.,** Conserved repeated sequences in vertebrate sex chromosomes, *Hum. Genet.,* 58, 46, 1981.

46. **Jones, K. W.,** Evolutionary conservation of sex specific DNA sequences, *Differentiation,* 23, S56, 1983.

47. **Lacroix, J.-C.,** Étude descriptive des chromosomes en écouvillon dans le genre *Pleurodeles* (Amphibien, urodèle), *Ann. Embryol. Morphog.,* 1, 179, 1986.

48. **Lacroix, J.-C.,** Variations expérimentales ou spontanées de la morphologie et de l'organisation des chromosomes en écouvillon dans le genre *Pleurodeles* (Amphibien, urodèle), *Ann. Embryol. Morphog.,* 1, 205, 1986.

49. **Lacroix, J.-C.,** Mise en évidence sur les chromosomes en écouvillon de *Pleurodeles poireti* Gervais, amphibien urodèle, d'une structure liée au sexe, identifiant le bivalent sexuel et marquant le chromosome W, *C. R. Acad. Sci. (Ser. D),* 271, 102, 1970.

50. **Bailly, S.,** Localisation et signification des zones Q observées sur les chromosomes mitotiques de l'amphibien *Pleurodeles waltlii* Michah, après coloration par la moutarde de quinacrine, *Chromosoma,* 54, 61, 1976.

51. **Labrousse, M., Guillemin, C., and Gallien, L.,** Mise en évidence, sur les chromosomes de l'amphibien *Pleurodeles waltlii* Michah, de secteurs d'affinité différente pour le colorant de Giemsa à pH 9, *C. R. Acad. Sci. (Ser. D),* 274, 1063, 1972.

52. **Baverstock, P. R., Adams, M., Polkinghorne, R. W., and Gelder, M.,** A sex-linked enzyme in birds — Z-chromosome conservation but no dosage compensation, *Nature (London),* 296, 763, 1982.

53. **Ferrier, V., Jaylet, A., Cayrol, C., Gasser, F., and Buisan, J.-J.,** Étude électrophorétique des peptidases érythrocytaires chez *Pleurodeles waltlii* (Amphibien Urodèle): mise en évidence d'une liaison avec le sexe, *C. R. Acad. Sci. (Ser. D),* 290, 571, 1980.

54. **Elinson, R. P.,** Inheritance and expression of a sex-linked enzyme in the frog, *Rana clamitans, Biochem. Genet.,* 21, 435, 1983.

55. **Wright, D. A. and Richards, C. M.,** Two sex-linked loci in the leopard frog, *Rana pipiens, Genetics,* 103, 249, 1983.

56. **Wright, D. A., Richards, C. M., Frost, J. S., Camozzi, A. M., and Kunz, B. J.,** Genetic mapping in amphibians, in *Isozymes: Current Topics in Biological and Medical Research,* Vol. 10, Alan R. Liss, New York, 1983, 287.

57. **Elinson, R. P.,** Genetic analysis of developmental arrest in an amphibian hybrid (*Rana catesbeiana, Rana clamitans*), *Dev. Biol.,* 81, 167, 1981.

58. **Ponse, K.,** Sur la digamétie du crapaud hermaphrodite, *Rev. Suisse Zool.,* 49, 185, 1942.

59. **Zaborski, P.,** Sur la constance de l'expression de l'antigène H-Y chez le sexe hétérogamétique de quelques Amphibiens et sur la mise en évidence d'un dimorphisme sexuel de l'expression de cet antigène chez l'Amphibien Anoure *Pelodytes punctatus* D., *C. R. Acad. Sci., (Ser. D),* 289, 1153, 1979.

60. **Witschi, E.,** Studies on sex differentiation and sex determination in amphibians. III. Rudimentary hermaphroditism and Y chromosome in *Rana temporaria, J. Exp. Zool.,* 54, 157, 1929.

61. **Schmid, M. and Haaf, T.,** unpublished results, 1987.

Chapter 5

EVOLUTION AND VARIETY OF SEX-DETERMINING MECHANISMS IN AMNIOTE VERTEBRATES

James J. Bull

TABLE OF CONTENTS

I. INTRODUCTION

It is especially appropriate to begin with reference to Susumu Ohno's first book on sex determination, which is in its 20th anniversary this year.[1] This book was characterized by bold extrapolations from the available data to suggestions of general principles and themes underlying vertebrate sex determination, and for this and other reasons, the book has served as an inspiration to our work on sex determination.

The theme of Ohno's that we address here is the *conservation* of mechanisms. Specifically, Ohno proposed that sex chromosome systems were highly conserved within each of three large vertebrate groups: mammals (male heterogamety), birds (female heterogamety), and snakes (female heterogamety). He went so far as to suggest that the sex chromosomes might even be homologous between these groups as well. We now suspect that Ohno was right about sex chromosome homologies within each of these groups, but that the sex chromosomes are not homologous between groups. More importantly, however, our understanding of vertebrate sex determination has been altered in one fundamental way over the last 20 years; sex-determining mechanisms are indeed conserved within groups identified by Ohno, but they are highly varied in other groups. Our appreciation for the diversity of mechanisms in vertebrates has been embellished mostly by the discovery of environmental sex determination (ESD), whereby the incubation temperature of the egg determines whether the embryo hatches as male or female. The discovery of ESD in amniotes (mammals, birds, reptiles) was essentially unprecedented 20 years ago, yet ESD is now known to occur in three of the four surviving lineages that arose from the stem reptiles (see below), and it has caused us to reevaluate our understanding of vertebrate sex determination. This paper addresses the evolution of amniote sex determination in view of these new discoveries. Readers wishing further details of the data and arguments presented in this chapter should consult Reference 2.

Terminology — The terminology surrounding work on sex determination is often confusing, and this paragraph explains some of the terms used in this chapter. A *sex-determining mechanism* is a description of the inherited basis of whether an individual develops as male or female. Genotypic sex determination (GSD), therefore, refers to the case in which genotypic differences lead to gender differences, whereas environmental sex determination refers to the case in which different experiences early in life lead to gender differences (genotypic differences having little or no role). A common form of GSD is heterogamety, labeled XX/XY (XX/XY is sometimes reserved for male heterogamety, ZZ/ZW for female heterogamety). In many animals and some plants, heterogamety is often accompanied by differences in size and shape between the chromosomes carrying the X and Y factors, in which case the term *heteromorphic sex chromosomes* is applied.

II. TWO TYPES OF SEX DETERMINATION

A. Characteristics and Taxonomic Distribution

Both GSD and ESD occur in amniotes. All known cases of GSD involve heteromorphic sex chromosomes. (Genetic evidence of heterogamety is available for some amphibians and fish in which heteromorphic sex chromosomes are unknown, but all cases of heterogamety in amniotes have either been identified because of heteromorphic sex chromosomes or are accompanied by sex chromosomes.) In the amniote mechanisms of ESD, incubation temperature of the embryo has the major effect on gender, although the moisture available to the egg has also been shown to have an influence.[3] Most credit for the discovery of ESD belongs to Claude Pieau, who in the 1970s published extensively on its characteristics in the European pond turtle and Greek tortoise.[4]

Both GSD and ESD are found together in all major lineages of amniotes except mammals (Table 1). This variety indicates that there has been considerable evolution of sex-determining

Table 1
SEX-DETERMINING MECHANISMS IN AMNIOTES

Lineage

Mammals

Male heterogamety with heteromorphic sex chromosomes occurs in the vast majority of species (in some cases as multiple sex chromosome derivatives). The X is homologous throughout placental and marsupial mammals, but this homology probably excludes monotremes. A few exceptional systems are known that deviate from the basic male heterogametic plan, but even these retain the ancestral X chromosome.

Archosaurs

Birds:

Female heterogamety with heteromorphic sex chromosomes is known in many species, with the Z chromosome apparently homologous; heteromorphic sex chromosomes are unknown in ratites (ostriches, emus).

Crocodilians:

Incubation temperature determines sex in all species studied. Sex chromosomes are unknown.

Squamates

Snakes:

Female heterogamety with heteromorphic sex chromosomes probably occurs in all colubrids, elapids, and viperids; sex chromosomes have been identified in one boid, but most boids and other "primitive" snakes lack detectable sex chromosomes. The Z chromosome is apparently homologous. ESD is unknown.

Lizards:

Male and female heterogamety occur together within two infraorders (Gekkota, Scincomorpha) and occur separately in a few other infraorders; these various sex chromosomes do not appear to be homologous. In addition, ESD occurs in two infraorders (Iguania and Gekkota).

Turtles

Species with ESD and species lacking ESD are known in both suborders of living turtles (side-neck and hidden-necks turtles), whereas sex chromosomes have been identified in just two genera of hidden-necked turtles. ESD is known from five families of hidden-neck turtles and is common in several of these families but is known in only one genus of side-neck turtles.

mechanisms in amniotes as a whole, and inspection on a finer level reveals that mechanisms are even highly variable within some groups. Thus, lizards (which share relatively recent common ancestry with snakes) have experienced multiple origins of ESD and of male heterogamety and/or female heterogamety; turtles likewise reveal considerable variety of mechanisms, although we know of many fewer examples of sex chromosomes in turtles than in lizards. Despite the fact that mechanisms are varied within some groups and between groups as well, it appears that mechanisms are highly conserved within birds, mammals, and snakes. It is interesting to note that, whereas 20 years ago it seemed that genotypic sex determination was ancestral to amniotes, no such conclusion is in evidence today.

The existence of ESD raises many interesting questions of natural history and evolution. The focus of this chapter is indeed the evolution of sex determination, but the ecology of sex ratio and sex determination is fundamental to this larger question, and some of the relevant details of ESD will therefore be presented here. Two basic points about ESD in reptiles are (1) the temperature effect is often so extreme as to cause 100% males at some temperatures and 100% female at others, and (2) temperature influences sex determination rather than causing differential mortality. One interesting finding is that the effect of temperature on gender is not the same among all species with ESD (Figure 1). Thus, in addition to species that lack ESD, some species develop as males at high temperatures and females at low temperatures, other species do the reverse, and some species develop as males at intermediate temperatures and as females at high and low extremes. These different forms of ESD do not strictly follow taxonomic boundaries.

When it has been studied in the field, ESD operates as a genuine sex-determining mechanism. In studies of crocodilians and turtles, various qualitative factors influencing nest sex ratio have been identified: nest position relative to shade, nest depth, season of nesting. Quantification of temperature effects on gender in the field are consistent with those from the laboratory.

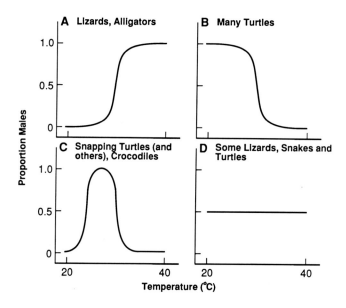

FIGURE 1.   Four patterns are observed between incubation temperature and hatchling sex ratio in reptiles. (A) Females develop at low temperatures, males at high ones, as known from two lizards, and some crocodilians. (B) Females develop at high temperatures, males at low ones, as in many turtles. (C) Males develop at intermediate temperatures, females develop at high and low extremes, as observed in several turtles and crocodiles. (D) The absence of temperature effects on gender, as in some turtles, squamates, and presumably all species with heteromorphic sex chromosomes. (From Bull, J. J., *Q. Rev. Biol.*, 55, 3, 1980. With permission.)

III. THE ANALYSIS OF VARIETY: WHAT, HOW, AND WHY

Having demonstrated that an interesting variety of mechanisms occurs in amniotes, we now turn to consider the evolution of this variety. Evolutionary biology is generally concerned with understanding variety, but there are several levels at which variety may be studied. At one level, we may try to understand the forces that caused organisms to diverge from the ancestral type (i.e., understand *why* the variety evolved), and such explanations are commonly phrased in terms of natural selection. At another level, we may investigate variety from the perspective of *how* the variety evolved, or what intermediate stages occurred between the ancestral form and present forms. Finally, a third level of investigating variety is the combinatorial level. Rather than consider processes that lead to variety (how and why), we may develop a framework to classify the variety and identify not only the types that exist, but also identify those that do not exist. All three levels of analysis may be applied to the study of sex determination.

A. The Observed Variety and Possible Variety

The objective of the combinatorial method is to identify the set of possible variety within which the observed variety occurs. Although this process of enumeration can sometimes be relatively straightforward, the variety revealed by nature is often a good place to begin considering the realm of possibilities. We have already noted that some amniotes possess mechanisms with complete genotypic control of sex determination, while others possess mechanisms with almost complete environmental control of sex. Parthenogenesis is also known in several squamates. Although this magnitude of variation is profound when compared to the magnitude of variety that was anticipated for amniotes 20 years ago, it is not so profound when compared to the diversity of mechanisms observed in a group such as fish, which includes GSD, ESD, various forms of parthenogenesis, simultaneous hermaphroditism, and sex change. Evolutionary biologists have acknowledged the absence of functional hermaphroditic species in amniotes, but there is no satisfying explanation.[5]

Employing the combinatorial approach at a finer level of analysis, one intriguing pattern does emerge. When ESD occurs in amniotes, temperature is the chief determinant of sex. In systems of ESD outside of vertebrates, temperature is not the determinant of sex, hence, gender could presumably be determined according to environmental factors other than temperature in reptiles. In considering the evolution of ESD in amniotes, our explanations must therefore account for the fact that only one cue in an endless variety of possible cues is used to determine gender.

B. The Transition Between ESD and GSD

A major evolutionary question within the present framework is *how* such profoundly different kinds of sex determination can evolve from a common ancestor. The smallest taxonomic group in which both GSD and ESD are known is the family level (Kinosternidae of turtles, Gekkonidae of lizards), so these transitions were not necessarily rapid. Yet geographic variation ranging from the absence of ESD to its presence occurs within one fish species; so in principle, the transition can occur rapidly.

Although sex-determining mechanisms are often regarded in a different light than are the visible phenotypes of an organism, their evolution is subject to many of the basic principles that apply to any evolutionary transition. For example, given the right mutations, an experimental geneticist would have no trouble producing a strain of ESD from a strain originally with XX/XY sex determination: a temperature-sensitive gene that allowed XX to be female at some temperatures and male at other temperatures could be selected so that all individuals of the strain were XX, gender being decided by temperature. Such temperature-sensitive sex transformer mutations/strains have in fact been isolated in *Drosophila,* in a mosquito, in a nematode, and in guppies (though not all of these mutations yield fertile transformations). The question, then, is whether the transition between ESD and GSD under natural conditions requires unusual circumstances and extreme selection (such as an experimenter can apply), or whether the transition can instead occur simply, without undue circumstances.

The evolution of ESD from GSD (or vice versa) under natural conditions can apparently occur without extreme selection or other complications, but proof of this conclusion requires a genetic model. In developing this argument, we assume that the proper kind of genetic variance is present, and the evidence cited above encourages us that this condition may often be satisfied. The evolution of ESD in a natural population must then obey certain rules: the sex ratio must remain largely unaltered during the transition, or extreme selection is required to produce the change. In view of this sex-ratio constraint, it is not obvious that the transition between ESD and GSD is unencumbered, because the sex ratio will depend on the level of ESD and on the frequencies of sex-determining genes, and both are changing during the transition. However, theoretical work to date has suggested that selection adjusts gene frequencies so that this constraint is automatically satisfied, and the transition between GSD and ESD is in fact straightforward (see below).

Consider a population in which the initial sex-determining mechanism is male heterogamety, XX/XY. The effect of ESD is introduced by assuming that an environmental factor is imposed so that some of the XX zygotes now develop as male rather than female. The genetic composition of the population changes from pure male heterogamety to:

Female	Male
XX	XY
	XX

As long as XY and XX males coexist, we say that this population contains a mixture of GSD and ESD: GSD is present because any individual with Y becomes male, whereas ESD is also present because XX sometimes develops as male and sometimes develops as female. At the moment we do not allow ESD to evolve (we simply imposed a given level of sex-reversing environmental effect on XX), but the magnitude of GSD is allowed to evolve through changes in the frequency of Y.

The evolution of Y depends on the sex ratio in the population: if more than 50% of the zygotes become male, Y is selected against; if less than 50% of the zygotes become male, Y is favored; if 50% of the zygotes become male, the frequency of Y does not change. The sex ratio, in turn, depends partly on the frequency of Y. If we start with 50% of the zygotes as XX and 50% as XY, an excess of males will result because of XY and some of the XX develop as male. However, some of the XX grow up to be male and they produce all daughters. Since half the zygotes of XY fathers are XX and all the zygotes of XX fathers are XX, more than half of the zygotes of the next generation will be XX. As long as the fraction of XX zygotes that develop as male remains constant from generation to generation (and does not exceed 50%), the frequency of XY zygotes automatically evolves toward the point that the sex ratio is $^1/_2$. At this point, GSD and ESD (XX and XY males) coexist indefinitely. Of course, if more than 50% of the XX zygotes develop as male, the sex ratio always exceeds $^1/_2$, but the Y is then eventually lost, and all traces of GSD disappear. By adjusting the magnitude of environmental effects on XX zygotes, therefore, we can generate equilibrium populations ranging from almost pure GSD (little environmental effect on XX) to pure ESD (Y lost).

We have described this process as though the environment alone imposes atypical sex determination (i.e., that some XX become male without a genetic basis), and this model may thus seem unrepresentative of evolution. Yet essentially, the same process can be shown to underly the more realistic but more complicated case in which ESD is allowed to coevolve with GSD — through genes that impart an environmental sensitivity to the X and/or Y. In general, therefore, studies of the transition between ESD and GSD have revealed that the sex ratio is selected toward its appropriate value, and that there are no intrinsic problems in changing from ESD to GSD or the reverse. Furthermore, mechanisms with any intermediate combination of genotypic and environmental effects may evolve and persist; there is no intrinsic tendency to evolve toward the extremes of pure GSD or pure ESD.

C. A Selective Advantage of ESD or GSD?

An interesting but difficult question concerns whether GSD or ESD has a selective advantage over the other mechanism. The previous section demonstrated that the mechanisms could coexist in a population at equilibrium, but those considerations were based on the effect of each mechanism on the population sex ratio. By studying models that specify greater details of these mechanisms, it is possible to uncover subtle selective differences between them. As we will see below, these selective differences may be embedded in the models used to describe the transition between GSD and ESD to understand how and why one mechanism may evolve from the other. In general, we may consider three classes of hypotheses that address the possibility of selective differences between ESD and GSD: (1) ESD is favored over GSD (under some conditions), (2) GSD is favored over ESD, and (3) neither mechanism has a relative advantage. Hypothesis (3) provides a suitable null hypothesis as it invokes no adaptationist principles. We thus proceed to propose alternatives to this null hypothesis for comparison.

1. Disadvantages of ESD

Early writers on the evolution of sex determination supposed that ESD was ancestral to but inferior to GSD; the assumption of inferiority was not explained, but presumably it was based on the assumption that ESD was more prone to produce intersexes than was GSD.

While such "physiological" differences may indeed exist between ESD and GSD, it is difficult to attach adaptationist significance to all but the most severe disturbances, and it may be more promising to look for other disadvantages of ESD. One such disadvantage is, in fact, evident on casual inspection: if sex is determined by environmental factors, the sex ratio may change from year to year strictly from changes in the physical environment. Thus, a cold year may greatly overproduce males in turtles and overproduce females in alligators. It has been shown, in fact, that such fluctuations tend to favor GSD, although these effects are weak in long-lived species where the fluctuations tend to cancel over the course of a generation. Nonetheless, the susceptibility of ESD to environmental fluctuations causing sex-ratio variation is a general and plausible disadvantage of ESD.

2. Advantages of ESD

Attempts have also been made to consider whether ESD might be superior to GSD under certain conditions. Charnov and Bull[6] proposed that ESD was, in fact, favored over GSD when three conditions occurred in the life history of a species. (1) The population inhabits a *patchy* environment: early in life all individuals enter a patch type that has a lasting effect on their fitness; the effect of patch type differs for males and females, such that some patches enhance male fitness more than female fitness, and other patches do the reverse. (2) Parents and offspring cannot chose patch type so as to preferentially put sons and daughters in different patches. (3) Individuals raised in different patch types mix freely before mating. These conditions favor ESD over GSD provided that sex is determined according to an appropriate environmental cue: ESD allows the individual to develop as the sex benefitting most under the patch type, whereas GSD would cause some individuals to develop as male in patches benefitting females and as male in patches benefitting females.

3. A Combination of Forces

What happens when we combine the factors that favor ESD with those that hinder ESD? Recent work[7] indicates that intermediate levels of ESD evolve, and the magnitude of environmental effects on gender observed at equilibrium simply depends on how strongly ESD vs. GSD is favored (e.g., suppose that temperature is the determinant of sex). When ESD is weakly selected relative to GSD, a mild form of ESD evolves so that temperature has only a weak effect on gender, and most temperatures produce both sexes (though not usually 50% males). When ESD is strongly selected, a strong form of ESD evolves, in which most temperatures produce either all females or all males.

D. A Formal Model

To formalize and combine some of these concepts, it is instructive to consider the following model that captures that Charnov-Bull paradigm within a framework for thinking about how the transition from GSD to ESD might occur. In the starting population, sex is determined strictly according to XX/XY (male heterogamety). Introduce into this population a rare, environment-sensitive allele at the sex-determining locus (labeled T); T may be a mutation of either the original X or Y, depending on the mode of action of this locus, but it causes TX to develop as male under some conditions and as female under other conditions; TY is always male. The composition of the population changes from pure XX/XY to:

Female	Male
XX	XY
TX	TX
	TY

ignoring the vanishingly rare TT genotypes. (As long as T is rare, TT can be ignored.) Whether one thinks of T as a modification of the X or of the Y, its importance is that TX is sometimes male, or sometimes female, due to its susceptibility to environmental influence.

To model the evolution of ESD is to understand whether the frequency of T increases or decreases following its introduction. In the spirit of the Charnov-Bull model, we allow the fitness of TX males and females to differ from the norm: TX males have fitness, w_m, relative to XY males, and TX female females have fitness, w_f, relative to XX females. A difference in fitness between TX and XY males (or between TX and XX females) would result, for example, if TX developed its gender in response to patch types that had different effects on male vs. female fitness. Since TY is always male, its fitness is the same as of XY males. The last addition to this model is the parameter P, the fraction of TX zygotes that develop as female. Omitting mathematical details, allele T increases when rare if,

$$Pw_f + (1 - P)w_m > 1 \qquad (1)$$

i.e., if the average fitness of an TX zygote exceeds that of the average non-T zygote. Clearly, if TX causes an individual to develop as the sex benefitting most according to patch type, its mean fitness is enhanced, and T will spread.

This result merely indicates that ESD will begin to evolve under certain conditions. The frequency of ESD is determined by the frequency of T, and as T becomes more common, our equations need to account for TT individuals, additional parameters need to be introduced, and a more extensive mathematical treatment is required. However, such studies have borne out the verbal predictions of the Charnov-Bull model. Thus, we see from this simple model how and why an XX/XY system could be converted into one of ESD.

IV. EMPIRICAL EVIDENCE FROM REPTILES

Above, we developed a theoretical framework for thinking about the evolution of ESD vs. GSD, so the next stage is to identify points of agreement and disagreement between observations and the models. Empirical evidence is sparse, however. There is no empirical evidence from amniotes on the transitions between ESD and GSD (or on any other possible transitions), and we have no studies of the genetics of sex determination in any reptile other than cytological observations of sex chromosomes.

At present, one of the most promising avenues for exploration concerns tests of the Charnov-Bull model. A clear prediction of this model is that *the environmental cue determining sex should have different consequences for males and females* (or should correlate with such effects). Our combinatorial perspective led us to realize that temperature is the chief determinant of sex across reptiles with ESD, hence, we anticipate that any effect of temperature consistent with Charnov-Bull should be fairly general to many reptiles rather than specific to individual species.

The Charnov-Bull model predicts that ESD evolves in response to environmental factors differentially influencing male vs. female *fitness*. Not surprisingly, variance in total fitness has not been measured for any reptile, and direct tests of this model are not forthcoming. Tests have, therefore, been confined to the elementary level of investigating whether incubation temperature has any lasting effects on individual phenotypes. If temperature is not found to have lasting phenotypic effects, it would be appropriate to reject the Charnov-Bull model, whereas if temperature does have lasting phenotypic effects, we must proceed carefully to understand whether these phenotypic effects likely translate into appropriate fitness consequences for males and females. It is encouraging that the first studies of this problem have identified major phenotypic effects of incubation temperature. A study of alligators revealed that incubation temperature influenced post-hatching growth rates (differently for males than for females),[8] and a recent study of leopard geckos revealed that incubation temperature influences hormone levels, and behavior of the adult.[9] This latter study also showed that the occasional females produced at male temperatures are effectively sterile (based on courtship behavior), hence, a possible drawback of ESD. Clearly, more studies of this nature are needed, but this avenue of exploration is one that offers considerable potential for further work.

V. CONCLUSIONS

A major accomplishment of the last 20 years toward understanding the evolution of amniote sex determination is the demonstration that various mechanisms of ESD and GSD coexist in reptiles. Twenty years ago, before the recognition that many reptiles have environmental sex determination, it was supposed that sex-determining mechanisms were highly conserved throughout amniotes. We now know that the mechanisms are conserved only within some groups (birds, mammals, snakes), yet are not conserved within others (lizards, turtles). A theoretical framework is in place to understand various aspects of evolution of this variety, but the framework has been in place for so short a time and reptiles are such poor subjects that little can be said in support of (or against) the paradigm.

At present our understanding of these various sex-determining mechanisms is confined largely to the level of inheritance (whether genotypic differences or environmental differences determine sex). The question remains, and the possibility is not inconsistent with the evolutionary perspective presented here, that these seemingly diverse mechanisms of ESD and GSD in amniotes are largely similar at the molecular level. We may hope that the next 20 years leaves us with an understanding at the molecular level that equals or surpasses the understanding we now enjoy at the level of the inheritance of sex.

REFERENCES

1. **Ohno, S.,** *Sex Chromosomes and Sex-linked Genes,* Springer-Verlag, Berlin, 1967.
2. **Bull, J. J.,** *Evolution of Sex Determining Mechanisms,* Benjamin/Cummings, Menlo Park, Calif., 1983.
3. **Gutzke, W. H. N. and Paukstis, G. L.,** Influence of the hydric environment on sexual differentiation of turtles, *J. Exp. Zool.,* 226, 467, 1983.
4. **Raynaud, A. and Pieau, C.,** Embryonic development of the genital system, in *Biology of the Reptilia,* Vol. 15, Gans, C. and Billet, F., Eds., John Wiley & Sons, New York, 1985, 149.
5. **Charnov, E. L.,** *The Theory of Sex Allocation,* Princeton University Press, Princeton, N.J., 1982.
6. **Charnov, E. L. and Bull, J. J.,** When is sex environmentally determined?, *Nature (London),* 266, 828, 1977.
7. **Bull, J. J.,** unpublished results, 1987.
8. **Joanen, T., McNease, L., and Ferguson, M. W. J.,** The effects of egg incubation temperature on post-hatching growth of American alligators, in *Wildlife Management: Crocodiles and Alligators,* Webb, G., Manolis, S., and Whitehead, P., Eds., Surrey Beatty and Sons, Chipping Norton, NSW, Australia, 1987, 303.
9. **Gutzke, W. H. N. and Crews, D. P.,** Embryonic temperature determines adult sexuality in a reptile, *Nature,* 332, 832, 1988.
10. **Bull, J. J.,** Sex determination in reptiles, *Q. Rev. Biol.,* 55, 3, 1980.

Part II: The Sex Chromosomes

Chapter 6

INACTIVATION PHENOMENA IN THE EVOLUTION AND FUNCTION OF SEX CHROMOSOMES

Kenneth W. Jones

TABLE OF CONTENTS

I. INTRODUCTION

The existence and unusual behavior of functionally specialized sex chromosomes present challenging problems in the fields of evolution and molecular genetics. Functional differentiation of the sex determinants resulting in genetic sex determination is achieved at the expense of incurring varying degrees of genetic loss, the most extreme example of which is the degenerate, functionally specialized, sex-determining chromosome such as the Y or W. Chromosomes also exhibit curious and bizarre sex-linked developmental behavior in some cases, ranging from X inactivation in mammals to the inactivation and degeneration of the entire paternal genome in some insects.

The conventional model for the specialization of sex chromosomes proposes that the slow accumulation of null mutations on the sex-determining chromosome occurs following the breakdown of crossing-over in the sex bivalent.[1] The resulting functional hemizygosity in the heterogametic sex is suggested to have led to the evolution of dosage-compensation systems.[2] Thus, X inactivation in mammals[3,4] and developmental equalization of the level of X-linked gene expression in the sexes in *Drosophila*[5] are both assumed to be primarily dosage compensation mechanisms.

This model, which has been discussed by previous authors,[1,6-9] however, does not explain certain observations on sex chromosomes in snakes, as will be discussed. It also fails to account for the fact, which has long puzzled geneticists, that Z chromosome inactivation and consequently Z-linked gene dosage compensation is absent in species with female heterogamety. Growing evidence of Z chromosome dosage effects in male homogamety[10,11] argues against an essential coupling between the evolution of a degenerate Y (or W) chromosome and the evolution of dosage compensation, and casts doubt on this aspect of the conventional theory. In acknowledgment of this difficulty, it has recently been suggested[12] that "dosage tolerance" evolved in female heterogamety. However, on the available evidence, it is equally valid to assume that the general phenomenon of sex-linked developmental chromosomal inactivation, of which X inactivation is only one example, has a completely different explanation. Such an assumption would seem particularly necessary in extreme cases, for example, those involving the inactivation of the paternal haploid genome in males of certain coccids. In the mealy bug, this results in functionally diploid females and haploid males,[13] and strongly suggests that sex-linked chromosomal developmental inactivation may have evolved for reasons other than restoring chromosome balance, as I will discuss.

II. EVOLUTION OF GENETIC SEX DETERMINATION

The mechanisms that initiated genetic control of previously indifferent sex alleles are believed to have been chromosomal inversions or mutations that suppressed crossing-over and genetically isolated the sex alleles.[14] Accepting this basic premise, the specific structure of the individual inversion concerned would have had considerable influence on the evolutionary outcome, depending upon whether it was confined to the sex alleles or disturbed the functions of other genes. Thus, in species in which the genotypes YY and WW are viable, as in Amphibia,[15] the inverted segments may have involved no other genes. By contrast, in species in which these genotypes are developmentally compromised, as, for example, in the Japanese medaka (*Oryzias latipes*)[16] and the guppy (*Poecilia reticulata*),[17,18] other genes appear to have been functionally compromised. In the medaka, the color gene, ruby, is sex linked and "non-crossover" YY genotypes are inviable. Simple inversions might account for this and similar limited heteromorphism of the sex bivalent involving heterochromatic segments but lacking true sex chromatin.[19,20] However, the mechanism of the origin of differential segments and their loss of function is a matter of debate. The classical model of sex chromosome specialization outlined above, and in particular, Müller's ratchet model,[21] has been criticized on the grounds that it is ineffective in accounting for species exhibiting limited degrees of chromosomal involvement and genetic loss.[12] This is because the model requires a differential segment (or a Y chromosome) of sufficiently large size for the overall mutation rate to reach a certain critical value.[22] Thus, the small differential segments, such as those found in the guppy and the medaka, are unlikely to have resulted from a ratchet mechanism, although there is evidence for loss of function within them. One suggested way out of this difficulty is the genetic hitchhiking model proposed by Rice[12] in which beneficial genes introduced into a differential segment carrying mildly deleterious mutations cause selection favoring the survival of that segment. Genetic hitchhiking is assumed to supplement the ratchet mechanism and to operate in conditions where this is ineffective. However, I wish to discuss a third model which offers a unified explanation for the origin of these inert differential segments, for entirely specialized sex chromosomes, and for sex-linked genomic inactivation. This model, which is based on observations of sex chromosomes in snakes, invokes specific classes of mutations resulting from inversions involving sex genes and affecting various levels of the hierarchy of chromosome control.

III. SEX CHROMOSOME EVOLUTION IN SNAKES

The conventional models of the evolution of sex chromosomes have been constructed mainly on the basis of observations in *Drosophila* and other species in which the process is relatively advanced. The sequence of events is clearer in snakes which exemplify a broad range of transitional stages.[23,24] These range from varieties with no discernible ZW bivalent, such as the relatively primitive nonpoisonous Boidae, to those in which the sex bivalent is highly differentiated, such as the highly evolved venomous Elapidae. Within this range, species of the family Colubridae have sex bivalents in which the Z and W chromosomes are morphologically virtually identical (homomorphic). Since differentiated ZW and XY sex chromosomes are thought to have evolved from a pair of homologous chromosomes, these appear to represent examples of a relatively early stage in sex chromosome evolution. It is interesting to note in passing that the Colubridae are also anatomically transitional between the Boidae and the Elapidae. Therefore, the degree of anatomical specialization also correlates with the extent of divergence within the sex bivalent, a point that I have discussed elsewhere.[25]

While there may be no gross morphological distinction between Z and W chromosomes in some colubrid species, chromosome banding,[26] labeling studies,[27] and *in situ* hybridization studies[28] reveal that the W chromosome, in contrast to the Z, is C-band positive and forms a somatic sex chromatin body[29] like the mammalian X chromosome. In fact, it appears that differentiated W chromosomes in all snake species are somatically inactivated whether hetero- or homomorphic. Thus, although there is ample evidence of the evolution of morphological divergence of Z and W in different species, the absence of examples that would indicate a gradual evolution of W chromosome inactivity is striking. The conclusion that inactivation was a relatively rapid phenomenon seems inescapable and at variance with the conventional model.

Similar observations of a W chromosome homomorphic with the Z and also completely heterochromatic have been reported in the lacertid turtle, *Takydromus sexlineatus*.[30] This is one of a relatively few species having morphologically identifiable sex chromosomes in an order which, for the most part, lacks them.

IV. REPEATED DNA SEQUENCES IN SEX BIVALENTS

The conclusion that inactivation was a relatively early and rapid event in sex chromosome evolution is supported by studies on the highly repeated simple nucleotide sequences of snakes. The species *Bungarus fasciatus* exhibits a female-specific satellite DNA (Bkm)[31] on isopycnic gradients, the sequences of which are evolutionarily highly conserved. It is significant that Bkm sequences detected by *in situ* hybridization are concentrated throughout the entire length of the W chromosome, whether in a homomorphic or heteromorphic sex bivalent, and essentially are not detectable on the Z chromosome. If homomorphic ZW sex bivalents represent an early stage in the evolution of chromosomal sex determination (CSD) and are somatically inert, it follows that both the cessation of crossing-over between Z and W and somatic inertness evolved early in the evolution of specialization. The most likely cause of this would seem to be an event which brought about W chromosome somatic inactivation. Since W inactivation persists into meiosis, it would have abolished crossing-over and caused the accumulation of highly repeated DNA on the genetically isolated W chromosome.

Since inversions or mutations suppressing crossing-over have been suggested to be implicated in the initiation of genetic control of sex, it is reasonable to suggest that the inactivation may have resulted from an inversion which caused a mutation affecting chromosomal behavior.

V. A MODEL TO ACCOUNT FOR SEX CHROMOSOME SPECIALIZATION

A model to account for such an initiating event has been proposed[32,33] and will be further elaborated here. This envisages that sex chromosome functional specialization originated as a result of the translocation of sex genes and their juxtaposition with genes (CE or control elements) that normally control gross chromosome conformation. This juxtaposition is suggested to have functionally subordinated the CE to the sex gene. I have referred to this hypothetical mechanism as chromosomal "hijacking".[32] A sex gene is defined broadly as a developmental regulatory gene the presence of which, or the expression of which at some stage, differs in the sexes. As a result of hijacking, sex-gene expression then determined the sexual phenotype and, incidentally, modified the condensation pattern of the sex chromosome. Expression of the sex gene is assumed to normalize chromosome conformation by permitting the simultaneous expression of the CE. Conversely, repression of the sex gene is assumed to bring about repression of the CE and to cause chromosome condensation and inactivation. As a result, the specialized sex chromosome exhibits allocyclic sex-linked behavior. The low probability of hijacking in evolutionary time explains the fact that not all closely related species, for example, among snakes, exhibit sex chromosome functional specialization.

According to the model, inactivation and reactivation of sex chromosomes in different tissues in each sex reflect sex-gene expression. For example, the expression of the X chromosome differs in female soma and germline where X inactivation is reversed or abolished. In contrast, X inactivation is absent in the male soma but occurs in the germline where one may reasonably assume that the expression of the major X-linked female sex gene concerned is repressed. It is interesting to note that Lifschytz and Lindsley[34] suggested from their studies in *Drosophila* that male germline X inactivation serves the purpose of inactivating genes whose activtities would be detrimental to spermatogenesis.

The permanent somatic inactivation of the Y and W chromosomes is interpreted to be an incidental consequence of the repressed state of their hijacking sex genes whose functions relate only to specific and limited germline development. Because these same genes are expressed discontinuously in the germline, they also interfere with the normal pairing conformation of the Y and W chromosomes, causing them to remain condensed during meiotic prophase. Consequently, pairing both of the Y and W was essentially abolished and they rapidly evolved substantial inertness. However, their sex genes, which according to the model control chromosome behavior, are expressed appropriately and are functionally conserved. Failure of pairing in the heterogametic sex is therefore interpreted in terms of the

Z and W and X and Y chromosomes carrying different hijacking sex genes, the different germline functions of which cause asynchronous chromosomal behavior.

VI. X INACTIVATION REVEALS THE PATTERN OF DEVELOPMENTAL GENE CONTROL

The early developmental plan of males and females is identical so that identical patterns of early somatic gene expression would be anticipated. From the fact that X chromosome inactivation coincides with the commencement of embryonic differentiaton, it may be suggested that the hijacking gene of the X chromosome is responsible for initiating development in male and female embryos. The pattern of X inactivation, which results in one X chromosome remaining active, indicates that the gene in question is expressed hemizygously in females as well as in males. The active allele is usually randomly and stably selected in each individual cell and is permanently required for somatic development. Where more than one X chromosome is present, only one is active. Therefore, the control of this hijacking gene incorporates a counting mechanism that may take account of the number of autosomal sets present. This latter conclusion is inferred from the reactivation behavior of the inactive X chromosome during meiosis.

In the mouse, reactivation of the inactive X chromosome occurs in the germline at about 12.5 days postcoitus when the germinal vesicle of diploid oocytes contains four times the haploid amount of DNA. However, XO females may be fertile so that this reactivation cannot be essential for germline development. It therefore seems likely that reactivation of the developmental gene which hijacked the X chromosome occurs in response to this increased ploidy.

Because both alleles become functional before meiotic prophase, X chromosome conformation is normalized such that when pachytene commences around day 14,[35] pairing occurs normally. Therefore, because of the continuous requirement for the expression of the X-linked female sex gene, and its response to X:autosome ratio, the hijacking of the X chromosome did not cause permanent genetic damage.

The somatically inactive W chromosome also decondenses (reactivates) during oogenesis,[28] indicating the involvement of its controlling sex gene in female germline differentiation. However, because this sex gene is responsible for differentiating between male and female germ cells, it must express only in the secondary oocytes. Accordingly, the W chromosome remains condensed and inactive in the germinal vesicle of the primary oocyte and becomes euchromatic only after reduction division.[28] Therefore, as in the case of the Y chromosome, W reactivation and decondensation occur too late to permit crossing-over with the Z chromosome.

One critical aspect of the model is supported by the fact that there is a *cis*-acting gene, *Xce (X-controlling element),* that controls X inactivation in somatic cells of mammalian females.[36] *Xce* is obviously a candidate for the hijacked CE proposed by the model. The model therefore predicts that a major gene, which initiates mammalian development, controls the function of *Xce* and may map in its immediate vicinity. This major gene is designated as a sex gene because, as indicated by the different patterns of X inactivation in males and females, its functions differ in the male and female germline.

VII. WHY Z CHROMOSOME INACTIVATION IS NEVER FOUND

The outstanding difference in sex chromosome evolution in the two major systems of heterogamety is the absence of Z chromosome inactivation. In principle, this could be explained by assuming that the remote chance of a hijacking mutation never happened to the Z. However, since it is hypothesized that the Y, W, and X chromosomes have sustained such mutations, this explanation is improbable. An alternative explanation is that hijacking of the Z chromosome is lethal. Lethality could be explained on the basis of the different distribution of the major sex genes in female, as compared with male, heterogamety. Thus, in XY male heterogamety, the genes respectively controlling male sex and heterogamety, both of which are somatically repressed, are carried by the Y chromosome and, by hijacking, have caused its somatic inactivation. In female heterogamety, however, the genes for maleness and heterogamety are carried separately — by the Z and W chromosomes, respectively. It therefore follows that both the W and Z chromosomes are at risk of becoming somatically inactivated as a result of being hijacked by sex genes that are limited in expression to the germline. Hijacking mutations on W chromosome are permissible and occur because its missing functions are supplied by the Z. However, according to this model, Z chromosome hijacking by its male-determining sex gene would result in both Z chromosomes inactivating somatically, and this would result in developmental nullisomy of the sex chromosomes in both sexes. For this reason, Z hijacking is not permissible and consequently, Z inactivation is never found.

VIII. THE HIJACK MODEL AND SEX DETERMINATION IN *DROSOPHILA*

The hijacking hypothesis explains the causal link between sex and X chromosome function as an incidental consequence of the mutation. This does not, of course, imply that sex chromosome inactivation is not adaptive and has no subsequent advantages.

The obligatory connection between sex-gene expression and X chromosome activity found in mammals is also evident in other systems. For example, major embryonic development in *Drosophila* is initiated by the gene *Sex-lethal, Sxl*, which is also fundamentally involved in setting the level of unrelated X-linked gene expression (reviewed in Chapter 3.[37]). Recent evidence also connects sex-gene expression with altered X chromosome function in *Cenorhabditis elegans*.[38] In this organism, a single X-linked gene, *scd-1*, controls both sex determination and levels of X-linked gene expression. Mutations in *scd-1* affect both masculinization (result in functional male development) and elevated levels of X-specific gene expression in XX animals. These data suggest that all of these systems in mammals, insects, and nematodes are similar and may be susceptible to explanation by the model under discussion.

Sex in *Drosophila* depends on the X:autosome ratio and is obligatively connected with the control of the level of X-linked gene expression in males and females. When female development is initiated, X-linked gene expression is halved in each X chromosome. Conversely, the level of X-linked gene expression is doubled when male development ensues.[39,40] Since a polytene X chromosome is constructed from several hundred strands each equivalent to one normal X chromosome, its behavior can be formally interpreted in terms of the individual response of each strand under the control of its sex gene, as envisaged by the hijack model. Thus, since in the eutherian mammalian female cell, one X chromosome inactivates at random, in a *Drosophila* cell with two polytene X chromosomes, on average, half of the constituent strands would be expected to inactivate. Thus, the female polytene X chromosome, instead of being totally inactivated as in mammals, is expressed at half the level of the single male X. Formally, therefore, the mechanism may involve sex genes controlling CEs competing independently for activating factors (L). The amount of L in cells of both sexes is sufficient to activate the number of strands equivalent to one polytene X and is determined by the number of sets of autosomes. XX female cells, therefore, will only contain enough L gene product to activate 50% of the constituent strands of each X chromosome. In male cells with one X chromosome, however, there is enough gene L product to activate 100% of the constituent strands, so the male X chromosome is transcribed at twice the level of the female X.

Consistent with a major conformational effect on the individual polytene strands, the chromatin structure of the single male X is less tightly packed and almost as broad as the paired homologs in females.[42] The male X also hybridizes much more efficiently *in situ* compared with the female X, or with the autosomes, consistent with a looser packing and consequently, greater accessibility to probes.[39]

As in mammals, all genes on the *Drosophila* X chromosome do not modulate their expression in response to the chromosomal sex of the cell. Therefore, the mechanism of inactivation once initiated must involve a class of secondary genes or target sequences adjacent to each individual-responding chromosomal domain, or perhaps each responding gene, on the chromosome.[41]

IX. GENES AND HIJACKING IN *DROSOPHILA*

In the light of current knowledge, the gene *Sxl*, which controls both sex and "dosage compensation",[39] is unlikely to be a candidate for the role of hijacking sex gene because it is "on" in females where the X is half expressed, and "off" in males where it is fully expressed. Thus, if we assume that repression of the hijacking sex gene (*S*) causes X inactivation, formally, *S* may control an X-linked repressor (*R*) of *Sxl*. Upon binding autosomal gene L product on a one-to-one basis, *S* would come "on", relax the constituent X chromosome polytene strand, and cause *R* to repress *Sxl* in *cis*. In XY:AA cells, gene *S* would become active on all constituent strands of the single X; therefore, all strands would relax and all *Sxl* gene copies would be "off", giving male development. Conversely, in XX:AA cells, only half of the copies of *S* would be "on" and therefore, only half of the strands would relax and half of the *Sxl* copies would be "on", giving female development.

The X:A ratio is important for male germline development in *Drosophila,* and the X chromosome is inactivated in this stage.[34] as in mammals. Therefore, on the basis of the model discussed above, *R* is repressed, so L must be unavailable; and L is perhaps bound preferentially by the Y chromosome. However, the role of *Sxl* is unclear in male germline development, so the model cannot be fully explored in this respect.

X. SEX-LINKED INACTIVATION OF GENOMES: A HIERARCHY OF CHROMOSOME CONTROL

A priori, it seems improbable that a gene with such fundamental effects as *Xce* would have arisen *de novo* on the mammalian X chromosome. More probably, it represents a genetic function carried by all chromosomes which has altered functionally on the sex chromosome, as envisaged by the model. It is therefore conceivable that the expression of normal, nonhijacked, CE genes is orchestrated in the mitotic cycle by master regulators (MCE, master chromosome elements) specific to the maternal and paternal genomes, respectively. Assuming that an MCE gene were to be hijacked by a major heterogametic sex gene, genomic inactivation would become sex linked. Developmental inactivation of the haploid somatic genome, known in certain insects such as mealy bugs,[43] would then ensue when this sex gene shut down during somatic development. In coccids, in both the Lecanoid and Comstockiella systems, embryos undergo total inactivation of the paternal somatic genome and become functionally haploid. The paternal chromosome set degenerates at meiosis. By contrast, the paternal and maternal genomes both remain euchromatic in females.

The levels of control of chromosome inactivation represented by CE and the postulated MCE may also include subordinate controlling elements (SCEs) downstream of the CEs which inactivate subdomains of the chromosome. In this case, hijacking in some ancestral species might have involved only such an SCE, thus causing only the corresponding segment of the sex chromosome to inactivate. This would explain the occurrence of differential segments in sex chromosomes which are not fully specialized. Thus, the hierarchy of control which almost certainly exists in chromosomes may risk various degrees of inactivation due to hijacking mutations ranging down from the whole parental genome to the individual chromosomal domain. Assuming a random probability of hijacking at the different control levels, the evolution of small differential segments will be favored over that of complete specialization. The evolution of paternal genomic inactivation would be the least probable because of the postulated singular nature of MCEs and the improbability of their being linked to major sex genes.

XI. HIJACKING MAINTAINS GENE BALANCE

Previous models for the evolution of specialized sex chromosomes have taken as their fundamental premise that the abrupt inactivation of the sex chromosome by a singular mutation would be lethal. Accordingly, they have gone to great lengths to attempt an explanation of how gene balance was maintained by prolonged and gradual selection for null alleles of small phenotypic effect. However, this assumption, which is based on the observed lethality of autosomal aneuploidy, may be unwarranted in the context of sex determination. Thus, the abrupt developmental inactivation of chromosomes under the control of major sex genes may, in fact, embody a mechanism which avoids resulting in a developmental gene imbalance. A clue to why sex-linked chromosomal inactivation, in general, is viable, is suggested from the pattern of mammalian somatic X inactivation. This is established at the inception of cellular differentiation in female somatic development;[44] and according to the hijack model, implies the hemizygous expression of the major X-linked sex genes which are, therefore, obviously dosage sensitive. Thus, major sex genes may be representative of a class of dosage-sensitive developmental genes. To facilitate discussion, we shall refer to this type of gene as class 1 and distinguish it from nondosage sensitive, or class 2, alleles which have been suggested to constitute the major proportion of the genome.[45]

From the pattern of X inactivation we can infer that all class 1 genes may be subject to a counting system which, by functionally excluding all but one allele, ensures hemizygous expression. By analogy with models for the control of X expression in *Drosophila*,[46,47] and consistent with X reactivation in 4C mammalian oocytes, the counting system may be suggested to be based on the number of sets of autosomes. Therefore, aneuploidy would probably disturb the balance of class 1 gene expression, causing inviability.

For normal development to occur, only one of each of any two class 1 alleles carried by a bivalent is required and is selected randomly and irreversibly. The developmental time of expression of a particular class 1 allele will differ between different genes. Therefore, as development proceeds, each one of a pair of homologous chromosome, complementarily, will express the class 1 alleles that are not expressed by its partner. On average, this means that each chromosome ultimately will be required to express only 50% of its class 1 alleles. Therefore, a mechanism that inactivates one of two homologous chromosomes at the very commencement of development will not create a dosage problem because normal development is sustained by expressing 100% of the class 1 alleles of the remaining chromosome.

The evidence from *Drosophila*[37,39] indicates that the sex genes initiate major embryonic development and control the expression of the X chromosome. Moreover, X inactivation in mammals precedes cellular differentiation.[44] Thus, by implication, the inactivation (or in *Drosophila*, the level of expression) of the X chromosome, under the control of the sex gene, precedes the expression of any embryonal class 1 genes.

It follows from this model that development is also sustainable if half of the genome is similarly inactivated at the commencement of development, as in the mealy bug. However, if chromosomal inactivation occurred later, as might happen if an autosome were hijacked by a later-acting class 1 gene, cellular and possibly organismal death would result. Sex chromosomes, therefore, may be unique in being able to sustain hijacking mutations in the absence of cellular lethality. However, one may speculate that chromosomal hijacking occurs continually in all chromosomes in evolutionary time and may have been adaptive in the evolution of patterns of developmental cell death which occurs as part of a normal developmental program.

The major sex genes involved in hijacking, therefore, may embody a mechanism which uniquely ensures that the abrupt developmental or evolutionary chromosomal inactivation, thereby caused, is not lethal. It follows that the imbalance problem which other models of sex chromosome evolution have sought to circumvent, more or less unsuccessfully, does not exist.

Hemizygosity, due to the evolution of chromosomal inactivation by hijacking, would expose recessive mutations which would result in increased selection pressure. However, this potential problem would be limited to class 2 genes because class 1 genes, by definition, are always selected for function in the hemizygous condition. This hemizygous mode of function would seem to be essential for most genes involved in controlling major patterns of development since it avoids the accumulation of functionally markedly different alleles which, because they control basic developmental patterns, would threaten the biological cohesion of a species. The unique exceptions to this rule are the sex genes, the alternative alleles of which ensure sexual dimorphism within species. However, such exceptions have evolved at the risk of losing an entire linkage group.

REFERENCES

1. **Müller, H.,** Genetic variability, twin hybrids and constant hybrids, in a case of balanced lethal factors, *Genetics,* 3, 422, 1918.
2. **Charlesworth, B.,** Model for the evolution of Y chromosomes and dosage compensation, *Proc. Natl. Acad. Sci. U.S.A.,* 75, 5618, 1978.
3. **Lyon, M.,** Gene action in the X chromosome of the mouse (*Mus musculus.* L.), *Nature (London),* 190, 372, 1961.
4. **Lyon, M.,** Evolution of X chromosome inactivation in mammals, *Nature (London),* 250, 651, 1974.
5. **Müller, H.,** Evidence for the precision of genetic adaptation, *Harvey Lect.,* 43, 165, 1950.
6. **Fisher, R. A.,** The sheltering of lethals, *Am. Nat.,* 69, 446, 1935.
7. **Hamilton, W. D.,** Extraordinary sex ratios, *Science,* 156, 477, 1967.
8. **Nei, M.,** Accumulation of nonfunctional genes on sheltered chromosomes, *Am. Nat.,* 104, 311, 1970.
9. **Lucchesi, J. C.,** Gene dosage compensation and the evolution of sex chromosomes, *Science,* 202, 711, 1978.
10. **Baverstock, P. R., Adams, M., Polkinghorn, R. W., and Gelder, M. A.,** sex-linked enzyme in birds Z chromosome conservation but no dosage compensation, *Nature (London),* 296, 763, 1984.
11. **Johnson, M. S. and Turner, R. G.,** Absence of dosage compensation for a sex-linked enzyme in butterflies (*Heliconius*), *Heredity,* 43, 71, 1979.
12. **Rice, W. R.,** Genetic hitchhiking and the evolution of reduced genetic activity of the Y sex chromosome, *Genetics,* 116, 161, 1987.
13. **Brown, S. W. and Chandra, H. S.,** Inactivation system of the mammalian X chromosome, *Proc. Natl. Acad. Sci. U.S.A.,* 70, 195, 1974.
14. **Ohno, S.,** Sex *Chromosomes and Sex-Linked Genes,* Springer-Verlag, Berlin, 1967.
15. **Humphrey, R. R.,** Sex determination in ambystomid salamander: a study of the progeny of females experimentally converted into males, *Am. J. Anat.,* 76, 33, 1945.
16. **Yamamoto, T.,** Sex differentiation, in *Fish Physiology III,* Hoar, W. S. and Randall, D. J., Eds., Academic Press, New York, 1969, 117.
17. **Winge, O. and Ditlevson, E.,** Color inheritance and sex determination, *Heredity,* 1, 65, 1947.
18. **Farr, J. A.,** Biased sex ratios in laboratory strains of guppies, *Lebistes reticulata, Heredity,* 47, 237, 1981.

19. **Manchino, G., Ragghanti, M., and Bucci-Innocenti, S.,** Cytotaxonomy and cytogenetics in European Newt species, in *The Reproductive Biology of Amphibians,* Taylor, D. H. and Guttman, S. I., Eds., Plenum Press, New York, 1977, 411.
20. **Schmid, M., Olert, J. and Klett, C.,** Chromosomal banding in amphibia. III. Sex chromosomes of *Triturus, Chromosoma,* 71, 29, 1979.
21. **Felsenstein, J.,** The evolutionary advantage of recombination, *Genetics,* 78, 737, 1974.
22. **Maynard-Smith, J.,** *The Evolution of Sex,* Cambridge University Press, New York, 1978.
23. **Beçak, W. and Beçak, M. L.,** Cytotaxonomy and chromosomal evolution in Serpentes, *Cytogenetics,* 8, 247, 1969.
24. **Singh, L.,** Evolution of karyotypes in snakes, *Chromosoma,* 38, 185, 1972.
25. **Jones, K. W.,** The evolution of sex chromosomes and their consequences for the evolutionary process, in *Chromosomes Today,* Bennett, M. D., Gropp, A., and Wolf, U., Eds., Allen and Unwin, London, 1984, 241.
26. **Ray-Chaudhuri, S. P., Singh, L., and Sharma, T.,** Sexual dimorphism in somatic interphase nuclei of snakes, *Cytogenetics,* 91, 410, 1971.
27. **Ray-Chaudhuri, S. P. and Singh, L.,** DNA replication pattern in sex chromosomes of snakes, *Nucleus (Calcutta),* 15, 200, 1972.
28. **Singh, L., Purdom, I. F., and Jones, K. W.,** Behaviour of sex chromosome-associated satellite DNAs in somatic and germ cells in snakes, *Chromosoma,* 71, 167, 1979.
29. **Singh, L. and Ray-Chaudhuri, S. P.,** Localization of C-band in the W sex chromosome of the common Indian krait *Bungarus caeruleus* (Schneider), *Nucleus (Calcutta,),* 18, 163, 1975.
30. **Olmo, E., Cobror, O., Morescalchi, A., and Odierna, G.,** Homomorphic sex chromosomes in the Lacertid lizard *Takydromus sexlineatus, Heredity,* 53 (Suppl 2), 457, 1984.
31. **Singh, L., Purdom, I. F., and Jones, K. W.,** Satellite DNA and evolution of sex chromosomes, *Chromosoma,* 59, 43, 1976.
32. **Jones, K. W.,** Evolution of sex chromosomes, in *Development in Mammals,* Vol. 5, Johnson, M. H., Ed., Elsevier, Amsterdam, 1983, 297.
33. **Jones, K. W.,** The evolution of sex chromosomes and chromosomal inactivation in reptiles and mammals, in *The Structure Development and Evolution of Reptiles,* Ferguson, M. W. J., Ed., Academic Press, New York, 1984, 315.
34. **Lifschytz, E. and Lindsley, D. L.,** The role of X chromosome inactivation during spermatogenesis, *Proc. Natl. Acad. Sci. U.S.A.,* 69, 182, 1972.
35. **Monk, M. and McLaren, A.,** X chromosome activity in foetal germ cells of the mouse, *J. Embryol. Exp. Morphol.,* 63, 75, 1981.
36. **Cattanach, B. M. and Isaacson, J. M.,** Controlling elements in the mouse X chromosome, *Genetics,* 57, 331, 1967.
37. **Cline, T. W.,** Exploring the role of the gene, Sex-Lethal, in the genetic programming of *Drosophila* sexual dimorphism, in *Evolutionary Mechanisms in Sex Determination,* Wachtel, S. S., Ed., CRC Press, Boca Raton, Fla., 1988, chap. 3.
38. **Villeneuve, A. M. and Meyer, B. J.,** Sdc-1: a link between sex determination and dosage compensation in *C. elegans, Cell,* 48, 25, 1987.
39. **Baker, B. S. and Belote, J. M.,** Sex determination and dosage compensation in *Drosophila melanogaster, Annu. Rev. Genet.,* 17, 345, 1983.
40. **Stewart, B. R. and Merriam, J. R.,** Regulation of gene activity by dosage compensation at the chromosome level in *Drosophila, Genetics,* 79, 635, 1975.
41. **Pardue, M-L., Lowenhaupt, K., Rich, A., and Nordheim, A.,** $(dC-dA)_n \cdot (dG-dT)_n$ sequences have evolutionarily conserved chromosomal locations in *Drosophila* with implications for roles in chromosome structure and function, *EMBO J.,* 6, 1781, 1987.
42. **Aronson, J. F., Rudkin, G. T., and Schultz, J.,** A comparison of the giant X chromosomes in male and female *Drosophila melanogaster* by cytophotography in the ultraviolet, *J. Histochem. Cytochem.,* 2, 458, 1954.
43. **Brown, S. W. and Chandra, H. S.,** Chromosome imprinting and the differential regulation of homologous chromosome, in *Cell Biology, Vol. 1,* Goldstein, L. and Prescott, D. M., Eds., Academic Press, New York, 1977, 109.
44. **Monk, M. and Harper, M. I.,** Sequential X chromosome inactivation coupled with cellular differentiation in early mouse embryos, *Nature (London),* 281, 311, 1979.
45. **Kaczer, H. and Burns, J.,** The molecular basis of dominance, *Genetics,* 97, 639, 1981.
46. **Gadagkar, R., Nanjundiah, V., Joshi, N., and Chandra, H. S.,** Dosage compensation and sex determination in *Drosophila:* mechanism of measurement of the X/A ratio, *J. Biosci.,* 4, 377, 1982.
47. **Chandra, H. S.,** Sex determination: a hypothesis based on noncoding DNA, *Proc. Natl. Acad. Sci. U.S.A.,* 82, 1165, 1985.

Chapter 7

A MOLECULAR ANALYSIS OF THE MOUSE Y CHROMOSOME

Colin E. Bishop, Christa Roberts, Jean Luc Michot, Claude M. Nagamine, Heinz Winking, Jean Louis Guenet, and Andreas Weith

TABLE OF CONTENTS

I. INTRODUCTION

The Y chromosome is the most specialized of mammalian chromosomes being involved almost exclusively in controlling primary sex determination and fertility. Under the dominant influence of the Y, the bipotent fetal gonad develops along the testicular pathway even when multiple copies of the X are present.[1,2] In the mouse, the process of testicular development, although undoubtedly initiated by the Y, involves a complex interaction with at least three autosomally located genes: *Tda-1, Tda-2,* and *Tas.*[3] At present, the way in which the Y controls this pathway is unknown. We favor the simplest explanation — that there is a primary sex-determining gene(s) (*Tdy*) regulatory in action and located on the Y. An alternative hypothesis has been put forward by Chandra[4] based on a passive Y chromosome containing only noncoding DNA.

The most powerful and direct method to analyze the role of the Y in this process would be to identify, map, and clone *Tdy.* This analysis has already started in the human with the analysis of numerous XX males and XY females by application of Y chromosome probes. In molecular terms, the human equivalent of *Tdy* (*TDF*) has been mapped to the short arm of the Y between the locus *MIC2* and locus *DXYS5* defined by random probe 47a.[5] Hence, it is proximal to the telomeric X-Y recombination (or pseudoautosomal) region, but distal to the centromere. In the mouse, an analysis of the *Sxr* mutation (see below) with a snake satellite DNA probe — banded krait minor (Bkm) — has led to the identification of the pericentric region of the Y as the putative sex-detemining region.[6] In this respect, it differs from the human sex determinant; it does not map near the telomeric pseudoautosomal region. Using a variety of molecular probes, the *Sxr* mutation, and a newly defined Y chromosome rearrangement (Y*), Eicher and colleagues[3,7] have been able to split the mouse Y into four functional regions. Region 1 contains the centromere; region 2, the Bkm-related sequences, *Tdy* and H-Y; region 3, a central region containing repeated viral sequences related to M720 and MuRVY; and region 4, a telomeric region involved in homologous pairing and recombination with the X (Figure 1). Keitges et al.[8] have presented evidence for the presence of X- and Y-linked genes for steroid sulfatase mapping to this region.

The present paper summarizes recent work in our laboratory involving isolation of DNA probes from the mouse Y chromosome and, in particular, from the pericentric region, and use of the probes to identify potential Y-located genes that may play a role in primary sex determination and/or fertility.

II. THE ORIGIN AND SEGREGATION OF THE *Sxr* MUTATION

The *Sxr (sex reversed)* mutation was discovered by Cattanach et al.[9] Males carrying this mutation were able to sire male mice with an apparent XX karyotype. Although the mutation segregated as an autosomal dominant, extensive linkage analysis failed to map the locus. In 1982, Singh and Jones[6] isolated minor satellite DNA sequences originally from a female-banded krait (Bkm) which hybridized strongly to the mouse Y. Using this probe, they were able to show that XX*Sxr* male mice did in fact carry in their genomes a small portion of the Y that presumably included *Tdy.* Subsequent *in situ* hybridization data showed that Bkm hybridized to the pericentric region of the normal Y and to both the pericentric and telomeric regions of the Y in XY*Sxr* carrier males. Unlike the situation in normal females, a concentration of Bkm could be found in the telomeric region of one X chromosome in XX*Sxr* males. This led to the proposed origin and segregation of *Sxr* as shown in Figure 1.[10-12] It was proposed that the pericentric sex-determining region of the Y was transposed from one chromatid to the telomere of the other, distal to the pseudoautosomal region. Hence, during male meiosis, this Sxr chromatin could be transferred to the distal region of the paternal X by recombination making the X function as a Y and thereby inducing testis formation in XX*Sxr* mice. As indicated, four types of progeny are produced: nonrecombinant XX females

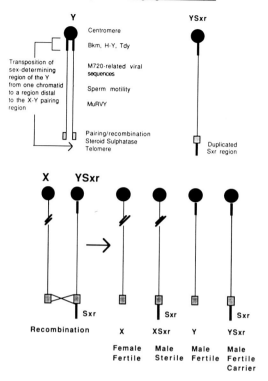

FIGURE 1. Diagram showing the proposed origin and segregation of *Sxr*.

and XY*Sxr* carriers, recombinant XX*Sxr* sterile males, and XY fertile males that have lost *Sxr* by recombination.

This heritable mutation offers significant potential for molecular analysis as XX*Sxr* mice carry a minute but critical portion of the Y bearing not only *Tdy* but also *H-Y* as defined by the cytotoxic T cell assay.[13] Recent evidence suggests that *H-Y* may be involved in spermatogenesis.[14,15] In addition, an aberrant Sxr type has been identified (designated *Sxr'*); it retains *Tdy* but no longer expresses *H-Y*,[16] and this represents another source of genetic material for analysis.

A. Isolation of DNA Sequences Derived from the Mouse Y Chromosome

1. Flow Sorting

The potential of random DNA probes for analyzing a similar, although generally sporadic, condition in the human — the XX male syndrome — has been proven. Hence, we decided to use this approach to analyze the mouse Y at the molecular level. We first constructed chromosome libraries highly enriched for the Y by flow sorting.[17] In this technique, isolated metaphase chromosomes are stained with ethidium bromide or other DNA fluorochromes and passed through a laser. The emitted fluorescence, which is proportional to DNA content, allows individual chromosomes to be separated by size. Preliminary flow karyotype analysis showed that although the Y could be separated from other chromosomes, it could not be resolved from chromosome 19. To circumvent this problem a male cell line homozygous for the Rb(9:19) robertsonian translocation was used. Due to its now increased size the 19 was removed from the Y peak (Figure 2). After collection of this peak, the DNA was extracted (approximately 70 ng representing 650,000 chromosomes) and half digested with EcoR 1 and half with Hind III. In this way, two complementary libraries representing many

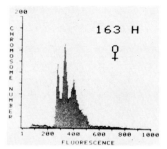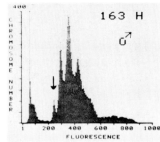

FIGURE 2.   Flow-sorting karyotype of the male mouse fibroblast line (163SV40/Pas) used for isolating the mouse Y chromosome. The Y peak is arrowed. Chromosome 19 which usually contaminates this peak has been removed by the use of the 9:19 robertsonian translocation.

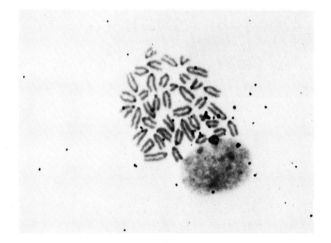

FIGURE 3.   *In situ* hybridization of the Bkm probe to metaphase chromosomes from a male mouse carrying the pericentric Y chromosome. The concentration of grains of the Y are localized to the short arm.

genome equivalents of the Y were constructed in phage lambda 1149. From these libraries a large number of Y-derived clones from all regions of the Y were obtained.

2. Microdissection and Microcloning

As the Sxr DNA represents only a small portion of the Y, a large number of clones must be screened to obtain sequences specific for this region. An alternative strategy is to directly microdissect the region of interest from metaphase chromosomes and microclone the picogram quantities of DNA obtained. This technique has been applied successfully to mouse chromosomes 17, X, and 1.[18-20] For this technique, it is necessary to have a marker chromosome that can be easily and unequivocally identified without staining. This is not feasible with the normal mouse Y. Recently, Winking[21] has reported finding a metacentric Y formed through a pericentric inversion. The short arm of this chromosome represents 25 to 30% of the Y. In addition, mice carrying this easily recognizable Y are fully fertile. In order to be sure that the putative sex-determining region was located on the short arm, *in situ* hybridization of Bkm which defines the region was carried out. As can be seen in Figure 3, an intense hybridization could be detected exclusively on the short arm of the Y. We then performed a microcloning experiment by directly dissecting this short arm from approxi-

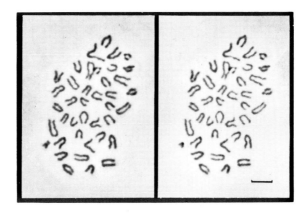

FIGURE 4. Microdissection of the short arm of the pericentric Y. Left panel: before microdissection, the Y can be seen at 8 o'clock (unstained preparations from embryonic yolk sac viewed under phase contrast microscopy). Right panel: same preparation after microdissection of the short arm.

mately 100 metaphase spreads as shown in Figure 4. The DNA (approximately 3 pg) was cloned into the immunity insertion vector NM1149, and 502 independent recombinants were obtained. The analysis of this bank is now underway; from 91 inserts tested, we have so far obtained four unique or low copy number probes mapping to the Sxr region.

B. Transcription from the Mouse Y Chromosome
1. Non-Sxr Transcription

Using pools of Y-derived probes taken from the flow-sorted Y libraries to probe Northern blots of testis mRNA, we identified a Y-specific sequence (pY353/B) that detected a specific transcript in the testis (Figure 5).[22] The genomic sequence itself is repeated about 250 times and our *in situ* hybridization studies suggest that it maps along the entire length of the Y. It does not react with XXSxr DNA, however, showing that it is absent from this region. The 1.3-kb transcript appears to be testis-specific as it could not be detected in RNA from liver, spleen, lung, heart, kidney, or brain. Sequence analysis of the cloned cDNA suggested that it was not retroviral in origin and that it contains a potential open reading frame of 760 bases. Although we have not yet been able to assign a function to the transcript, it is possible that it might play a role in sperm motility.

2. Transcription from the Sxr Region

In an effort to identify transcription from this region, we probed our flow-sorted Y libraries at low density with a mixture of total mouse DNA, Bkm, M720, pY353/B, and a mixture of anonymous Y-repeated sequences. The negative plaques were then replica-plated and screened with cDNA probes synthesized from male testis, male liver, and female liver. Five clones that were positive with testis RNA but negative with liver RNA were identified. By probing genomic blots with these clones, we identified pCRY8/B. Y8 is a 2.2-kb EcoR 1 fragment; shown in Figure 6 (left panel), it reacts with male DNA from the inbred C57BL/6 mouse strain but not with female DNA. This indicates that the bands are Y located. Further, all the bands can be seen to react with XXSxr DNA indicating that they are contained within this critical region. Examination of DNA from a family-segregating Sxr (Figure 6, right panel) reveals that the probe reacts with sequences from XYSxr, XY, and XXSxr males, but not with sequences from the normal XX female. An increased intensity of hybridization can be seen on the XYSxr DNA as compared to XY or XXSxr DNA, showing that the

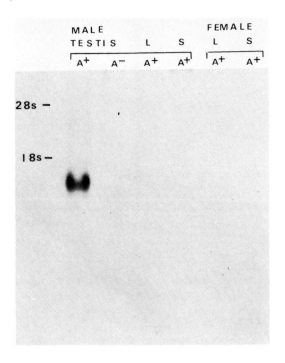

FIGURE 5. Northern blot analysis of the probe pY353/
B. An approximately 1.3-kb transcript can be detected in
the testis but not in male or female liver or spleen.

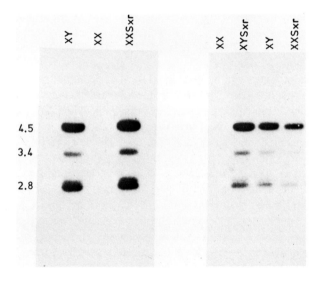

FIGURE 6. Southern blot analysis (Taq 1 restricted) using Sxr-specific
probe pYCR8/B. Left panel: Four Y-specific bands can be detected in
male (XY) but not female (XX) C57BL/6 DNA. These same bands can
be seen in sex-reversed male (XX*Sxr*) DNA showing them to be contained
within the Sxr region. Right panel: reaction of the probe with DNA from
a family segregating Sxr (129Sv/Pas N4).

L   T

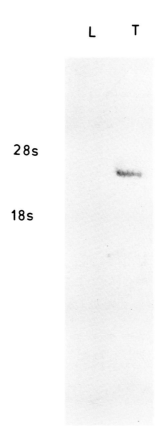

FIGURE 7. Northern blot analysis of poly A⁺RNA from adult male liver
(L) and testis (T) (strain C57BL/6) using probe pYCR8/B. A transcript of
approximately 3.5 kb can be seen in the testis but not in the liver.

sequence probably has been duplicated in XY*Sxr* cells. Careful dosage studies have confirmed this result and are consistent with the proposed origin of Sxr by duplication.

As this sequence was initially identified with cDNA synthesized from testis mRNA, it should be transcribed. Preliminary data obtained by probing Northern blots of mRNA from testis and male and female liver indicate that pCRY8/B detects a single-band transcript of approximately 3.5 kb in the testis but not in the liver (Figure 7). At present we are trying to clone the corresponding cDNA and to analyze further this potentially important result.

C. Identification of a Functional Y-Located Gene for *Steroid Sulphatase (STS)* and Linkage to *Sxr*

Evidence has recently been put forward by Keitges et al.[8] for X and Y linkage of the gene for steroid sulfatase in the mouse. This was based on the apparently paradoxical situation that showed that *STS* was definitely X-linked although it segregated as if it were an autosomal gene. They postulated that if there were functional X- and Y-linked genes in the pseudoautosomal region, the alleles could recombine during male meiosis and mimic an autosomal segregation pattern. As the observed crossover frequently, with respect to sex, was 50%, the genes should map very close to the telomere. In the Y*Sxr* chromosome, one copy of the Sxr region is situated distal to the X-Y recombination (or pseudoautosomal) region. Hence, on this chromosome these two interesting regions must be physically very close, and *Sxr* itself can be used as a marker for the telomere. We therefore set up crosses using the steroid-sulfatase-deficient mouse strain C3H/An and *Sxr* to test the hypothesis. At the same time,

Table 1
LINKAGE ANALYSIS OF *STEROID SULFATASE (STS)* AND *Sxr*

Number of mice N = 47	Phenotype	Presence of entire Y chromosome		Presence of Sxr region		Steroid sulfatase	Genotype
		p80	p353	pY8	pC66		
11	F	−	−	−	−	−	XX
11	M	−	−	+	+	+	XX*Sxr*
15	M	+	+	+	+	−	XY or XY*Sxr*
10	M	+	+	+	+	+	XY*Sxr* or XY

Note: Analysis of *STS* was performed using ^{3}H dehydroepiandrosterone sulfate on spleen or liver samples. Phenotype was based on examination of external genitalia. Taq 1-restricted DNA from spleen was used for Southern blot analysis. Presence of the Y chromosome was detected using repeated (non-Sxr located) Y chromosome probes pY80/B and pY353/B, thus distinguishing normal and XX*Sxr* males. The presence of the Sxr region was detected using Sxr-specific probes pCRY8/B and pYC66/B.

we were able to analyze the progeny of the cross with our Y-derived DNA probes to look for new rearrangements involving this region of the Y.

The following cross was set up: $STS^-/STS^- \times (STS^-/YSxr)$. If there is a telomeric Y-located *STS* gene linked to *Sxr*, four types of progeny should result: female XX ($STS-/-$), male XX*Sxr* ($STS-/+$), male XY ($STS-/-$), and male XY*Sxr* ($STS-/+$) (Figure 1). The results of analysis of 47 backcross mice are presented in Table 1. Of 11 XX female mice, all were STS negative, and of 11 XX*Sxr* males, all were STS positive. In order to distinguish XY males from XY*Sxr*-carrier males, we selected ten mice at random for progeny testing. Preliminary results show that the STS positive males segregate *Sxr* as predicted, whereas the STS negative males do not. These results clearly demonstrate that the Y carries a fully functional *STS* gene. Because it is tightly linked to *Sxr*, the *STS* gene must be located in the distal X-Y pairing region. To date, one exceptional male mouse which is STS positive but which does carry *Sxr* has been found. The simplest explanation is that despite its distal location, *STS* and *Sxr* can occasionally be separated by recombination, and that *Sxr* must be more distal than *STS* on the Y. Experiments are currently under way to assess the frequency. Careful DNA analysis suggests that it is routinely possible to distinguish XY males from XY*Sxr* carrier males by dosage with Sxr probe pYCR8/B. Hence, the tedious process of progeny testing can probably be avoided in the future. Finally, no obvious rearrangements were found at the DNA level using Y-derived repeated (non-Sxr located) probes pY353/B or pY80/B or Sxr-located probes pCR8/B or pYC66/B.

D. Polymorphism in the Sxr Region

We have previously reported that by use of the Y-specific repeated probe pY353/B, the Y chromosomes of the European semi-species *Mus musculus musculus* and *M. musculus domesticus* could be distinguished on the basis of RFLP (restriction fragment-length polymorphisms). Further analysis of DNA from common laboratory strains showed that despite presence of domesticus-type mitochondrial DNA, only SJL and AKR/J carried a domesticus-derived Y chromosome.[23,24] The other strains tested, including C57BL/6, carried a musculus-derived Y. We have subsequently confirmed this observation with numerous Y-derived probes both repeated and single copy.[24] When the domesticus Y chromosome is placed on the C57BL/6 background, XY sex reversal results, generating XY females and hermaphrodites (reviewed by Eicher and Washburn[3]). Those authors suggest that this may result from

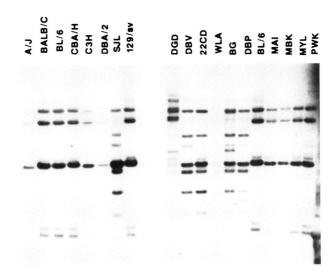

FIGURE 8. Southern blot analysis (Msp1 restricted) of laboratory and wild-derived mouse strains using probe pYCR8/B. Left panel: a clear polymorphism can be seen between laboratory strain SJL/Pas which carries the domesticus type Y and the other strains which all carry a musculus-derived Y. Right panel: no polymorphisms can be seen within the musculus mouse group (MAI, MBK, MYL, PWK, and C57BL/6) although extensive polymorphisms can be detected within the domesticus group (DGD, DBV, 22CD, WLA, BG, and DBP). The weak signal seen in WLA was due to underloading.

an improper interaction between different musculus and domesticus-derived alleles of the Y-located *Tdy* and autosomally-located *Tda-1* and *Tas* genes. Recently, Nagamine and Koo[25] presented evidence that the X-Y pairing region may be involved in *Tda-1* inherited sex reversal. In addition, they reported finding different degrees of fetal sex reversal when the Y chromosomes from SJL and AKR were placed on the C57BL/6 background.[26] They suggested that although both SJL and AKR carry a domesticus type Y, they may carry different alleles at *Tdy* or at other genes involved in the testis differentiation pathway.

We recently used our Sxr-specific probe pCRY8/B to investigate the polymorphisms that might exist in this region of the Y. Such a study might throw some light on these complex XY sex-reversed phenomena. Figure 8 shows that when Y8 was used to probe Msp1-digested DNA from laboratory inbred strains, a musculus/domesticus RFLP could be seen with only the SJL strain carrying the domesticus type Y. When different domesticus and musculus mice originating from several locations in Europe were tested (for the origins of these mice see Bishop et al.[23] and Bonhomme and Guenet[27]), we were unable to detect any RFLP within the musculus group, but within the domesticus mice six haplotypes could be distinguished. A more detailed analysis using more enzymes, mice, and probes is now under way. These preliminary results indicate, however, that the Sxr region of the Y is evolving at a surprisingly rapid rate, and may well correlate with the type and degree of inherited XY sex reversal associated with the domesticus Y chromosome.

E. Prospectives

The isolation of unique or low copy number DNA sequences that detect specific DNA fragments within the Sxr sex-determining region of the mouse makes possible a direct analysis of this region at the molecular level. Furthermore, as the Sxr region is linked to the pseudoautosomal region in XY*Sxr* and in XX*Sxr* mice, and because the isolation of DNA from this region has already been reported,[28] the simultaneous analysis of both regions becomes possible. The development of pulsed field gel electrophoresis[29,30] has allowed long-range mapping of DNA over distances well in excess of a million base pairs.[31,32] The application of this technique to an analysis of the Sxr region of the mouse will be an important first step and is under way in our laboratory. Among the questions we want to answer are how big is the region, and how much of it is made up of the GATA-GACA tandem repeats? To date, we have isolated five Sxr clones from our flow-sorted and microcloned libraries. These probes range in size from 0.6 to 2.2 kb. Some of them detect multiple fragments in the Sxr region which, on the basis of Southern blot analysis, appear not to be contiguous. Hence, it may be possible to clone a significant portion of the region simply by using them to isolate large genomic fragments from an XX*Sxr* cosmid library.

One of the main goals is to isolate functional genes from this region. Our preliminary data show that pYCR8/B detects a 3.5-kb transcript in the testis originating from this region but further analysis will have to await the isolation of the cDNA. Further transcripts can be sought by using isolated cosmids to probe northern blots of testis RNA. This approach has its problems for although H-Y has been shown to be ubiquitously expressed, *Tdy* may not be expressed in the adult testis. Possibly, expression for a relatively short time during embryogenesis is all that is required to initiate testis development. Perhaps a more effective approach would be to look for sequence conservation between man and mouse, because such sequences often represent expressed genes.[33] Sxr-located cosmids that are also cross-reactive with the short arm of the human Y and present in the genome of XX (Y$^+$) males would surely be good candidates. McLaren et al.[16] have described an Sxr' mouse that does not express H-Y. Although it is unknown if this arose through a small deletion or point mutation which would be difficult to detect, it would certainly be worthwhile examining this mouse with Sxr-specific probes. It is intriguing that the region appears to be highly polymorphic and must be evolving at a very rapid rate. Further study may well reveal a correlation between the Y haplotypes detected in the wild-derived strains and their ability to produce sex reversal when bred to laboratory strains.

In conclusion, the isolation of numerous DNA probes from the Sxr region of the mouse Y chromosome represents a significant advance in the study of primary sex determination and fertility. The use of these tools, in conjunction with recent advances such as pulsed field gel electrophoresis and transgenic mice, should now allow rapid progress in this area of research.

III. SUMMARY

The mouse Y chromosome plays a fundamental role in the control of primary sex determination and fertility. Genetic and molecular biological evidence has shown that the majority of these functions are encoded by a minute piece of the Y (the Sxr region) which has arisen through duplication of the pericentric region of the normal Y. The present article described the isolation of random Y chromosome probes and their use to investigate this Sxr region at the molecular level. Total mouse Y chromosome libraries were constructed from flow-sorted material and an Sxr regional library was constructed after specific microdissection and cloning. Transcription has been detected in the testis using Sxr-specific and non-Sxr-located genomic probes taken from these libraries. In addition, we have been able to confirm the presence of an active steroid sulfatase gene on the mouse Y. This gene is located in the distal portion of the pseudoautosomal region and is tightly linked to *Sxr*. Finally, by use of an Sxr-specific probe we can define multiple Y chromosome haplotypes in the mouse showing that the region is evolving very rapidly.

ACKNOWLEDGMENTS

We acknowledge the continuous interest and support of Drs. M. Fellous, P. Avner, and C. Rijnders during the course of this work. We thank Dr. E. Eicher for supplying the M720 probe, and Dr. Jean Weissenbach for reading the manuscript.

REFERENCES

1. **Jacobs, P. A. and Strong, J. A.,** A case of human intersexuality having a possible XXY sex-determining mechanism, *Nature (London),* 183, 302, 1959.
2. **Russell, L. B. and Chu, E. H.,** An XXY male in the mouse, *Proc. Natl. Acad. Sci. U.S.A.,* 47, 571, 1961.
3. **Eicher, E. and Washburn, L. L.,** Genetic control of primary sex determination in mice, *Annu. Rev. Genet.,* 20, 327, 1986.
4. **Chandra, H. S.,** Sex-determination: an hypothesis based on noncoding DNA, *Proc. Natl. Acad. Sci. U.S.A.,* 82, 1165, 1985.
5. **Vergnaud, G., Page, D. C., Simmler, M. C., Brown, L., Rouyer, F., Noel, B., Botstein, D., de la Chapelle, A., and Weissenbach, J.,** A deletion map of the human Y chromosome based on DNA hybridization, *Am. J. Hum. Genet.,* 38, 109, 1986.
6. **Singh, L. and Jones, K. W.,** Sex reversal in the mouse is caused by a recurrent nonreciprocal crossover involving the X and an aberrant Y chromosome, *Cell,* 28, 205, 1982.
7. **Eicher, E., Phillips, S. J., and Washburn, L. L.,** The use of molecular probes and chromosomal rearrangements to partition the mouse Y chromosome into functional regions, in *Recombinant DNA and Medical Genetics,* Messer, A. and Porter, I. H., Eds., Academic Press, New York, 1983, 57.
8. **Keitges, E., Rivest, M., Siniscalco, M., and Gartler, S. M.,** X linkage of steroid sulphatase in the mouse and evidence for a functional Y linked allele, *Nature (London),* 315, 226, 1985.
9. **Cattanach, B. M., Pollard, C. E., and Hawkes, S. G.,** Sex-reversed mice: XX and XO males, *Cytogenetics,* 10, 318, 1971.
10. **Eicher, E. M.,** Primary sex determining genes in mice, in *Prospects for Sexing Mammalian Sperm,* Amann, R. P. and Seidel, G. E., Jr., Eds., Colorado Association University Press, Boulder, 1982, 121.
11. **Hansmann, I.,** Sex-reversal in the mouse, *Cell,* 30, 331, 1982.
12. **Burgoyne, P. S.,** Genetic homology and crossing over in the X and Y chromosomes of mammals, *Hum. Genet.,* 61, 85, 1982.
13. **Simpson, E., Edwards, P., Wachtel, S., McLaren, A., and Chandler, P.,** H-Y antigen in Sxr mice detected by H-2 restricted cytotoxic T cells, *Immunogenetics,* 13, 355, 1981.
14. **Burgoyne, P. S., Levy, E. R., and McLaren, A.,** Spermatogenetic failure in male mice lacking H-Y antigen, *Nature (London),* 320, 170, 1986.
15. **Levy, E. R. and Burgoyne, P. S.,** The fate of XO germ cells in the testis of XO/XY and XO/XY/XYY mouse mosaics: evidence for a spermatogenesis gene on the mouse Y chromosome, *Cytogenet. Cell Genet.,* 42, 208, 1986.
16. **McLaren, A., Simpson, E., Tomonari, K., Chandler, P., and Hogg, H.,** Male sexual differentiation in mice lacking H-Y antigen, *Nature (London),* 312, 552, 1984.
17. **Baron, B., Metezeau, P., Hatat, D., Roberts, C., Goldberg, M., and Bishop, C. E.,** Cloning of DNA libraries from mouse Y chromosomes purified by flow cytometry, *Somatic Cell Mol. Genet.,* 12, 289, 1986.
18. **Rohme, D., Fox, H., Hermann, B., Frischauf, A.-M., Edstron, J. E., Mains, P., Silver, L., and Lehrach, H.,** Molecular clones of the mouse t complex derived from microdissected metaphase chromosomes, *Cell,* 36, 783, 1984.
19. **Fisher, E. M., Cavanna, S. S., and Brown, S. D.,** Microdissection and microcloning of the mouse X chromosome, *Proc. Natl. Acad. Sci. U.S.A.,* 82, 5846, 1985.
20. **Weith, A., Winking, H., Brackmann, B., Baldyreff, B., and Traut, W.,** Microclones from a mouse germ line HSR detect amplification and complex rearrangements of DNA sequences, *EMBO J.,* in press, 1988.
21. **Winking, H.,** personal communication, 1987.
22. **Bishop, C. E. and Hatat, D.,** Molecular cloning and sequence analysis of a mouse Y chromosome transcript expressed in the testis, *Nucleic Acids Res.,* in press, 1988.

23. **Bishop, C. E., Boursot, P., Baron, B., Bonhomme, F., and Hatat, D.,** Most classical *Mus musculus domesticus* laboratory mouse strains carry a *Mus musculus musculus* Y chromosome, *Nature (London)*, 315, 70, 1985.

24. **Bishop, C. E.,** unpublished results, 1987.

25. **Nagamine, C. M. and Koo, G.,** Evidence that the X-Y pairing/recombination region in the mouse may be involved in Tda-1 inherited sex-reversal, in *Genetic Markers of Sex Determination*, Plenum Press, New York, in press, 1988.

26. **Nagamine, C. M., Taketo, T., and Koo, G.,** Studies on the genetics of Tda-1 XY sex reversal in the mouse, *Differentiation*, in press, 1988.

27. **Bonhomme, F. and Guenet, J.-L.,** The wild house mouse and its relatives, in *Genetic Variants and Strains of the Laboratory Mouse*, Lyon, M. F. and Searle, A. G., Eds., Oxford University Press, New York, in press, 1988.

28. **Harbers, K., Soriano, P., Müller, U., and Jaenisch, R.,** High frequency of unequal recombination in pseudoautosomal region shown by proviral insertion in transgenic mice, *Nature (London)*, 324, 682, 1986.

29. **Carle, G. F. and Olson, M. V.,** Separation of chromosomal DNA molecules from yeast by orthagonal-field-alteration gel electrophoresis, *Nucleic Acids Res.*, 12, 5647, 1984.

30. **Schwartz, D. A. and Cantor, C. R.,** Separation of yeast chromosome sized DNA's by pulsed field gradient gel electrophoresis, *Cell*, 37, 67, 1984.

31. **Brown, W. R. A. and Bird, A. P.,** Long range restriction site mapping of mammalian DNA, *Nature (London)*, 322, 477, 1986.

32. **Van Omen, G. J. B., Verkerk, J. M. H., Hofker, M. H., Monaco, A. P., Kunkel, L. M., Ray, P., Worton, R., Wieringa, B., Bakker, E., and Pearson, P. L.,** A physical map of 4 million bp around the Duchenne dystrophy gene on the human X chromosome, *Cell*, 47, 499, 1986.

33. **Monaco, A. P., Neve, R. L., Colletti-Feener, C., Bertelson, C. J., Kurmit, D. M., and Kunkel, L. M.,** Isolation of candidate cDNA's for portions of the Duchenne muscular dystrophy gene, *Nature (London)*, 323, 646, 1986.

Chapter 8

MOLECULAR BIOLOGY OF THE HUMAN Y CHROMOSOME

Ulrich Müller

TABLE OF CONTENTS

I. INTRODUCTION

The human Y chromosome can be divided into two portions by cytogenetic criteria. One, the euchromatic portion, comprises the Y chromosome short arm and proximal long arm and can be subdivided into several segments by chromosome banding.[1] The second, or heterochromatic, portion consists of the quinacrine-positive,[2] heterochromatic, late-replicating material[3] that makes up the distal part of the Y chromosome long arm and the centromeric region. The occasional absence of the entire quinacrine-positive long arm in normal fertile males[4] suggests that this portion does not contain functional genes. Y-chromosomal genes appear to be confined to the euchromatic portion. These include putative testis-determining genes on the Y chromosome short arm,[5] and genes required for normal spermatogenesis on the proximal long arm.[6]

This chapter reviews the current knowledge of the structure of the Y chromosome at the molecular level and summarizes current approaches toward the isolation of gene(s) that might be involved in testicular determination/differentiation.

II. Y HETEROCHROMATIC PORTION

A. Distal Yq (Yq12)

Differential hybridization experiments using male DNA and an excess of female DNA allowed the isolation of repeated DNA sequences that were primarily present in male DNA.[7] These sequences were assigned to Yq by deletion mapping.[8] Similarly, restriction enzyme cleavage of DNA of both sexes with Hae III revealed two repeated sequence-derived sets of fragments of 3.4 kb and 2.1 kb in male but not in female DNA.[9] By use of the two Hae III restriction fragments or subclones thereof as hybridization probes, the repeats were assigned to Yq.[10] Although the major portion of the repeated sequences is confined to the heterochromatic portion (Yq12), a small percentage of the 3.4-kb fragments may spread into the euchromatic portion of Yq.[11] The 2.1-kb repeat appears to be located most distally on the Y long arm, close to the tip.[12] Both Hae III fragments are tandemly repeated and are related to human satellite III.

There are approximately 4000 copies of the 3.4-kb repeat and about 2000 copies of the 2.1-kb repeat.[11] The Y-chromosomal satellite DNA of Yq may comprise 60 to 70% of the total Y-chromosomal DNA content.[11] Since there appears to be no functional genes in the heterochromatic portion of the Y chromosome, the Yq repeats either may be "selfish"[13] or "junk" DNA, or may have some structural, but as yet unknown, function. The notion that deletions of Yq may result in meiotic arrest[14] suggests a function of distal Yq in meiosis.

B. Y Centromere

The centromeres of all chromosomes contain alphoid satellite DNAs. Different chromosomes may differ, however, in the composition of their centromeric satellites. Thus, a subset of alphoid satellites was found, confined primarily to the X chromosome.[15,16] Y-chromosomal centromeric alphoid repeats also have been detected.[17] On screening human Y-derived cosmid clones with a DNA probe characteristic of human X-chromosomal alphoid repeats, Wolfe et al.[17] isolated several positive cosmids. A 5.5-kbp (kilobase pair) EcoR1 fragment isolated from one of the cosmids was specific for Y-chromosomal repeated sequences. These sequences were confined to the Y centromere; they were repeated approximately 100-fold and exhibited about 70% sequence homology with X-chromosomal centromeric repeats. (For further details on the molecular structure of the Y centromere, see Tyler-Smith[18].)

III. Y EUCHROMATIC PORTION

A. Pseudoautosomal Region

The distal short arms of the nonhomologous human sex chromosomes form a synaptonemal complex during meiosis I. Crossing-over occurs between the pairing segments.[19] These observations have led to the suggestion that at least partial homology exists between distal Xp and Yp.[20] Replication kinetics studies on prophasic human X and Y chromosomes support this notion. An early replicating segment was detected on the distal short arm of both the X and the Y chromsomes.[21] This might reflect homologous regions on the distal short arms of the sex chromosomes.

Recent molecular findings substantiate the circumstantial cytogenetic evidence of sequence homology between distal Xp and Yp. The distal short arms of both sex chromosomes were found to contain highly polymorphic DNA sequences.[22-25] Pedigree studies using these polymorphic sequences did not reveal sex linkage. The DNA sequences of this region recombine frequently. This indicates their homology on distal Xp and Yp and assigns them to the "pseudoautosomal region"[26] of the sex chromosomes. Interestingly, the recombination frequency in the terminal regions of Xp and Yp (male meiosis) is tenfold greater than it is in the same regions of the X chromosomes in female meiosis.[24] The hypervariaibility of the pseudoautosomal region could be ascribed to differences in the copy number of arrays of short tandemly repeated DNA sequences ("minisatellites").[27] One structural gene was assigned to this region of Xp and Yp.[28,29] This gene(s) codes for 12E7, a cell-surface antigen of unknown function. The regulation of its expression is not yet fully understood.

B. Yp and Proximal Yq

The Y-chromosomal regions between the centromere and pseudoautosomal region on the short arm and between the centromere and Yq12 on the long arm are hypothesized to contain functional genes. These include the testis-determining gene(s) on Yp[5] and the gene(s) required for normal spermatogenesis on proximal Yq.[6] Probing with Y-chromosomal DNA sequences ("probes") of DNA from patients with disorders of gonadal differentiation allows the delineation of the testis-determining locus. Similarly, Y-chromosomal probes can be used for deletion mapping of the putative spermatogenesis locus on Yq in male individuals with disorders of spermatogenesis and aberrations of Yq. Although many Y-chromosomal DNA sequences have been isolated from proximal Yq, a deletion map of this region with respect to the spermatogenesis locus is not yet available. The following discussion focuses on the putative testis-determining locus and reviews recent experiments toward the eventual isolation of the testis-determining gene(s).

C. Testis-Determining Locus

Y-chromosomal DNA probes can be isolated from recombinant phage DNA libraries constructed from flow-sorted human Y chromosomes,[30,31] or from libraries constructed from human/rodent somatic cell hybrids retaining the Y as the only human chromosome.[32-34] In addition, Y-chromosomal DNA sequences may be obtained by the random screening of genomic DNA libraries constructed from total male human DNA. The mapping of the Y-chromosomal DNA probes in relation to the testis-determining locus is accomplished by their hybridization with DNA from XX males and XY females.

XX males are phenotypically male but infertile individuals with a 46,XX[35] or — very rarely — with a 47,XXX karyotype.[36] Cytogenetic findings in certain XX males revealed Y-chromosomal material on the tip of the short arm of one of their X chromosomes.[37-39] Molecular studies using Y-chromosomal DNA sequences as hybridization probes have now revealed the presence of Y-chromosomal DNA in most XX males.[40-43] Application of in situ hybridization to Y DNA-positive XX males[44-46] and to a 47,XXX male[47] demonstrated the presence of the Y DNA in these individuals on the tip of the short arm of one X chromosome. These findings lend strong support to the hypothesis of Ferguson-Smith,[48] suggesting that an unequal X/Y interchange during paternal meiosis results in the insertion of Y-chromosomal DNA (including testis-determining genes) in a 46,XX zygote.

46,XY females are phenotypic females with streak gonads and a 46,XY karyotype (XY gonadal dysgenesis or Swyer's syndrome[49,50]). Although most 46,XY females have a cy-togenetically normal Y chromosome, a deletion of the Y short arm was detected in some individuals.[51-54] Unlike patients with "classic" Swyer's syndrome, however, these individuals had Turner stigmata in addition to gonadal dysgenesis. Molecular studies revealed absence of Y-chromosomal DNA sequences in a 46,XY female with no cytogenetically detectable Y-chromosomal aberration[41,42] and in two patients with cytogenetically detectable deletions of Yp.[30,53,54]

The molecular findings in XX males and in XY females have allowed the construction of a deletion map of Yp with respect to the testis-determining locus.[41] This map is based on the finding that XX males differ in the amount of Y DNA present (Table 1). Assuming that most Y-chromosomal translocations into the DNA of XX males are contiguous, the Y-DNA sequences in XX males reflect their position on Yp. Accordingly, DNA sequences present in most XX males appear to be closest to the testis-determining locus; those found in some XX males only are more distal.

Figure 1 gives a tentative map of Yp with respect to the testis-determining locus.[55] Probe pDP34,[56] assigned to Yp by *in situ* hybridization,[57] did not hybridize with any of our XX males. Therefore, DNA sequences hybridizing with this probe are closest to the centromere. Probes Y-156, Y-182, Y-190, Y-223a, and Y-219, each hybridize with Y-specific repeated DNA sequences and with the same set of XX males. Probes located more distally on Yp hybridize with single-copy Y-chromosomal DNA sequences; some of them hybridize in addition to an autosomal single-copy sequence.

The data suggest that there is a bulk of Y-specific repeated DNA sequences between the pDP34 sequences (that have X homology) and the Y-specific single-copy sequences on Yp. Our data and those of others[43] further indicate that the testis-determining locus is distal on Yp, yet proximal to the pseudoautosomal region. The map of Figure 1 is supported by findings in the 46,XY females tested; DNA sequences that appear to be closest to the testis-determining locus were missing in all three 46,XY females with deletions of Yp.[30,41,42]

Probes Y-286 and Y-280 of Figure 1, that appear to be close to testis-determining gene(s), are being used as "starting points" for "chromosome walking" and pulsed field gel electrophoresis.

The goal of these experiments is the isolation of evolutionarily conserved Y-chromosomal DNA sequences which are likely to be parts of functional gene(s). Expression of such sequences will be studied in various (embryonic) tissues. This may eventually allow the isolation of candidates for the testis-determining/differentiating gene(s).

IV. SUMMARY

The human Y chromosome is divided into heterochromatic and euchromatic portions. The former comprises the distal long arm (Yq12) and the centromere; the latter, the short arm and the proximal long arm. The heterochromatic portion is composed mainly of highly reiterated satellite DNA. Functional genes appear to be confined to the Y euchromatin. These genes include the putative testis-determining gene(s) on Yp. Hybridization of Y-chromosomal DNA probes with DNA from XX males and XY females allows the construction of a deletion map of Yp with respect to the testis-determining locus. This locus is assigned to distal Yp, close to the pseudoautosomal region.

Table 1
HYBRIDIZATION OF Y-SPECIFIC DNA PROBES WITH DNA FROM SIXTEEN 46,XX MALES AND A 47,XXX MALE[55]

Probe	46,XX males[55]																47,XXX male
	460	462	548	756	775	693	547	#11	(p22.33; p11.2)	GM2626	GM2670	102	510	GM1189	481	385	
pDP 34	−	−	−	−	−	−	−	−	n.t.[a]	−	−	−	−	n.t.	n.t.	n.t.	n.t.
Y-190	+	+	+	+	+	+	+	+	+	−	−	−	−	−	−	−	+
Y-156	+	+	+	+	+	+	+	+	+	−	−	−	−	−	−	−	+
Y-182	+	+	+	+	+	+	+	+	+	−	−	−	−	−	−	−	+
Y-223a	+	+	+	+	+	+	+	+	+	−	−	−	−	−	−	−	+
Y-219	+	+	+	+	+	+	+	+	+	−	−	−	−	−	−	−	+
Y-227	+	+	+	+	+	+	+	+	+	+	+	−	−	−	−	−	+
Y-228	+	+	+	+	+	+	+	+	+	+	+	−	−	−	−	−	+
Y-280	+	+	+	+	+	+	+	+	+	+	+	+	+	−	−	−	+
Y-286	+	+	+	+	+	+	+	+	+	+	+	+	+	−	−	−	+
Y-198	−	−	−	−	−	−	−	−	−	−	−	+	−	−	−	−	−
Y-253	−	−	−	−	−	−	−	−	−	−	−	+	−	−	−	−	−
Y-202	−	−	−	−	−	−	−	−	−	−	−	+	−	−	−	−	−
Y-216a	−	−	−	−	−	−	−	−	−	−	−	−	−	−	−	−	−
Y-221	−	−	−	−	−	−	−	−	−	−	−	−	−	−	−	−	−
Y-294	−	−	−	−	−	−	−	−	−	−	−	−	−	−	−	−	−
Y-214	−	−	−	−	−	−	−	−	−	−	−	−	−	−	−	−	−
Y-157a	−	−	−	−	−	−	−	−	−	−	−	−	−	−	−	−	−

Note: Probes Y-198; Y-253; Y-202; Y-216a; Y-221; Y-294; Y-214; and Y-157a were assigned to Yq11.21→Yq11.23 in patients with 46,X,del(Y)(pter→q11.23:) and 45,X/46,X,del(Y)(pter→q11.21:) karyotypes.

a Not tested.

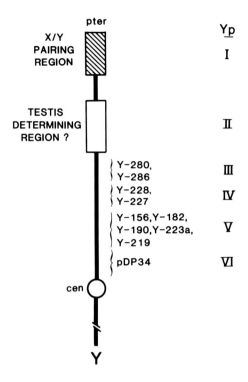

FIGURE 1.   Schematic diagram of the relative arrangement of Y-chromosomal DNA sequences hybridizing with various probes in relation to the testis-determining locus on Yp. Distances given in the map merely reflect the order of the probes and not physical or genetic distances of the probes from the testis-determing locus. For further details, see text. (From Müller, U. et al., *Nucleic Acids Res.*, 14, 6489, 1986. With permission.)

ACKNOWLEDGMENTS

The author is indebted to Dr. Samuel A. Latt for critically reading the manuscript and for many stimulating discussions. This article was supported by grant HD18658 from the NIH. The author was a recipient of a Heisenberg fellowship from the Deutsche Forschungsgemeinschaft.

REFERENCES

1. **Francke, U.,** High resolution ideograms of trypsin banded human chromosomes, *Cytogenet. Cell Genet.,* 31, 24, 1981.
2. **Zech, L.,** Investigation of metaphase chromosomes with DNA-binding fluorochromes, *Exp. Cell Res.,* 58, 463, 1969.
3. **Schempp, W. and Müller, U.,** High resolution replication patterns of the human Y chromosome. Intra- and interindividual variation, *Chromosoma,* 86, 229, 1982.
4. **Verma, R. S., Dosik, H., Scharf, T., and Lubs, H. A.,** Length heteromorphisms of fluorescent (f) and non-fluorescent (nf) segments of human Y chromosome: classification, frequencies, and incidence in normal Caucasians, *J. Med. Genet.,* 15, 277, 1978.
5. **Zuffardi, O., Maraschio, P., Lo Curto, F., Müller, U., Giarola, A., and Perotti, L.,** The role of Yp in sex determination: new evidence from X/Y translocations, *Am. J. Med. Genet.,* 12, 175, 1982.
6. **Tiepolo, L. and Zuffardi, O.,** Localization of factors controlling spermatogenesis in the nonfluorescence portion of the human Y chromosome long arm, *Hum. Genet.,* 34, 119, 1976.

7. **Kunkel, L. M., Smith, K. D., and Boyer, S. H.,** Human Y chromosome specific reiterated DNA, *Science,* 191, 1189, 1976.

8. **Kunkel, L. M., Smith, K. D., Boyer, S. H., Borgaonkar, D. S., Wachtel, S. S., Miller, O. J., Breg, W. R., Jones, H. W., and Rary, J. M.,** Analysis of human Y chromosome specific reiterated DNA in chromosome variants, *Proc. Natl. Acad. Sci. U.S.A.,* 74, 1245, 1977.

9. **Cooke, H. J.,** Repeated sequences specific of human males, *Nature (London),* 262, 182, 1976.

10. **McKay, R. D. G., Bobrow, M., and Cooke, H. J.,** The identification of a repeated DNA sequence involved in the karyotype polymorphism of the human Y chromosome, *Cytogenet. Cell Genet.,* 21, 19, 1978.

11. **Cooke, H. J., Fantes, J., and Green, D.,** Structure and evolution of human Y chromosome DNA, *Differentiation,* 23, S48, 1983.

12. **Schmidtke, J. and Schmid, M.,** Regional assignment of a 2.1 kb repetitive sequence to the distal part of the human Y heterochromatin, *Hum. Genet.,* 55, 255, 1980.

13. **Orgel, L. E. and Crick, F. H. C.,** Selfish DNA: the ultimate parasite, *Nature (London),* 284, 604, 1980.

14. **Münke, M., de Martinville, B., Lieber, E., and Francke, U.,** Minute chromosomes replacing the Y chromosome carry Y-specific sequences by restriction fragment analysis and in situ hybridization, *Am. J. Med. Genet.,* 22, 361, 1985.

15. **Yang, T. P., Hansen, S. K., Oishi, K. K., Ryder, O. A., and Hamkalo, B. A.,** Characterization of a cloned repetitive DNA sequence concentrated on the human X chromosome, *Proc. Natl. Acad. Sci. U.S.A.,* 79, 6593, 1982.

16. **Willard, H., Smith, K., and Sutherland, J.,** Isolation and characterization of a major tandem repeat family from the human X chromosome, *Nucleic Acids Res.,* 11, 2017, 1983.

17. **Wolfe, J., Darling, S. M., Erickson, R. P., Craig, I. W., Buckle, V. J., Rigby, P. W. J., Willard, H. F., and Goodfellow, P. N.,** Isolation and characterization of an alphoid centromeric repeat family from the human Y chromosome, *J. Mol. Biol.,* 182, 477, 1985.

18. **Tyler-Smith, C.,** Structure of repeated sequences in the centromeric region of the human Y chromosome, *Development,* 101, 93, 1987.

19. **Solari, A. J.,** Synaptonemal complexes and associated structures in microspread human spermatocytes, *Chromosoma,* 81, 315, 1980.

20. **Polani, P. E.,** Pairing of X and Y chromosomes, non-inactivation of X-linked genes, and the maleness factor, *Hum. Genet.,* 60, 207, 1982.

21. **Müller, U. and Schempp, W.,** Homologous early replication patterns of the distal short arms of prometaphasic X and Y chromosomes, *Hum. Genet.,* 60, 274, 1982.

22. **Cooke, H. J., Brown, W. R. A., and Rappold, G. A.,** Hypervariable telomeric sequences from the human sex chromosomes are pseudoautosomal, *Nature (London),* 317, 687, 1985.

23. **Simmler, M. C., Rouyer, F., Vergnaud, G., Nystrom-Lahti, M., Ngo, K. Y., de la Chapelle, A., and Weissenbach, J.,** Pseudoautosomal DNA sequences in the pairing region of the human sex chromosomes, *Nature (London),* 317, 692, 1985.

24. **Rouyer, F., Simmler, M. C., Johnsson, C., Vergnaud, G., Cooke, H., and Weissenbach, J.,** A gradient of sex linkage in the pseudoautosomal region of the human sex chromosomes, *Nature (London),* 319, 291, 1986.

25. **Goodfellow, P. J., Darling, S. M., Thomas, N. S., and Goodfellow, P. N.,** A pseudoautosomal gene in man, *Science,* 234, 740, 1986.

26. **Burgoyne, P. S.,** Genetic homology and crossing-over in the X and Y chromosomes of mammals, *Hum. Genet.,* 61, 85, 1982.

27. **Simmler, M. C., Johnsson, C., Petit, C., Rouyer, F., Vergnaud, G., and Weissenbach, J.,** Two highly polymorphic minisatellites from the pseudoautosomal region of the human sex chromosomes, *EMBO J.,* 6, 963, 1987.

28. **Goodfellow, P.,** Expression of the 12E7 antigen is controlled independently by genes on the human X and Y chromosomes, *Differentiation,* 23, S35, 1983.

29. **Darling, S. M., Banting, G. S., Pym, B., Wolfe, J., and Goodfellow, P. N.,** Cloning an expressed gene shared by the human sex chromosomes, *Proc. Natl. Acad. Sci. U.S.A.,* 83, 135, 1986.

30. **Müller, U., Lalande, M., Disteche, C. M., and Latt, S. A.,** Construction, analysis, and application to 46,XY gonadal dysgenesis of a recombinant phage DNA library from flow-sorted human Y chromosomes, *Cytometry,* 7, 418, 1986.

31. **Van Dilla, M. A., Deaven, L. L., Albright, K. L., Allen, N. A., Aubuchon, M. R., Bartholdi, M. F., Brown, N. C., Campbell, E. W., Carrano, A. V., Clark, L. M., Cram, L. S., Crawford, B. D., Fuscoe, J. C., Gray, J. W., Hildebrand, C. E., Jackson, P. J., Jett, J. H., Longmire, J. L., Lozes, C. R., Luedemann, M. L., Martin, J. C., McNinch, J. S., Meincke, L. J., Mendelsohn, M. L., Meyne, J., Moyzis, R. K., Munk, A. C., Perlman, J., Peters, D. C., Silva, A. J., and Trask, B. J.,** Human chromosome-specific DNA libraries: construction and availability, *Biotechnology,* 4, 537, 1986.

32. **Bishop, C. E., Guellaen, G., Geldwerth, D., Voss, R., Fellous, M., and Weissenbach, J.,** Single-copy DNA sequences specific for the human Y chromosome, *Nature (London)*, 303, 831, 1983.

33. **Bishop, C. E., Guellaen, G., Geldwerth, D., Fellous, M., and Weissenbach, J.,** Extensive sequence homologies between Y and other human chromosomes, *J. Mol. Biol.*, 173, 403, 1984.

34. **Wolfe, J., Erickson, R. P., Rigby, P. W. J., and Goodfellow, P. N.,** Cosmid clones derived from both euchromatic and heterochromatic regions of the human Y chromosome, *EMBO J.*, 3, 1997, 1984.

35. **de la Chapelle, A.,** The etiology of maleness in XX men, *Hum. Genet.*, 58, 105, 1981.

36. **Bigozzi, V., Simoni, G., Montali, E., Dalpra, L., Rossella, F., Piazzini, M., and Borghi, A.,** 47,XXX chromosome constitution in a male, *J. Med. Genet.*, 17, 62, 1980.

37. **Madan, K. and Walker, S.,** Possible evidence for Xp+ in an XX male, *Lancet*, 1, 1223, 1974.

38. **Wachtel, S. S., Koo, G. C., Breg, W. R., Thaler, H. T., Dillard, G. M., Rosenthal, I. M., Dosik, H., Gerald, P. S., Salenger, P., New, M., Lieber, E., and Miller, O. J.,** Serologic detection of a Y-linked gene in XX males and XX true hermaphrodites, *N. Engl. J. Med.*, 295, 750, 1976.

39. **Magenis, R. E., Webb, M. J., McKean, R. S., Tomar, D., Allen, L. J., Kammer, H., Van Dyke, D. L., and Lovrien, E.,** Translocation (X; Y) (p22.3; p11.2) in XX males: etiology of male phenotype, *Hum. Genet.*, 62, 271, 1982.

40. **Affara, N. A., Ferguson-Smith, M. A., Tolmie, J., Kwok, K., Mitchell, M., Jamieson, D., Cooke, A., and Florentin, L.,** Variable transfer of Y-specific sequences in XX males, *Nucleic Acids Res.*, 14, 5375, 1986.

41. **Müller, U., Donlon, T., Schmid, M., Fitch, N., Richer, C. L., Lalande, M., and Latt, S. A.,** Deletion mapping of the testis determining locus with DNA probes in 46,XX males and in 46,XY and 46,X,dic(Y) females, *Nucleic Acids Res.*, 14, 6489, 1986.

42. **Müller, U., Lalande, M., Donlon, T., and Latt, S. A.,** Moderately repeated DNA sequences specific for the short arm of the human Y chromosome are present in XX males and reduced in copy number in an XY female, *Nucleic Acids Res.*, 14, 1325, 1986.

43. **Vergnaud, G., Page, D. C., Simmler, M. C., Brown, L., Rouyer, F., Noel, B., Botstein, D., de la Chapelle, A., and Weissenbach, J.,** A deletion map of the human Y chromosome based on DNA hybridization, *Am. J. Hum. Genet.*, 38, 109, 1986.

44. **Magenis, R. E., Tomar, D., Sheehy, R., Fellous, M., Bishop, C., and Casanova, M.,** Y short arm material translocated to distal X short arm in XX males: evidence from in situ hybridization of Y-specific single copy DNA probes, *Am. J. Hum. Genet.*, 36, 102S, 1984.

45. **Andersson, M., Page, D. C., and de la Chapelle, A.,** Chromosome Y-specific DNA is transferred to the short arm of X chromosome in human XX males, *Science*, 233, 786, 1986.

46. **Magenis, R. E., Casanova, M., Fellous, M., Olson, S., and Sheehy, R.,** Further cytologic evidence for Xp-Yp translocation in XX males using *in situ* hybridization with Y-derived probe, *Hum. Genet.*, 75, 228, 1987.

47. **Müller, U., Latt, S. A., and Donlon, T.,** Y-specific DNA sequences in male patients with 46,XX and 47,XXX karyotypes, *Am. J. Med. Genet.*, 28, 393, 1987.

48. **Ferguson-Smith, M. A.,** X-Y chromosomal interchange in the aetiology of true hermaphroditism and of XX Klinefelter's syndrome, *Lancet*, 2, 475, 1966.

49. **Swyer, G. I. M.,** Male pseudohermaphroditism: a hitherto undescribed form, *Br. Med. J.*, 2, 709, 1955.

50. **Simpson, J. L., Blagowidow, N., and Martin, A. O.,** XY gonadal dysgenesis: genetic heterogeneity based upon clinical observation, H-Y antigen status, and segregation analysis, *Hum. Genet.*, 58, 91, 1981.

51. **Rosenfeld, R. G., Luzzatti, L., Hintz, R. L., Miller, O. J., Koo, G. D., and Wachtel, S. S.,** Sexual and somatic determinants of the human Y chromosome. studies in a 46,XYp− phenotypic female, *Am. J. Hum. Genet.*, 31, 458, 1979.

52. **Magenis, R. E., Tochen, M. L., Holahan, K. P., Carey, T., Allen, L., and Brown, M. G.,** Turner syndrome resulting from partial deletion of Y chromosome short arm: localization of male determinants, *J. Pediatr.*, 105, 916, 1984.

53. **Disteche, C., Saal, H., Friedman, C., Thuline, H., and Sybert, V.,** Female and male patients with Y chromosome abnormalities studied by a prometaphase analysis and by Y-specific DNA probes, *Am. J. Hum. Genet.*, 36 (Suppl.), 91S, 1984.

54. **Disteche, C. M., Casanova, M., Saal, H., Friedman, C., Sybert, V., Graham, J., Thuline, H., Page, D. C., and Fellous, M.,** Small deletions of the short arm of the Y chromosome in 46,XY females, *Proc. Natl. Acad. Sci. U.S.A.*, 83, 7841, 1986.

55. **Müller, U.,** Mapping of testis determining locus on Yp by the molecular genetic analysis of XX males and XY females, *Development*, 101, 51, 1987.

56. **Page, D. C., de Martinville, B., Barker, D., Wyman, A., White, R., Francke, U., and Botstein, D.,** Single-copy sequence hybridizes to polymorphic and homologous loci on human X and Y chromosomes, *Proc. Natl. Acad. Sci. U.S.A.*, 79, 2352, 1982.

57. **Page, D. C., Harper, M., Love, J., and Botstein, D.,** Occurrence of a transposition from the X-chromosome long arm to the Y-chromosome short arm during human evolution, *Nature (London)*, 311, 119, 1984.

Chapter 9

THE PSEUDOAUTOSOMAL REGION OF MAN

Peter N. Goodfellow and Paul J. Goodfellow

TABLE OF CONTENTS

I. INTRODUCTION

The sex-determination decision in mammals is based on the segregation of sex chromosomes during male meiosis. Male gametes that receive a Y chromosome will produce male offspring; X chromosome-bearing male gametes produce females. For autosomes, correct meiotic segregation of homologs depends on chromosome pairing, and indirect evidence suggests an additional requirement for at least one chiasma between each paired chromosome arm. The mammalian sex chromosomes are very different in size and genetic content, and this precludes extensive homologous pairing and chiasma formation. In many species, during meiosis, the sex chromosomes form a specialized structure known as the sex vesicle, and the precise physical arrangement of the sex chromosomes is not easy to discern.[1] However, half a century ago Koller and Darlington[2] described the cytogenetic association of rat sex chromosomes and laid the foundation of our present understanding of sex chromosome pairing and segregation. Those authors suggested that the sex chromosomes of mammals are composed of two distinct parts: the first part is sex chromosome-specific and is responsible for sex determination (Y chromosome) and the observed sex-limited traits (X chromosome); the second part is shared by the sex chromosomes and is responsible for meiotic pairing. It was predicted that genes within the X-Y shared part would recombine and fail to show complete sex linkage. Subsequently, the term "pseudoautosomal" was introduced to describe the inheritance of postulated genes and sequences that were exchanged between the sex chromosomes.[3]

Formal proof of the existence of the pseudoautosomal region proved difficult to obtain.[4] In the mouse, an inherited form of female sex reversal (Sxr) was found to be due to the presumptive duplication and translocation of the sex determining region from a pericenter to a telomeric location on the Y chromosome.[5,6] In this new location, the testis-determining gene, *Tdy*, is inherited psuedoautosomally.[5,6] More recently, evidence has been presented for the pseudoautosomal inheritance of *STS*[7] and an integrated viral sequence.[8]

In man, the existence of the pseudoautosomal region was indicated by the gene *MIC2*, shared by the X and Y chromosomes. However, the conclusive proof was provided by application of the techniques of molecular genetics. In this review, we consider these recent results and conclude that the predictions made by Koller and Darlington[2] are correct.

II. THE SEX CHROMOSOME PAIRING REGION IN MAN

The human sex chromosomes pair during male meiosis.[9] The pairing occurs between the short arms of the X and Y chromosomes,[10] and the extent of pairing varies extensively between different meiotic spreads.[11,12] This variation may be due to nonspecific promiscuous association of unpaired chromosomes occurring late in meiosis,[13] and this makes it difficult to estimate the size of the homologous pairing region. However, in a detailed study, Chandley et al.[11] found that the minimum pairing corresponded to about 11% of the Y chromosome. Direct measurement of human chromosomes suggests that the Y chromosome consists of about 5×10^7 bp (base pairs) of DNA.[14] Assuming that the minimum pairing observed reflects the homologous pairing, the latter region is approximately 5×10^6 bp in length.

III. MOLECULAR CLONING OF SEQUENCES DERIVED FROM THE PSEUDOAUTOSOMAL REGION

Over 250 different sequences have been cloned from the human Y chromosome.[15] The majority of these sequences have been isolated either from human-rodent somatic cell hybrids or from flow-sorted chromosomes. In the first technique, cosmid libraries were constructed with DNA derived from rodent-human somatic cell hybrids which retained the human Y chromosome as the only human genetic contribution.[16-19] Y-derived clones were recognized by screening the libraries with total human DNA or cloned human-specific repeats. In the second technique, phage lambda libraries were constructed with DNA isolated from flow-sorted human Y chromosomes.[20]

The Y-derived sequences have been used to construct a map of the Y chromosome by utilizing *in situ* hybridization, Y chromosome structural abnormalities, and the DNA from XX males.[21,22] Among the sequences isolated at random from the Y chromosome, one group was found to be derived from the X and Y chromosomes within the pairing region. Three of these sequences have been studied in detail:

1. *DXYS14*. This sequence maps to a chromosomal location immediately adjacent to the telomere. As might be predicted, by analogy with the telomeres of lower eukaryotes, probes recognizing this sequence define a large number of polymorphisms due to duplications and deletions of repeat sequences within the telomeric region.[18]
2. *DXYS15*. This locus is associated with a minisatellite that is unusual because it has a high AT content. Variations in the minisatellite make this locus highly polymorphic.[23,24]
3. *DXYS17*. This locus is also highly polymorphic because of an associated minisatellite. The *DXYS15* and *DXYS17* minisatellites are not related.

It may not be coincidental that several of the pseudoautosomal loci are associated with minisatellites; Weissenbach et al.[25] have suggested that their repeat sequences may help promote the high levels of male recombination seen in the pseudoautosomal region.

A fourth pseudoautosomal marker was provided by cloning the structural gene, *MIC2*,[26] which encodes a cell-surface antigen recognized by the monoclonal antibody 12E7.[27] The cDNA and genomic clones corresponding to *MIC2* recognize multiple polymorphisms.[26,28] Recently, several new pseudoautosomal loci have been described.[25]

IV. RECOMBINATION OF PSEUDOAUTOSOMAL SEQUENCES IN MALE MEIOSIS

Formal proof of the existence of the pseudoautosomal region in man was obtained by studying the inheritance of *DXYS14*, *DXYS15*, and *DXYS17*. These three markers freely recombine with each other and with the X- and Y-specific parts of the sex chromosomes.[29,23] Subsequently, *MIC2* was also shown to be pseudoautosomal. However, this marker recombines only rarely between the sex chromosomes and shows close linkage to sex chromosome-specific sequences.[28]

Based on the observed recombination, a map of the human pseudoautosomal region can be constructed (Figure 1). This map has several notable features. First, approximately 50% recombination occurs between *DXYS14* and *MIC2*. Second, there is a gradient of recombination. Third, the distances between the markers are additive. This suggests that a single obligate recombination event occurs in male meiosis. Consistent with this interpretation, double recombination events have not been observed.[25,28,29]

V. RECOMBINATION OF PSEUDOAUTOSOMAL SEQUENCES IN FEMALE MEIOSIS

In contrast to the high levels of recombination seen in male meiosis between *MIC2* and *DXYS14*, only 7% recombination is seen between the same markers in female meiosis. Recently, we extended the meiotic map of the X chromosome short arm by looking for linkage between *MIC2* and other markers on the distal part of the human X chromosome. These studies suggest that previous studies had overestimated the distances between markers in this region.[30]

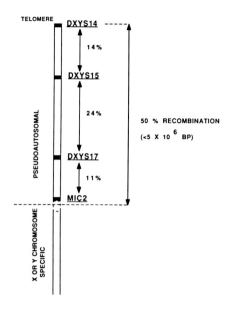

FIGURE 1. The pseudoautosomal region of man. The recombination fractions for male meiosis are taken from References 25 and 28.

VI. THE PHYSICAL SIZE OF THE PSEUDOAUTOSOMAL REGION

There are four indirect estimates of the size of the pseudoautosomal region. As discussed above, the minimum pairing region seen in male meiosis consists of about 5×10^6 bp. Based on sequence representation frequency in Y-derived libraries, Weissenbach et al.[25] have suggested that the pseudoautosomal region consists of about 3×10^6 bp. This estimate is only approximate and could be seriously biased by DNA-cloning artifacts. The third estimate is based on a deleted X chromosome originally found in association with chondrodysplasia punctata.[31] This X chromosome is deleted for all the pseudoautosomal sequences and sequences such as DXS31 which are known to be inherited in an X-linked manner and are absent from the Y chromosome short arm.[32] Cytogenetic analysis of the deleted chromosome implies the deletion is no bigger than 5×10^6 bp. The fourth estimate comes from the analysis by flow cytometry of the X chromosomes in XX males. In 12 out of 20 XX males analyzed, one X of the pair had gained about 6×10^6 bp of extra DNA. If it is assumed that XX males are generated by terminal exchange between the X and Y chromosomes, the extra DNA should include the testis-determining gene and the pseudoautosomal region.[33] Unfortunately, neither the amount of DNA deleted from the X chromosome during the exchange nor the amount of transferred nonpseudoautosomal Y DNA can be easily estimated.

Despite the caveats, the estimates of the size of the pseudoautosomal region are all consistent with a figure of around 5×10^6 bp. This implies that recombination in male meiosis occurs at a frequency ten times higher than the genomic average. The rates in females approximate closely to the genomic average.[34] The absence of double recombinants in males may mean that recombinational interference is proportional to physical size and not to meiotic distance.

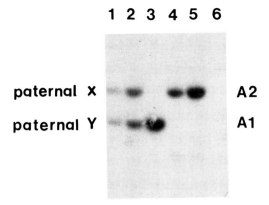

1 2 3 4 5 6

paternal X ······ A2

paternal Y ······ A1

FIGURE 2. Analysis of inheritance of *MIC2* in PPXXM, an XX male who failed to inherit Y chromosome-derived sequences. Somatic cell hybrids were made between cells from JP (father of the XX male) and HPRT⁻ hamster cells. The hybrids segregate the paternal X chromosome and from those the X-linked *MIC2* allele can be determined. 5 µg of Taq1 digest DNA per track. (1) JP — father, (2) PP — XX male son, (3) BP — mother, (4) JP3.3 human/hamster hybrid retaining JP's (father's) X chromosome, (5) JP3.4 an independent human/hamster hybrid retaining JP's (father's) X chromosome, (6) hamster control, A1 = 2.5 kb, A2 = 3.2 kb, paternal X: paternal X chromosome allele, and paternal Y: paternal Y chromosome allele. Hybridization with the *MIC2* genomic sequence p19B.[28]

VII. THE PSEUDOAUTOSOMAL REGION AND THE GENERATION OF XX MALES

The isolation of Y-derived sequences has allowed the partial solution of a conundrum in the biology of sex determination. One in 20,000 individuals with a male phenotype does not have a Y chromosome.[35] About 80% of these XX males do have Y-derived sequences, and different XX males have inherited different amounts of X-derived material.[22,32,35] Assuming that the Y-derived sequences have been transferred to the X chromosome by a simple terminal exchange, it is possible to construct a map of the Y chromosome including a location for *TDF* (testis-determining factor). The terminal exchange model allows two predictions. First, Y-derived sequences will be present at the tip of the short arm of the paternal X chromosome. This has been confirmed by *in situ* hybridization with Y-derived probes.[37,38] Second, XX males would be expected to inherit the pseudoautosomal region from the Y chromosome. This also has been confirmed. Page et al.[39] described two families in which an XX male inherited his father's Y-chromosomal pseudoautosomal region, and Petit et al.[40] described the same result in six families. The latter authors also showed that in three families, the XX males had not inherited the Y pseudoautosomal region. Recently, we found a similar family (Figure 2). The phasing of the father's pseudoautosomal loci was achieved by segregating his X and Y chromosomes in somatic cell hybrids. The XX male had inherited his father's X chromosome *MIC2* locus. In all four families where the XX males had failed to inherit the Y chromosome pseudoautosomal region, they also had failed to inherit any detectable Y sequences. This result supports suggestions that some XX males may be caused by mutations elsewhere in the genome.

VIII. THE LIMIT OF THE PSEUDOAUTOSOMAL REGION

In several of the models proposed for the Y chromosome, the pairing region is composed of two parts: a strictly homologous region and a partly homologous region.[3] The latter region

Table 1

**THE RELATIONSHIP BETWEEN
RECOMBINATION FRACTION AND
ALLELIC DIFFERENTIATION BETWEEN X
AND Y CHROMOSOMES IN THE
PSEUDOAUTOSOMAL REGION**

r	JXY/JXX
10^{-2}	1.000
10^{-3}	0.999
10^{-4}	0.989
10^{-5}	0.877
10^{-6}	0.404
10^{-7}	0.063

Note: r is the recombination fraction between a pseudoautosomal locus and the sex chromosome-specific region. JXY/JXX is the allelic similarity at the pseudoautosomal locus between the X and Y chromosomes compared with allelic similarity between different X chromosomes. It has been assumed that the mutation rate is 10^{-6} and that the population size is 20,000 individuals. Full details of the calculations can be found in Reference 41. From Bengtsson, B. O. and Goodfellow, P. N., *Ann. Hum. Genet.*, 51, 57, 1987. With permission.

would be sufficiently homologous for pairing, but would be insufficiently homologous for recombination. This model of the pairing region is unlikely to be correct in humans. In a recent theoretical paper, Bengtsson and Goodfellow[41] posed the following question: "What degree of exchange between the sex chromosomes is required to maintain full sequence identity between the X and Y chromosomes?" The result of one calculation to answer this question is presented in Table 1. Qualitatively, it can be stated that as long as the recombination rate is higher than the mutation rate the sequences on the X and Y chromosomes will not diverge. If XX males are generated by X-Y terminal exchange with a frequency of about 1 in 20,000,[35] recombination between the X and Y chromosome distal to sex-determining locus, *TDF*, will occur at a frequency greater than 1 in 20,000. This frequency should be sufficient to maintain total sequence identity between the X and Y sequences as long as identity of these sequences does not affect fertility or viability. We predict that adjacent to the pseudoautosomal region on the Y chromosome, an expressed gene will be found. This gene may be *TDF*.

IX. GENES IN THE PSEUDOAUTOSOMAL REGION

The only well-defined pseudoautosomal gene in man is *MIC2*[28,42,43] which encodes a cell-surface antigen of wide tissue distribution and unknown function.[27,44] *MIC2* escapes inactivation on the inactive X chromosome, thereby maintaining male-female gene dosage equivalence.[45] Genomic sequences and cDNA corresponding to *MIC2* each have been cloned.[26,28] Amino acid sequences deduced from the cDNA clones are consistent with a cell-surface molecule with an external C-terminus.[46] Neither the nucleic acid sequence nor the deduced amino acid sequence shows any homology to any previously described sequence. The *MIC2* probes react strongly with primate DNA, but fail to react, or react only weakly, with rodent DNA even at very reduced stringency.[47]

Although several hundred people have been screened, we have not found an individual who fails to express the antigenic product of the *MIC2* locus. However, one human-tumor-derived cell line, LY65, fails to produce *MIC2* mRNA or protein, suggesting that mutation at *MIC2* is unlikely to be cell lethal.[48]

Recently, we presented evidence for the existence of a second pseudoautosomal gene that regulates, in *cis*, the expression of *MIC2* on red blood cells and the X-linked red blood cell locus *XG*.[49,50] This postulated locus, *XGR* (*XG regulator*), has not been shown formally to be distinct from *MIC2*.

As the function of the *MIC2* gene is presently not understood, it is not possible to address the question of whether a pseudoautosomal location has functional implications for a gene.

X. THE PSEUDOAUTOSOMAL REGION IN OTHER MAMMALS

It is assumed that all eutherian mammals use a similar mechanism for ensuring correct segregation of sex chromosomes. However, it should be pointed out that metatherians have very reduced Y chromsomes which do not pair with the X chromosome in male meiosis, although correct sex chromosome segregation must occur in these animals.

Apart from cytogenetic observation, direct evidence for the existence of the pseudoautosomal region is lacking in all mammalian species except for mouse and man. In the mouse, three loci, *STS*, *Sxr*, and *Mov15* show pseudoautosomal inheritance. However, of these three loci, only *STS* is normally pseudoautosomal. *Sxr* is associated with the duplication and terminal translocation of the pericentric region of the mouse Y chromosome.[5,6,51] This results in female sex reversal in XX animals that have inherited *Sxr* by pseudoautosomal exchange between the X and Y chromosome, due to the presence of *Tdy* within the Sxr region. *Sxr* is exchanged with a high frequency between the mouse X and Y chromosomes, although this recombination may not always occur between strictly homologous sequences as deletions of Bkm (banded krait minor satellite DNA) sequences associated with *Sxr* may occur during meiosis.[18] A similar deletion of sequences has been recorded for the integrated virus sequence *Mov15*, however, unlike *Sxr* and *STS*, *Mov15* recombines at a frequency substantially less than 50%.[8]

The steroid sulfatase gene *STS*, is the only locus which is known to be pseudoautosomal in the wild-type mouse. This locus in humans is X linked, but is located just below the pseudoautosomal region,[31,52] prompting suggestions that *STS* was once pseudoautosomal in the human lineage.[53] The presence of a Y-located pseudogene on the long arm of the human Y chromosome[54] may be consistent with a Y chromosome pericentric inversion that has disrupted the human pseudoautosomal region. It is possible that this inversion is related to another major difference between the human and mouse Y chromosomes: in humans, the *TDF* gene lies just below the pseudoautosomal region; in the mouse, the equivalent gene, *Tdy*, is thought to map close to the centromere. These observations raise the possibility that the structure of the pseudoautosomal region may be fundamentally different in humans and rodents.

XI. EVOLUTION OF THE PSEUDOAUTOSOMAL REGION

Previously, we suggested that the origin of the pseudoautosomal region may be distinct from the origin of the rest of the sex chromosomes.[55] Unfortunately, this question cannot be addressed directly because of the lack of information about the pseudoautosomal regions in species other than human or mouse. Obtaining this information may be difficult as the human pseudoautosomal sequences isolated so far display very little sequence conservation outside of the primates.

XII. CONCLUSIONS

The predictions made by Koller and Darlington[2] have been proven correct. Questions still to be answered include the evolution of the pseudoautosomal region, the nature of the sequences at the extreme of the pseudoautosomal region, the function of pseudoautosomal genes, and how high levels of recombination are promoted at the molecular level in so small a region.

XIII. SUMMARY

Molecular analysis has confirmed predictions, made 50 years previously, that the Y chromosome of eutherian mammals is composed of two distinct regions: one region shared by the X and Y chromosomes and one region specific for the Y chromosome. The X-Y shared region is responsible for meiotic pairing and segregation of the sex chromosomes in male meiosis, and the Y-specific region encodes the male sex-determining gene.

The shared region has been described as pseudoautosomal because sequences in this region exchange in male meiosis and consequently fail to show complete sex linkage. In this review, we have compared the physical and genetic size of the pseudoautosomal region, and we have described *MIC2*, a human pseudoautosomal gene. Eventually, studies on the pseudoautosomal region should provide important clues as to how mammals regulate recombination rates and also should point to the evolutionary origin of the mammalian sex chromosomes.

ACKNOWLEDGMENTS

We thank our colleagues in the Laboratory of Human Molecular Genetics for helpful discussion and for sharing in many of the experiments described in this review. We also thank Mrs. C. Middlemiss for editorial assistance.

REFERENCES

1. **Solari, A. J.,** The behaviour of the X-Y pair in mammals, *Int. Res. Cytol.,* 38, 273, 1974.
2. **Koller, P. C. and Darlington, C. D.,** The genetical and mechanical properties of the sex chromosomes. I. *Rattus novegicus, J. Genet.,* 29, 159, 1934.
3. **Burgoyne, P. S.,** Genetic homology and crossing over in the X and Y chromosome of mammals, *Hum. Genet.,* 61, 85, 1982.
4. **Haldane, J. B. S.,** A search for incomplete sex-linkage in man, *Ann. Eugenic.,* 7, 28, 1936.
5. **Evans, E. P., Burtenshaw, M., and Cattanach, B. M.,** Meiotic crossing over between X and Y chromosomes of male mice carrying the sex reversing (*Sxr*) factor, *Nature (London),* 300, 443, 1982.
6. **Singh, L. and Jones, K. W.,** Sex reversal in the mouse (*Mus musculus*) is caused by a recurrent nonreciprocal crossover involving the X and an aberrant Y chromosome, *Cell,* 28, 205, 1982.
7. **Keitges, E., Rivest, M., Siniscalco, M., and Gartler, S. M.,** X-linkage of steroid sulphatase in the mouse is evidence for a functional Y-linked allele, *Nature (London),* 315, 226, 1985.
8. **Harbers, K., Soriano, P., Muller, U., and Jaenisch, P.,** High frequency of unequal recombination in pseudoautosomal region shown by proviral insertion in transgenic mouse, *Nature (London),* 324, 682, 1986.
9. **Pearson, P. L. and Bobrow, M.,** Definitive evidence for the short arm of the Y chromosome associating with the X during meiosis in the human male, *Nature (London),* 226, 959, 1970.
10. **Chen, A. and Falek, A.,** Cytological evidence for the association of the short arms of the X and Y in male meiosis, *Nature (London),* 232, 555, 1971.
11. **Chandley, A. C., Goetz, P., Hargreave, T. B., Joseph, A. M., and Speed, R. M.,** On the nature and extent of the X-Y pairing at meiotic prophase in man, *Cytogenet. Cell Genet.,* 38, 241, 1984.

12. **Moses, M. J., Counce, S. J., and Paulson, D. F.,** Synaptoneal complex complement of man in spreads of spermatocytes with details of the sex chromosome pair, *Science,* 187, 363, 1975.
13. **Rassmussen, S. W. and Holm, P. B.,** Mechanics of meiosis, *Hereditas,* 93, 187, 1980.
14. **Mendelsohn, M. L., Mayall, B. H., Bogart, E., Moore, D. H., and Perry, B. H.,** DNA content and DNA based centromeric index of the 24 human chromosomes, *Science,* 179, 1126, 1973.
15. **Goodfellow, P. N., Wolfe, J., and Craig, I. W.,** The mammalian Y chromosome: molecular search for the sex determining gene, *Development,* 101 (Suppl.), 1987.
16. **Bishop, C. E., Guellaen, G., Geldwerth, D., Voss, R., Fellous, M., and Weissenbach, J.,** Single copy DNA sequences specific for the human Y chromosome, *Nature (London),* 303, 831, 1983.
17. **Burk, R. D., Ma, P., and Smith, K. D.,** Characterisation and evolution of a single-copy sequence from the human Y chromosome, *Mol. Cell Biol.,* 5, 576, 1985.
18. **Cooke, H. J., Brown, W. R. A., and Rappold, G. A.,** Hypervariable telomeric sequences from the human sex chromosomes are pseudoautosomal, *Nature (London),* 317, 687, 1985.
19. **Wolfe, J., Erickson, R. P., Rigby, P. W. J., and Goodfellow, P. N.,** Cosmid clones derived from both euchromatic and heterochromatic regions of the human Y chromosome, *EMBO J.,* 3, 1997, 1984.
20. **Müller, C. R., Davies, K. E., Cremer, C., Rappold, G., Gray, J. W., and Ropers, H.-H.,** Cloning of genomic sequences from the human Y chromosome after purification by dual beam sorting, *Hum. Genet.,* 64, 110, 1983.
21. **Goodfellow, P. N., Davies, K. E., and Ropers, H.-H.,** Report of the Committee on the Genetic Constitution of the X and Y chromosomes, *Cytogenet. Cell Genet.,* 40, 296, 1985.
22. **Vergnaud, G., Page, D. C., Simmler, M-C., Brown, L., Rouyer, F., Noel, B., Botstein, D., de la Chapelle, A., and Weissenbach, J.,** A deletion map of the human Y chromosome based on DNA hybridization, *Am. J. Hum. Genet.,* 38, 109, 1986.
23. **Simmler, M-C., Rouyer, F., Vergnaud, G., Nystrom-Lahti, M., Ngo, K. Y., de la Chapelle, A., and Weissenbach, J.,** Pseudoautosomal DNA sequences in the pairing region of the human sex chromosome, *Nature (London),* 317, 692, 1985.
24. **Simmler, M-C., Johnsson, C., Petit, C., Rouyer, F., Vergnaud, G., and Weissenbach, J.,** Two highly polymorphic minisatellites from the pseudoautosomal region of the human sex chromosomes, *EMBO J.,* 6, 963, 1987.
25. **Weissenbach, J., Levilliers, J., Petit, C., Rouyer, F., and Simmler, M-C.,** Normal and abnormal exchanges between the human X and Y chromosomes, *Development,* 101 (Suppl.), 1987.
26. **Darling, S. M., Banting, G. S., Pym, B., Wolfe, J., and Goodfellow, P. N.,** Cloning an expressed gene shared by the human sex chromosomes, *Proc. Natl. Acad. Sci. U.S.A.,* 83, 135, 1986.
27. **Goodfellow, P. N.,** Expression of the 12E7 antigen is controlled independently by genes on the human X and Y chromosomes, *Differentiation,* 23, 535, 1983.
28. **Goodfellow, P. J., Darling, S. M., Thomas, N. S., and Goodfellow, P. N.,** A pseudoautosomal gene in man, *Science,* 234, 740, 1986.
29. **Rouyer, F., Simmler, M.-C., Johnsson, C., Vergnaud, G., Cooke, H. J., and Weissenbach, J.,** A gradient of sex linkage in the pseudoautosomal region of the human sex chromosomes, *Nature (London),* 319, 291, 1986.
30. **Goodfellow, P. J., Harper, P., and Goodfellow, P. N.,** unpublished data, 1987.
31. **Curry, C. J. R., Magenis, R. E., Brown, M., Lauman, J. T., Tsa, J. T., O'Langue, P., Goodfellow, P. N., Mohandas, T., Bergner, E. A., and Shapiro, L. M.,** Inherited chondrodysplasia punctata due to a deletion of the terminal short arm of an X chromosomes, *N. Engl. J. Med.,* 311, 1010, 1984.
32. **Mondello, C., Ropers, H.-H., Craig, I. W., and Goodfellow, P. N.,** Physical mapping of genes and sequences at the end of the human X chromosome short arm, *Ann. Hum. Genet.,* 51, 137, 1987.
33. **Ferguson-Smith, M. A., Affara, N. A., and Magenis, R. E.,** Ordering of Y specific sequences by deletion mapping and by analysis of X-Y interchange males and females, *Development,* 101 (Suppl.), 1987.
34. **Renwick, J. H.,** Progress in mapping human autosomes, *Br. Med. Bull.,* 25, 65, 1986.
35. **de la Chapelle, A.,** The etiology of maleness in XX men, *Hum. Genet.,* 58, 105, 1981.
36. **Guellaen, G., Casanova, M., Bishop, C., Gelwerth, D., Audre, G., Fellous, M., and Weissenbach, J.,** Human XX males with Y single copy DNA fragments, *Nature (London),* 307, 172, 1984.
37. **Andersson, M., Page, D. C., and de la Chapelle, A.,** Chromosome Y-specific DNA is transferred to the short arm of the X chromosome in XX males, *Science,* 233, 786, 1986.
38. **Buckle, V., Boyd, Y., Fraser, N., Craig, I. W., Goodfellow, P. N., and Wolf, J.,** Localization of Y chromosomal sequences in normal and XX males, *Cytogenet. Cell Genet.,* 40, 593, 1985.
39. **Page, D. C., Brown, L., Pollack, J., Biecker, K., de la Chapelle, A., Disteche, C., and McGillivray, B.,** Sex reversal and the male-determining function of the human Y chromosome, *Development,* 101 (Suppl.), 1987.
40. **Petit, C., de la Chapelle, A., Levilliers, J., Castillo, S., Noel, B., and Weissenbach, J.,** An abnormal terminal X-Y interchange accounts for most but not all cases of human XX maleness, *Cell,* 49, 595, 1987.

41. **Bengtsson, B. O. and Goodfellow, P. N.,** The effect of recombination between the X and Y chromosomes of mammals, *Ann. Hum. Genet.*, 51, 57, 1987.

42. **Buckle, V., Mondello, C., Darling, S., Craig, I. W., and Goodfellow, P. N.,** Homologous expressed genes in the human sex chromosome pairing region, *Nature (London)*, 317, 739, 1985.

43. **Goodfellow, P. N., Banting, G., Sheer, D., Ropers, H.-H., Caine, A., Ferguson-Smith, M. A., Povey, S., and Voss, R.,** Genetic evidence that a Y-linked gene in man is homologous to a gene on the X chromosome, *Nature (London)*, 302, 346, 1983.

44. **Pym, B., Banting, G. S., Katz, F., Darling, S., and Goodfellow, P.,** Five independent monoclonal antibodies recognise the products of the human *MIC2* loci: genetic and biochemical analysis using DNA transfectants, submitted, 1986.

45. **Goodfellow, P. N., Pym, B., Mohandas, T., and Shapiro, L. J.,** The cell surface antigen locus *MIC2X* escapes X-inactivation, *Am. J. Hum. Genet.*, 36, 777, 1984.

46. **Darling, S. M., Goodfellow, P. J., Pym, B., Banting, G. S., Pritchard, C., and Goodfellow, P. N.,** Molecular genetics of *MIC2*: a gene shared by the human X and Y chromosomes, *Cold Spring Harbor. Symp. Quant. Biol.*, 51, 205, 1986.

47. **Goodfellow, P. N. and Goodfellow, P. J.,** unpublished data, 1987.

48. **Banting, G.,** personal communication, 1987.

49. **Goodfellow, P. J., Pritchard, C., Tippett, P., and Goodfellow, P. N.,** Recombination between the X and Y chromosomes at the *YG* locus: implications for the relationship between *MIC2*, *XG* and *YG*, *Ann. Hum. Genet.*, 51, 161, 1987.

50. **Goodfellow, P. N. and Tippett, P.,** A human polymorphism related to XG blood groups, *Nature (London)*, 289, 404, 1981.

51. **Epplen, J.,** Molecular biological analysis of *Sxr* (abstract), *Development*, 101 (Suppl.), 1987.

52. **Geller, R. L., Shapiro, L. J., and Mohandas, T. K.,** Fine mapping of the distal short arm of the human X chromosome using X/Y translocations, *Am. J. Hum. Genet.*, 38, 884, 1986.

53. **Craig, I. W. and Tolley, E.,** Steroid sulphatase and the conservation of mammalian X chromosomes, *Trends Genet.*, 20, 201, 1986.

54. **Craig, I. W.,** personal communication, 1987.

55. **Goodfellow, P. J., Darling, S., Pym, B., Pritchard, C., and Goodfellow, P. N.,** Homologies between the sex chromosomes of man, *Int. Congr. Hum. Genet.*, in press, 1988.

Part III: Male Antigens

Chapter 10

H-Y ANTIGEN: DOGMAS AND ENIGMAS

Ulf H. Wiberg

TABLE OF CONTENTS

I. INTRODUCTION

As the name implies, the H-Y antigen is an immunological entity. This antigen was first described by Eichwald and Silmser[1] as a male-specific transplantation antigen present in male skin grafts of certain inbred strains of mice. Later studies, demonstrating that the male-specific antigen was lacking in spleen cells from XO and XX female mice, but present in spleen cells by XXY and XY male mice,[2] suggested Y-chromosomal linkage of the genes responsible for the expression of the antigen. This early evidence of Y-linkage caused Billingham and Silvers[3] to introduce the term "H-Y" (histocompatibility-Y) antigen. In this article we use the term H-Yt to refer to the male-specific antigen as defined by transplantation.

In addition to skin and spleen, the H-Yt has been found in almost all male tissues investigated.[4] Moreover, it has been demonstrated to be male-specific in the rat, hamster, pig, rabbit, and human, and in the newt and platyfish, each of which has an XX/XY mechanism of sex determination. However, in the chicken in which the female is hetero-gametic (ZZ/ZW mechanism), a sex-specific transplantation antigen (H-W) is found in the female.[4] Thus, there is evidence for evolutionary conservation of sex-specific transplantation antigens (H-Yt and H-Wt) in the vertebrate subphylum.

By definition, the H-Yt is a weak or minor histocompatibility (nonmajor histocompatibility complex, non-MHC) antigen. Operationally, this means that graft rejection caused by H-Yt alone is slower than that caused by MHC antigens. Because H-Yt is a minor histocompatibility antigen, typing for H-Yt was originally performed within inbred strains. In 1973, Silvers and Yang[5] sensitized female mice with xenogeneic male cells of certain rat strains, and subsequently grafted the female mice with syngeneic male skin. In this way they were able to demonstrate the presence of H-Yt on male, but not female, rat cells. This demonstrated homology of H-Yt of mouse and rat, and suggested that it was possible, by means of priming inbred female mice with xenogeneic male cells, to assay for the H-Yt in noninbred species as well. This indirect H-Y transplantation assay seems to have been forgotten for several years, and it was not until 1985 that Wiberg and colleagues revived it, thereby demonstrating the presence of H-Yt in the human[6] and wood lemming.[7-8]

In 1971, Goldberg et al.[9] found that sera from inbred female mice became cytotoxic for male target cells in the presence of complement after grafting with syngeneic male skin or spleen cells. Cytotoxicity could be absorbed out with cells from mammalian males but not females.[4] Thus, male skin and spleen carry an antigen or group of antigens that induces an anti-male antibody in immunized females. At present, it is not known whether this serologically detectable male-specific antigen is the same as the antigen detected by skin grafting. We shall discuss this in detail below. For the sake of clarity, the serological antigen will be referred to as the H-Ys antigen in this chapter.[10]

In addition to skin grafting and serological typing, another method is used to assay for the H-Y antigen(s). According to that method, female mice are primed with syngeneic male cells in vivo. The spleen cells of the primed females are then restimulated in vitro by exposure to irradiated syngeneic male spleen cells. The female spleen cells, now twice immunized, are added to test suspensions containig ^{51}Cr-labeled male spleen cells, and specific lysis of the latter is determined by quantification of chromium release. This cell-mediated cytotoxicity (CMC) test was first described by Goldberg et al.[11] With regard to the H-Y antigen, the female effector cells in the CMC assay are referred to as H-Y-specific cytotoxic T lymphocytes (CTLs). Male-specific antigen(s) detected in the CMC assay will be referred to here as H-Yc because there is no direct (i.e., molecular) evidence of identity of H-Yc and H-Yt.

II. H-Y IMMUNOLOGY AND ISSUES

As hinted in the Introduction, there is no consensus as to the number of existing H-Y antigens. This seems to be due in part to complexity of the immunological factors affecting the different assay systems. In other words, if a certain subject is typed H-Y positive with one assay system and H-Y negative with another, this could merely indicate restriction of the latter assay, rather than the existence of supernumerary H-Y antigens (although it does not exclude the latter alternative). For example, human XX males carrying certain DNA sequences of the Y chromosome short arm (Yp) but lacking Y chromosome long arm (Yq) sequences have been found to type H-Yc negative and XY human females lacking part of Yp (46,XYp⁻) but carrying most or all of Yq type H-Yc positive.[12] In contrast, all XX males so far tested type positive for the H-Ys, and the few 46,XYp⁻ females investigated are reported to be negative, or possibly weakly positive, for the H-Ys.[4] The question arises whether this should be regarded as evidence that H-Yc and H-Ys are different antigens. Before trying to answer, it may be useful to briefly review current knowledge of the mechanisms of immune responses by the different cell types of the immune system.

A. Interaction Between Cells of the Immune System

Although T and B lymphocytes both are activated via antigen-specific receptors, there are two main differences in the way activation occurs. First, with certain exceptions (some polymers can trigger antibody synthesis by B cells alone), activation of B cells is dependent on T cells. Second, T cells recognize antigens and MHC molecules simultaneously; in general, they cannot recognize antigen alone. (There is no strong evidence that recognition by B cells is MHC dependent.) In other words, B cells need help to become capable of producing antibodies.[13-16]

Similarly, CTLs, the effector cells in the MHC-restricted H-Y-specific CMC assay (responsible for an H-Yc answer), differentiate from precursor cytotoxic T cells (T_{CP}) only if they receive help.[17] It is known that a subpopulation of T cells, T helper cells (T_H), are the ones governing this crucial function. However, the mechanisms of help to B and T_{CP} by T_H cells are different, and should be briefly outlined here.

B. B Cell/T_H Cell Interactions

There are two main classes of MHC molecules: class I antigens (HLA-A and HLA-B in the human, and H-2K, H-2D and H-2L in the mouse), and class II antigens (HLA-D in the human, and H-2A and H-2E in the mouse). The genes for the class II antigens are called *Ir* genes, and their gene products are called Ia molecules.[17]

B cells recognize and take up antigen by specific Ig receptors. Macrophages may also recognize and take up antigen, but their recognition is nonspecific. After uptake, the antigen (comprising, for example, the epitopes A, B, C, D, and E) is internalized by phagocytosis and processed by the B cell. Processing includes digestion with cathepsin-like proteases into amphipathic segments, and one such segment is chosen and presented on the cell surface together with Ia (class II) antigen. For example, a certain B cell presents the molecular complex Ia + the A epitope on the cell surface. This complex is then recognized by a T_H cell carrying the proper T cell receptor (TCR). After binding between TCR and Ia + A, the T_H cell delivers a signal to the B cell which then differentiates into a plasma cell and produces specific anti-A-antibodies. Simultaneously, the B cell secretes interleukin-1 which stimulates differentiation of the T_H cell into a blast cell; the blast cell secretes interleukin-2 (see the next section). Most likely, a B cell population is able to recognize, internalize, process, and present all of the epitopes A to E of the stated example.[16] Moreover, the signal by the stimulated T_H cell is specific for the antigen-presenting B cell, although nonspecific T cell replacing factors[17] are secreted in addition; these stimulate proliferation of other B cells. Thus, B cell presentation of antigen to T_H cells comprises a complete epitope (A to E) repertoire.

C. Macrophage/T_H Cell/T_{CP} Cell Interactions

In contrast to the interaction between B and T_H cells, which is class II restricted, differentiation of T_C cells is class I restricted. Macrophages nonspecifically recognize antigen (e.g., epitopes A to E). They phagocytize, process, and present an epitope bound to class I molecules. A cytotoxic T precursor (T_{CP}) cell, with the proper TCR, recognizes this complex of specific epitope (e.g., B) + class I molecule and, with the help of interleukin-2 secreted by T_H cells (see the foregoing section), the T_{CP} cell differentiates into a T_{CP} blast cell which then further differentiates into a T_C cell which in turn specifically recognizes the "B" epitope.[17]

It is now clear that T cells and B cells recognize different antigenic sites,[16] and that the T cells recognize a limited number of antigenic sites (e.g., A, B, E), whereas antibodies are directed at multiple sites probably involving the entire surface of the antigen (A to E). It is also clear that different classes of T cells (e.g., suppressor and helper T cells) recognize different, nonoverlapping epitopes.[17,18]

Given an antigen with epitopes A to E, one may conclude that antibodies are synthesized against all of the epitopes (i.e., complete B cell response), whereas cytotoxic T cells may evolve with reactivity for only some of the epitopes (e.g., A, B, E).

Thus, certain subjects that type positive for the H-Ys antigen (B cell response), may type H-Yc negative when tested in the CMC assay (T_C cell response). Indeed, it is well known that H-Y of the mouse is not detected by T_C cells when associated with certain class I haplotypes (e.g., H-2^k), whereas strong T_C responses are induced when H-Y is associated with certain other class I haplotypes (e.g., H-2^b).[19,20]

This seems to imply that H-Y is physically associated with MHC class I molecules, in keeping with Ohno's proposal[21] that H-Y of nongonadal cells is bound to β_2-microglobulin of the class I molecule. If so, it is easy to imagine that certain combinations of H-Y + class I are detected by T_C cells while certain others are not. Furthermore, it is believed that only proteins are able to initiate T cell responses in any significant way, while all other structures activate predominantly B lymphocytes.[17] Thus, H-Ys antibodies directed mainly against the sugar residues of H-Y could explain H-Ys positive and H-Yc negative typing in certain cases (there is evidence that H-Y is a glycoprotein[22-23]).

D. Mechanisms Involving Graft Rejection In Vivo

Several different mechanisms interact to destroy foreign (nonself) grafts. The actual killing of graft cells can be mediated by effector cells such as T_C cells, killer (K) cells and activated macrophages, or by soluble factors such as lymphotoxin and antibodies + complement.[17] As is the case for the T_C cells of the CMC assay, T_C cells involved in graft rejection in vivo are subject to MHC class I restriction. Thus, some inbred strains of mice are responder strains with respect to H-Y while some other strains are not.[19]

The fact that macrophages, K cells, and B cells (antibodies + complement) are among the effector cells responsible for graft rejection indicates that, when testing for the H-Yt using the *indirect* transplantation assay, a complete epitope repertoire is likely to be involved in the recognition, processing, presentation, and response to H-Y. Thus, activated macrophages and/or B cells may present epitope + Ia to T_H cells, and antibodies secreted by the helped B cells may coat the foreign cells and induce antibody-dependent cell-mediated killing by K cells.[17] Therefore, it seems that the indirect H-Yt assay resembles the H-Ys assays more closely than it does the CMC (H-Yc) assay in regard to the complete epitope repertoire response. In fact, in all cases where the subjects under investigation have been typed for both the H-Ys and the H-Yt, there is complete concordance.[10] As an additional point, xenogeneic priming of C57BL/6 mice leads to an H-Yt response,[6-8] but not to a CMC (H-Yc) response.[24]

We may now return to the question raised above: do H-Ys positive, H-Yc negative XX men and H-Ys negative, H-Yc positive 46,XYp$^-$ women suggest occurrence of alternative H-Y antigens? It may be argued that there is no strong evidence for more than one H-Y antigen. We shall return to this question, and present a model for the genetic regulation and biochemical modification of the H-Y antigen after this discussion of the genetics of the H-Y antigen.

III. H-Y GENETICS AND PROBLEMS

There is now evidence[25] that genes coding for the H-Ys of the human are on chromosome 6p. Lau and colleagues[25] isolated a 1-kb mRNA using cDNA clones originating from a mouse testicular cDNA library. By immunoscreening with H-Ys antibodies, they found the mRNA to be expressed (1) in the testis of the human and a number of other mammals, (2) in pre- and postmeiotic rat testes, and (3) in Sertoli cell cultures. Furthermore, they detected trace amounts of this mRNA in the female, but there was a 1000-fold higher dosage of it in the male.[26]

Since minute amounts were detected in the female, these investigators now call the antigen MEA, for "male-enhanced antigen". The autosomal location of the gene for MEA (H-Ys), combined with earlier genetic and immunological evidence on the H-Ys,[10] suggest that its expression is regulated by other genes located on the Y and X chromosomes.[27] Although not yet shown, for evolutionary considerations, one would expect the H-Ys gene(s) to be autosomal in other mammals as well, including the mouse.

There is now evidence for a gene(s) on the human Yq controlling H-Yc activity. Thus, 46,XX males carrying certain Yp sequences but lacking Yq sequences type H-Yc negative, whereas 46,XYp⁻ women type H-Yc positive.[12] It is not known whether this H-Yc gene(s) on Yq is structural or regulatory; this question will be addressed below.

Autosomal location of the human H-Ys gene(s) and regulation of that gene by X- and Y-linked genes is compatible with the findings of H-Ys in 45,X (Turner's syndrome) females,[28] 46,XX true hermaphrodites lacking Y-specific DNA sequences,[29] 46,XX males,[4] and with the description of H-Ys negative or weakly positive 46,XYp⁻ females. It is also compatible with H-Ys and H-Yt positive X*Y, X*X, X*O, and XO female wood lemmings.[7,8,30] (Wood lemmings have not been typed for the H-Yc.) This will be shown in the model below.

In the mouse, a segment located rather close to the centromere on the normal Y chromosome is responsible for H-Yc and H-Yt[31,32] and for the development of testes.[33] This segment may be duplicated and transposed to the distal end of the Y chromosome. In the latter position, it may cross over to the X chromosome during male meiosis, thus giving rise to an XX sex-reversed mouse. The translocated piece of the Y chromosome is known as the Sxr (sex-reversed) fragment. Sex-reversed XX*Sxr* mice have testes (albeit small and infertile) and type H-Yt, H-Ys, and H-Yc positive.[34,35] Thus, in addition to genetic information essential for development of the testis, the Sxr fragment contains information for expression of H-Y antigen.

A mutation in the Sxr region was recently described.[31] This mutation, termed *Sxr'*, does not prevent testicular development in affected animals, but the spleen and skin of adult Sxr' mice type negative for the H-Yc[31-36] and H-Yt;[32,37] Sxr' mice have not yet been tested for the H-Ys antigen. Thus, it seems that Y-linked H-Y controlling genes are present and active in the Sxr fragment, but lost or silenced (inactive) in the Sxr' fragment.

IV. A MODEL FOR THE GENETIC REGULATION AND BIOCHEMICAL MODIFICATION OF H-Y ANTIGEN

As discussed above, discordant H-Y typing, with different assays for H-Y, does not necessarily indicate the existence of more than one male-specific antigen. We share the view of Ohno[38] (which is supported by the available immunological evidence reviewed above) that it is likely that T_H cells and T_C (T_{CP}) cells recognize alternative epitopes of the same antigen, and thereby generate confusion.

These postulates are required to render the following model valid. (1) Structural and

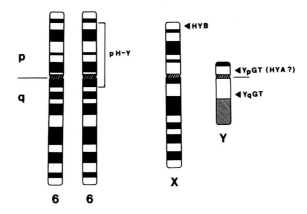

FIGURE 1.   Location of H-Y structural and regulatory genes according to the proposed model. The human karyotype is used for illustration. The gene for pH-Y (precursor H-Y) is thought to be synonymous with the *MEA* gene of Lau et al.,[25] and is located in the region 6p23—6q12.[26] *HYB* is the X-linked repressor gene which escapes X-inactivation;[42] thus, a double dosage of the HYB gene product is present in the normal XX female (complete repression of *pH-Y*), whereas the single gene dosage allows for expression of pH-Y in the XO female (and XY male). The Yp glycosyltransferase (GT) gene may or may not be identical to the *HYA* gene.[42] The *YqGT* gene is supposed to be the equivalent of the *H-Yc* gene, and is lacking in the 46,X,t(X;Yp) cases of Simpson et al.,[12] for example. p: short arm, and q: long arm of the chromosomes, respectively. See text and Table 1.

regulatory genes for H-Y are chromosomally similarly located in mammals. (2) H-Y antibodies are directed against protein and, preferentially, carbohydrate epitopes of the H-Y glycoprotein. (3) H-Y-specific MHC-restricted cytotoxic T lymphocytes, detecting and defining the H-Yc, are raised exclusively against protein epitopes of the H-Y glycoprotein. (4) The MEA gene product of chromosome 6 of the human is the same as the unglycosylated part of the H-Ys glycoprotein.

A. The Model
The human karyotype is chosen in order to visualize the model.

1. Mammalian autosomal genes code for a precursor protein, pH-Y.[39]
2. The pH-Y molecule is structurally modified by gene products coded for by genes on Yp and Yq. These gene products are likely to be glycosyltransferases (GTs) (and/or glycosidases) involved in terminal glycosylation.[39,40]
3. The YpGT, when present alone, glycosylates the pH-Y such that the protein parts of the latter become "hidden" and are not recognized by T_C cells or antibodies directed against protein epitopes of the pH-Y.
4. The YqGT modifies the Yp-glycosylated pH-Y such that protein epitopes are exposed and coexisting with carbohydrate epitopes.
5. When the YqGT acts alone, little or no carbohydrate epitope is present.
6. Finally, the *pH-Y* gene is under repression by X-chromosomally coded gene products.[28,41]

The location of the genes postulated in the model are synoptically illustrated in Figure 1.

Table 1

PREDICTIONS[a] AS TO H-Y PHENOTYPE IN CERTAIN CASES OF ABNORMAL SEX CHROMOSOME CONSTITUTION IN THE HUMAN

| | Genes present[b] | | | | H-Y type | | | |
Karyotype	pH-Y	HYB	YpGT	YqGT	H-Yt	H-Ys	H-Yc	Ref.
46,XX	Yes	Yes (x2)[c]	No	No	−	−	−	4
46,XY	Yes	Yes (x1)[c]	Yes	Yes	+	+	+	4
45,X	Yes	Yes (x1)[c]	No	No	+	−(±)[d]	−[e]	6, 28, 43
46,XYp⁻	Yes	Yes (x1)[c]	No	Yes	−(±)[d,e]	−(±)[d]	+[f]	4, 12
46,X,t(X;Yp)	Yes	Yes (x?)[c]	Yes	No	+[e]	+	−	4, 12
47,XX,t(X;Yp)	Yes	Yes (x?)[c]	Yes	No	+	+[e]	−[e]	44

[a] Based on the model for genetic regulation and biochemical modification of H-Y.
[b] Compare with text and Figure 1.
[c] Dosage of the HYB genes present.
[d] (±), Weakly positive.
[e] Predicted H-Y typing.
[f] All other + and − signs are based on experiments.

B. Predictions

1. The normal XY male should type positive and the XX female negative for H-Yt, H-Ys, and H-Yc.
2. Individuals carrying the YpGT gene but lacking the YqGT gene should type positive for H-Yt and H-Ys, but negative for H-Yc.
3. Individuals carrying the YqGT but lacking the YpGT gene should type negative or weakly positive for the H-Yt and the H-Ys, and positive for the H-Yc.

The suggestion in the model of more than a single enzyme modifying the pH-Y is in line with current glycobiology; the carbohydrate moieties of glycoconjugates change in structure during development and differentiation.[40] Moreover, the sugar part of glycoproteins exists in covalent association with the protein part. This makes it more likely that carbohydrate epitopes, too, are presented to T_H cells (in association with Ia) by B cells or macrophages after cytoplasmatic processing of antigen by the macrophages. Finally, it is clear that a change in a single sugar residue will affect not only the composition but the overall conformation of a carbohydrate, and thus, will affect specific recognition of a glycoprotein.[40]

How does the proposed model fit with the experimental data concerning the H-Y states of various subjects (Table 1)? First, it explains the discrepant findings of H-Ys positive[4] but H-Yc negative[12] 46,XX males, and H-Yc positive[12] 46,XYp⁻ females; the latter may be weakly H-Ys positive for the same reasons that true XO females are weakly positive (see below). Second, 45,X females with Turner's syndrome may type H-Ys and H-Yt positive,[6,28] and H-Yc positive because of hidden XO/XY mosaicism.[45] True XO females should type H-Yc negative. Because of weak reaction of anti-H-Y antibodies with the protein epitopes of the pH-Y, true XO females may type weakly H-Ys positive and, most likely, H-Yt positive as well. Weak reaction of anti-H-Y antibodies with the pH-Y of XO females, furthermore, would explain earlier reports of reduced H-Ys in these patients.[28] Expression of pH-Y in XO females (and XY males) is allowed for because of a single copy of the X-linked repressor gene (HYB) product[28,41] (Figure 1).

Next, all 46,XX true hermaphrodites investigated have been found to lack Y-specific DNA sequences[29,46] and to type H-Ys positive.[4,29] (XX true hermaphrodites of the human have not been tested for the H-Yt and H-Yc, but the model would require that they type positive for the former and negative for the latter; see below.) There is evidence that XX true hermaphroditism in mammals may be inherited as an autosomal recessive (goat), or recessive or dominant trait (dog); and these subjects, too, type H-Ys positive.[4] In these cases, therefore, H-Ys positivity is not readily explained by insufficient product of the X-linked repressor gene. Instead, a mutation at the repressor site of the autosomal pH-Y gene is likely as suggested by Waibel et al.[29]

A recently described[47] 47,XXX male provided an interesting test of the model. We used cultured fibroblasts from skin and testes of the XXX male to sensitize B6 female mice. The skin and testis fibroblasts each strongly sensitized the B6 females to subsequent grafts of syngeneic male skin. Thus, the XXX cells carried the H-Yt.[44]

There is evidence that two of the X chromosomes of the XXX male are of maternal origin, and the third, of paternal origin.[47] Moreover, the paternally derived X was shown, by *in situ* hybridization, to carry Yp sequences, but there was no evidence for the presence of Yq material.[47,48] Because of the lack of Yq sequences in skin and testis of the XXX male, the model would require that these tissue types be H-Yc negative. With regard to the H-Ys he should type positive, as do XX males (see above).

The model would also explain the H-Y findings in the wood lemming. As already mentioned, X*Y, X*O, X*X, and XO female wood lemmings (and normal XY males) are H-Ys and H-Yt positive.[7,8,30] These findings are readily explained by postulating a deletion or mutation of the presumptive X-linked repressor gene on the X chromosome, and indeed, there is evidence for cytogenetic rearrangements on the short arm of this chromosome.[49] H-Yt and H-Ys positive XO wood lemmings are explained in the same way as XO human females. Since wood lemmings are not inbred, it may be difficult to test them for the H-Yc in an allogeneic test system. Furthermore, the recent finding that xenogeneic H-Y positive cells of the human do not stimulate a CTL (H-Yc) response in inbred C57BL/6 mice[24] seems to limit the chances for using a similar system for H-Yc typing in the wood lemming.

In the mouse, one would predict that genes homologous to those coding for YpGT and YqGT in the human should be closely linked on the acrocentric Y chromosome; these should be present in the Sxr, but absent in the Sxr′ fragments because XY and Sxr male mice type H-Yc positive and H-Yt positive, whereas Sxr′ male mice type negative for these epitopes. Since adult Sxr′ mice type negative for the H-Yt, the model would require that they type H-Ys negative as well.

V. H-Y ANTIGEN AND THE TESTIS

We have so far avoided discussion of the possible biological function(s) of H-Y antigen(s). Currently this subject is a matter for controversy.

Ohno and Wachtel[50,51] postulated that the H-Y antigen is the product of the mammalian testis-determining gene, and indeed, there is a large body of experimental data that seem to support this hypothesis.[4,38,51] On the other hand, the fact that skin and spleen of adult Sxr′ male mice type negative for H-Yc and H-Yt seems to falsify the hypothesis. If fetal tissues (in particular testis) of Sxr′ mice should turn out to lack the H-Ys, we should have strong evidence against the hypothesis.

XSxrO mice have spermatogenesis and are H-Yt, H-Ys, and H-Yc positive, whereas XSxr′O mice are aspermatic and H-Yt and H-Yc negative (see earlier discussion). These facts led to the hypothesis by Burgoyne et al.[36] that the H-Yc is the product of a Y-linked spermatogenesis gene. The circumstantial evidence seems to favor this hypothesis.

The prediction by Ohno[52] of a gonad-specific receptor for the H-Y, and later evidence of its existence,[53,54] are compatible with both of the hypotheses mentioned. (It has been proposed that the receptor is X linked.[30]) There is no conclusive evidence which of the cell types of the testis carry the receptor. According to the hypothesis of Ohno[52] and Wachtel,[4] it is the Sertoli cells (there is experimental evidence for this[55]), but this does not exclude presence of the receptor on germ cells (except for the early diploid stages[56]) since these (i.e., all or part of them) are H-Y antigen positive. Presence of the receptor on ovarian cells[53,54] (whether on somatic cells, or germ cells, or both), would likewise be compatible with both hypotheses because presence of H-Y on sperm and its receptor on the ovum could indicate a function of H-Y in sperm-egg recognition, for example. It should be stressed that this latter suggestion is only speculation; its aim is to demonstrate that there may be many possible roles for a male-specific cell surface antigen such as H-Y should the above-mentioned hypotheses be falsified.

In fact, H-Y antigens may exist as a family of differentially glycosylated proteins (see Chapter 13).[57] If so, each H-Y family member could have a specific function. In particular, Edelman[58] has shown that cell adhesion molecules (CAMs) involved in morphogenesis undergo remarkable sequences of expression, changes in concentration, and changes in localization at the cell surface. Also, CAMs appear in a definite spatiotemporal order during development and histogenesis, and chemical modulation of different forms of CAMs involves glycosylation. These characteristics could apply to H-Y molecules as well.

In any event, there must be a major testis-determining gene on the mammalian X chromosome. This notion is based on studies in the human[59] and, in particular, on data obtained in the wood lemming. In the latter species, there are two morphologically distinguishable X chromosomes.[49,60] One of these represents the normal, wild-type X, and the other, a mutant chromosome, designated X*. When the X* chromosome is present together with a normal Y chromosome, the X*Y wood lemming becomes a normal fertile female.[60] Aberrant X*XY wood lemmings occur as males, females, or hermaphrodites.[61] We have shown that if the X* chromosome is X inactivated in such X*XY lemmings, they develop into males, whereas inactivation of the wild-type X causes female development.[62] This is strong evidence for a gene involved in testis determination on the wood lemming X; evidently this gene is deleted or mutated on the X* chromosome. One might postulate the existence of a similar gene on the mammalian X chromosome in general.[63]

VI. EPILOGUE

Whatever the function(s) of H-Y antigen(s), it remains a truth that the falsification, not the corroboration, of hypotheses is the driving force of science. Similarly, the enigmas, not the dogmas, make science a fascinating occupation.

ACKNOWLEDGMENTS

This work was supported by grants from the *Deutsche Forschungsgemeinschaft*. I thank Prof. U. Wolf for reading the manuscript and for discussions.

REFERENCES

1. **Eichwald, E. J. and Silmser, C. R.,** Untitled communication, *Transpl. Bull.,* 2, 148, 1955.
2. **Celada, F. and Welshons, W. J.,** An immunogenetic analysis of the male antigen in mice utilizing animals with an exceptional chromosome constitution, *Genetics,* 48, 139, 1963.
3. **Billingham, R. E. and Silvers, W. K.,** Studies on tolerance of the Y chromosome antigen in mice, *J. Immunol.,* 85, 14, 1960.
4. **Wachtel, S. S.,** *H-Y Antigen and the Biology of Sex Determination,* Grune & Stratton, New York, 1983.
5. **Silvers, W. K. and Yang, S.-L.,** Male specific antigen: its homology in mice and rats, *Science,* 181, 570, 1973.
6. **Wiberg, U. H.,** H-Y transplantation antigen in human XO females, *Hum. Genet.,* 69, 15, 1985.
7. **Wiberg, U. H. and Fredga, K.,** The H-Y transplantation antigen is present in XO and X*X female wood lemmings (*Myopus schisticolor*), *Immunogenetics,* 22, 495, 1985.

8. **Wiberg, U. H. and Günther, E.,** Female wood lemmings with the mutant X*-chromosome carry the H-Y transplantation antigen, *Immunogenetics,* 21, 91, 1985.

9. **Goldberg, E. H., Boyse, E. A., Bennett, D., Scheid, M., and Carswell, E. A.,** Serological demonstration of H-Y (male) antigen on mouse sperm, *Nature (London),* 232, 478, 1971.

10. **Wiberg, U. H.,** Facts and considerations about sex-specific antigens, *Hum. Genet.,* 76, 207, 1987.

11. **Goldberg, E. H., Shen, F., and Tokuda, S.,** Detection of H-Y (male) antigen on mouse lymph node cells by the cell to cell cytotoxicity test, *Transplantation,* 15, 334, 1973.

12. **Simpson, E., Chandler, P., Goulmy, E., Disteche, C. M., Ferguson-Smith, M. A., and Page, D. C.,** Separation of the genetic loci for the H-Y antigen and testis determination on human Y chromosome, *Nature (London),* 326, 876, 1987.

13. **Lanzavecchia, A.,** Antigen specific interaction between T and B cells, *Nature (London),* 314, 537, 1985.

14. **Howard, J. C.,** Immunological help at last, *Nature (London),* 314, 494, 1985.

15. **Marrack, P.,** New insights into antigen recognition, *Science,* 235, 1311, 1987.

16. **Smith, J. A. and Rose, G. D.,** Immune recognition of proteins: conclusions, dilemmas and enigmas, *Bio. Essays,* 6, 112, 1987.

17. **Klein, J.,** *Immunology: The Science of Self-Nonself Discrimination,* John Wiley & Sons, New York, 1982, 129.

18. **Gammon, G., Dunn, K., Shastri, N., Oki, A., Wilbur, S., and Sercarz, E. E.,** Neonatal T-cell tolerance to minimal immunogenic peptides is caused by clonal inactivation, *Nature (London),* 319, 413, 1986.

19. **Hurme, H., Chandler, P. R., Hetherington, C. M., and Simpson, E.,** Cytotoxic T-cell responses to H-Y: correlation with the rejection of syngeneic male skin grafts, *J. Exp. Med.,* 147, 768, 1978.

20. **Simpson, E.,** The role of H-Y as a minor transplantation antigen, *Immunol. Today,* 3, 97, 1982.

21. **Ohno, S.,** The original function of MHC antigens as the general plasma membrane anchorage site of organogenesis-directing proteins, *Immunol. Rev.,* 33, 59, 1977.

22. **Shapiro, M. and Erickson, R. P.,** Evidence that the serological determinant of H-Y antigen is carbohydrate, *Nature (London),* 290, 503, 1981.

23. **Shapiro, M. and Goldberg, E. H.,** Analysis of a serological determinant of H-Y antigen: evidence for carbohydrate specificity using an H-Y specific monoclonal antibody, *J. Immunogenet.,* 11, 209, 1984.

24. **Braun, A. and Cleve, H.,** The assay for cell mediated lympholysis in inbred mice is not suitable for H-Y antigen determination of human cells, *Hum. Genet.,* in press.

25. **Lau, Y.-F., Chan, K., Kan, Y. W., and Goldberg, E.,** Isolation of a male-specific and conserved gene using an anti-H-Y antibody, *Am. J. Hum. Genet.,* Suppl., 39, A142, 1986.

26. **Lau, Y.-F.,** personal communication, 1987.

27. **Wolf, U.,** Studies on H-Y antigen in disorders of sexual development, in *Actualités Gynécologiques,* 11ᵉ Serie, Netter, A. and Gorius, A., Eds., Masson, Paris, 1980, 96.

28. **Wolf, U., Fraccaro, M., Mayerová, A., Hecht, T., Zuffardi, O., and Hameister, H.,** Turner syndrome patients are H-Y positive, *Hum. Genet.,* 54, 315, 1980.

29. **Waibel, F., Scherer, G., Fraccaro, M., Hustinx, T. W. J., Weissenbach, J., Wieland, J., Mayerová, A., Back, E., and Wolf, U.,** Absence of Y-specific DNA sequences in human 46,XX true hermaphrodites and in 45,X mixed gonadal dysgenesis, *Hum. Genet.,* in press.

30. **Wiberg, U., Mayerová, A., Müller, U., Fredga, K., and Wolf, U.,** X-linked genes of the H-Y antigen system in the wood lemming (*Myopus schisticolor*), *Hum. Genet.,* 60, 163, 1982.

31. **McLaren, A., Simpson, E., Tomonari, K., Chandler, P., and Hogg, H.,** Male sexual differentiation in mice lacking H-Y antigen, *Nature (London),* 312, 552, 1984.

32. **Simpson, E., Chandler, P., Hunt, R., Hogg, H., Tomonari, K., and McLaren, A.,** H-Y status of X/X *Sxr'* male mice: *in vivo* tests, *Immunology,* 57, 345, 1986.

33. **Cattanach, B. M., Pollard, C. E., and Hawkes, S. G.,** Sex-reversed mice: XX and XO males, *Cytogenetics,* 10, 318, 1971.

34. **Bennett, D., Mathieson, B. J., Scheid, M., Yanagisawa, K., Boyse, E. A., Wachtel, S. S., and Cattanach, B. M.,** Serological evidence for H-Y antigen in *Sxr,* XX sex-reversed phenotypic males, *Nature (London),* 265, 255, 1977.

35. **Simpson, E., Edwards, P., Wachtel, S., McLaren, A., and Chandler, P.,** H-Y antigen in *Sxr* mice detected by H-2 restricted cytotoxic T cells, *Immunogenetics,* 13, 355, 1981.

36. **Burgoyne, P. S., Levy, E. R., and McLaren, A.,** Spermatogeneic failure in male mice lacking H-Y antigen, *Nature (London),* 320, 170, 1986.

37. **Wiberg, U. H. and Lattermann, U.,** Syngeneic male graft rejection by B6 female mice primed with spleen and testes of *Sxr* and *Sxr'* mice, *Expl. Clin. Immunogenet.,* 4, 167, 1987.

38. **Ohno, S.,** Of testis and H-Y antigen, comment to Wiberg, U.H.; Facts and considerations about sex-specific antigens, *Hum. Genet.,* 76, 215, 1987.

39. **Polani, P. E. and Adinolfi, M.,** The H-Y antigen and its functions: a review and a hypothesis, *J. Immunogenet.,* 10, 85, 1983.

40. **Stanley, P.,** Glycosylation mutants and the function of mammalian carbohydrates, *TIG,* 3, 77, 1987.
41. **Wolf, U., Fraccaro, M., Mayerová, A., Hecht, T., Maraschio, P., and Hameister, H.,** A gene controlling H-Y antigen on the X chromosome. Tentative assignment by deletion mapping to Xp223, *Hum. Genet.,* 54, 149, 1980.
42. Human Gene Mapping 6 (1981): Sixth International Workshop on Human Gene Mapping, *Cytogenet. Cell Genet.,* 32, 179, 1982.
43. **Haseltine, F. P., DePonte, K. K., Breg, W. R., and Genel, M.,** Presence of H-Y antigen in patients with Ullrich-Turner syndrome and X chromosome rearrangements, *Am. J. Med. Genet.,* 11, 97, 1982.
44. **Wiberg, U.,** unpublished data, 1987.
45. **Müller, U., Donlon, T. A., Kunkel, S. M., Lalande, M., and Latt, S. A.,** Y-190, a DNA probe for the sensitive detection of Y-derived marker chromosomes and mosaicism, *Hum. Genet.,* 75, 109, 1987.
46. **Wiberg, U. H. and Scherer, G.,** Evidence for the presence of testicular tissue and *Sxs* antigen in the absence of Y-derived sequences, *Development,* in press, 1987.
47. **Annerén, G., Andersson, M., Page, D. C., Brown, L. G., Berg, M., Läckgren, G., Gustavson, K.-H., and de la Chapelle, A.,** XXX male resulting from paternal X-Y interchange and maternal X-X nondisjunction, *Am. J. Med. Genet.,* in press, 1987.
48. **Müller, U., Latt, S. A., and Donlon, T.,** Y-specific DNA sequences in male patients with 46,XX and 47,XXX karyotypes, *Am. J. Med. Genet.,* in press, 1987.
49. **Herbst, E. W., Fredga, K., Frank F., Winking, H., and Gropp, A.,** Cytological identification of two X-chromosome types in the wood lemming (*Myopus schisticolor*), *Chromosoma,* 69, 185, 1978.
50. **Wachtel, S. S., Ohno, S., Koo, G. C., and Boyse, E. A.,** Possible role for H-Y antigen in the primary determination of sex, *Nature (London),* 257, 235, 1975.
51. **Ohno, S.,** *Major Sex Determining Genes,* Springer-Verlag, Berlin, 1979.
52. **Ohno, S.,** Major regulatory genes for mammalian sexual development, *Cell,* 7, 315, 1976.
53. **Müller, U., Aschmoneit, I., Zenzes, M. T., and Wolf, U.,** Binding studies of H-Y antigen: indication for a gonad specific receptor, *Hum. Genet.,* 43, 151, 1978.
54. **Müller, U., Wolf, U., Siebers, J. W., and Günther, E.,** Evidence for a gonad-specific receptor for H-Y antigen: binding of exogenous H-Y antigen to gonadal cells is independent on β_2-microglobulin, *Cell,* 17, 331, 1979.
55. **Müller, U.,** Testis-determining H-Y antigen and the induction of the hCG receptor, in *Chorionic Gonadotropin,* Segal, S. S., Ed., Plenum Press, New York, 1980, 371.
56. **Zenzes, M. T., Müller, U., Aschmoneit, I., and Wolf, U.,** Studies on H-Y antigen in different cell fractions of the testis during pubescence, *Hum. Genet.,* 45, 297, 1978.
57. **Bradley, M. P. and Heslop, B. F.,** Recent development in the serology and biochemistry of testicular sex-specific (H-Y) antigens, in *Evolutionary Mechanisms in Sex Determination,* Wachtel, S., Ed., CRC Press, Boca Raton, Fla., 1988, chap. 13.
58. **Edelman, G. M.,** Cell adhesion and the molecular processes of morphogenesis, *Ann. Rev. Biochem.,* 54, 135, 1985.
59. **Wolf, U.,** XY gonadel dysgenesis and the H-Y antigen: report on 12 cases, *Hum. Genet.,* 47, 269, 1979.
60. **Fredga, K., Gropp, A., Winking, H., and Frank, F.,** Fertile XX- and XY-type females in the wood lemming, *Nature (London),* 261, 225, 1976.
61. **Winking, H., Gropp, A., and Fredga, K.,** Sex determination and phenotype in wood lemmings with XXY and related karyotypic anomalies, *Hum. Genet.,* 58, 98, 1981.
62. **Schempp, W., Wiberg, U., and Fredga, K.,** Correlation between sexual phenotype and X-inactivation pattern in the X*XY wood lemming, *Cytogenet. Cell Genet.,* 39, 30, 1985.
63. **Ohno, S.,** *Sex Chromosomes and Sex-Linked Genes,* Springer-Verlag, Berlin, 1967.

Chapter 11

SEROLOGIC H-Y ANTIGEN IN THE HOMOGAMETIC SEX

Ulrich Wolf and Cecilia Ebensperger

TABLE OF CONTENTS

I. INTRODUCTION

Anti-H-Y antiserum is usually raised in female laboratory rodents by immunization with syngeneic male cells. Immunization is expected to occur if the homogametic female lacks H-Y antigen which, in turn, is male specific in mammals.

Testing of H-Y antigen in various species of nonmammalian vertebrates using mammalian-derived anti-H-Y antiserum revealed cross-reaction in the heterogametic but not in the homogametic sex.[1] Thus, it can be generalized that in the XY/XX mechanism of sex determination the male types H-Y positive, while in the ZZ/ZW mechanism the female types H-Y positive.

We have provided evidence that, under certain conditions, H-Y antigen also occurs in the homogametic sex. This allows for the conclusion that the determinant gene is shared by both sexes. Consequently, the sex specificity of H-Y antigen as normally found, must be due to differential genetic control in the two sexes.

The examples to be discussed here are (1) *Coris julis,* a teleostean fish with the XY/XX mechanism, (2) the chicken with the ZZ/ZW mechanism, and (3) human XX true hermaphrodites, as a representative of mammals.

Following a proposal by Wiberg,[2] serologic H-Y antigen will be called "Sxs antigen" for "serologic sex-specific antigen", in this chapter. There are two reasons for this new designation: first, it avoids confusion with the H-Y transplantation antigen; and second, it takes into account the specificity of this factor not only for the male sex in mammals, but for the heterogametic sex in general, within the vertebrate subphylum.

II. *CORIS JULIS*

The wrasse is a diandric protogynous Labrid fish living in the Mediterranean. This species includes primary males which never change sex and which resemble the rather inconspicuous females in their external appearance, and secondary males which result from physiological sex inversion of females and which develop a characteristic color pattern. It has been shown that sex inversion can also be induced experimentally by administration of androgens to functional females.

In cooperation with Prof. R. Reinboth (Mainz, W. Germany), we recently determined Sxs antigen in gonads and nongonadal tissues of these different sexual types of *Coris julis.*[3] It turned out that the males, whether primary or secondary, were Sxs positive in all tissues studied (gonads, spleen, muscle, brain). In contrast, the respective female tissues were Sxs negative. Administration of testosterone to females results in degeneration of oocytes and formation of testicular tubules; external changes in color pattern also occur. These experimentally sex-inversed females proved to be Sxs positive in gonads and nongonadal tissues.

These findings suggest that androgens are responsible, directly or indirectly, not only for the sex inversion as such, but also for the occurrence of Sxs antigen in the homogametic female. If this is true, the physiological sex change and the presence of Sxs antigen in secondary males may also be due to the influence of androgens.

III. *GALLUS DOMESTICUS*

In the chicken, Sxs antigen is present in the heterogametic female but absent in the normal male. It is well known that estrogens, administered to the embryonic male chicken, cause the formation of an ovotestis.[4] In early experiments, we detected Sxs in the sex-inversed male gonad on day 13 of embryonic life after treatment of 4-day-old embryos with estradiol in vivo. In contrast, nongonadal tissues of the treated males lacked the Sxs antigen.[5]

From these findings, it appeared that estradiol effectuates, by whatever mechanism, the occurrence of Sxs antigen in the treated male. We therefore asked if the female embryo lacks Sxs antigen at the indifferent stage of gonad development and before estrogens are produced physiologically. We approached that question in cooperation with Prof. U. Drews (Tübingen, W. Germany) by studying various tissues of chicken embryos from the age of $5^{1}/_{2}$ days onward. At day 6, the embryonic gonad is still indifferent histologically; the sexes can first be distinguished around $6^{1}/_{2}$ days. Testing for Sxs antigen revealed that, at day 6, individual embryos as well as pooled organs (gonads, mesonephros, liver) each were indeed negative. At $6^{1}/_{2}$ days, Sxs activity was first detectable in pooled embryos, though not yet at full titer of the female control. At day 7, the sex difference was clear-cut; male organs (gonads and nongonadal tissues) were Sxs negative and female organs were Sxs positive, similar to the control samples from older animals.

Thus, in the female, Sxs antigen appears at the stage when gonadal differentiation starts. Interestingly, at about the same developmental stage, estrogens become detectable in the female embryo.

To further study the association between estrogens and Sxs antigen, we repeated the sex-inversion experiments, again in cooperation with Prof. Drews, but this time under well-defined in vitro conditions using the following experimental design. Gonads and mesonephros, respectively, of 8-day old chicken embryos were pooled separately for each sex, dissociated proteolytically into single-cell suspensions and brought into rotation culture. Parts of the suspensions were treated with estradiol. After 3 days of culture, the resulting aggregates were tested for Sxs antigen and the gonadal aggregates examined histologically. After the estradiol treatment, the testicular cultures developed a distinct ovarian-like organization, and the ovarian cultures showed an intensification of the normal female organization. Sxs antigen was absent in the male control aggregates of testicular and mesonephric tissues, but present in the female aggregates. Administration of estradiol caused the testicular aggregates to become Sxs positive, while the treated male mesonephric cultures remained Sxs negative, similar to the controls. There was no difference between treated and untreated female cultures; all were Sxs positive. Thus, our former results obtained in vivo were confirmed in these histologically well-defined in vitro experiments.

These findings allow for the conclusion that in the ZZ/ZW system of the chicken, estrogens are responsible for the occurrence of Sxs antigen in the homogametic male sex. This effect, however, appears to be restricted to the gonads, as was the case in our in vivo experiments.

IV. HUMAN XX TRUE HERMAPHRODITES

In humans, the occurrence of XX males and XX true hermaphrodites is a well-known phenomenon. From the point of view of this paper, the question arises whether this represents sex inversion of the homogametic female sex. With respect to XX males, it can now be stated that this is not true, at least in the majority of cases. On searching for DNA sequences derived from the Y chromosome in the genome of XX males, one finds that most cases react positively with one or another Y-specific DNA probe.[6] By means of in situ hybridization with these probes, as performed in a number of cases, evidence was obtained that the XX male condition originates from X/Y interchange.[7]

In our laboratory, we have investigated 11 XX males using six different DNA probes specific for various parts of the Y chromosome. In each case a positive signal for at least one of these probes was obtained. Seven of these cases were also studied by the technique of in situ hybridization and in each, Y-specific DNA could be assigned to the distal short arm of one of the X chromosomes.[8] It may be concluded that XX males positive for Y-specific DNA are not the result of a gene mutation causing sex inversion of the homogametic sex. Rather they represent pseudohomogametic individuals resulting from X/Y interchange (theoretically they could also originate from another translocation mechanism introducing a segment of the Y chromosome into their genome).

Our findings in XX true hermaphrodites contrast sharply with our findings in XX males. We recently studied seven cases of 46,XX true hermaphroditism, confirmed by gonadal histology — including two siblings — and an additional case with 45,X mixed gonadal dysgenesis.[9] We were not able to demonstrate presence of Y-specific DNA in any of these cases. In particular, the Y-derived DNA probe Fr 80-II, which gave a positive signal in all of the 11 XX males, did not react with the DNA of any of the hermaphrodites. As far as we are aware, other investigators also have failed to detect Y-DNA in this condition.

Based on these findings, we start from the assumption that in the cases of XX true hermaphroditism studied so far, Y-specific DNA is indeed lacking, and that the relevant individuals may be considered homogametic.

A larger number of XX males including our own cases have been examined for Sxs antigen, and they were consistently found to be Sxs positive.[1] Interestingly, XX true hermaphrodites are also Sxs positive in all cases studied,[1] and we have also shown this explicitly in the cases mentioned here.[9] It appears, however, that they have a somewhat reduced Sxs titer as compared to male controls.

If we take the situation as it stands now, it can be stated (1) that testicular morphogenesis can take place in the absence of DNA from the Y chromosome, and (2) that Sxs antigen can be present in the absence of Y-DNA. If this is true, it follows that the gene for Sxs antigen is shared by both sexes in the human, and that it can be expressed in the homogametic sex.

It is not intended within the scope of this article, to discuss the genetic reason for the occurrence of XX true hermaphroditism. It may be said only that a ''constitutive'' mutation seems to be responsible for the expression of the *Sxs* gene which otherwise is repressed in the female.

V. GENERAL CONCLUSIONS

The occurrence of Sxs antigen in the homogametic sex of various vertebrate species, irrespective of the mechanism of sex determination (XY/XX or ZZ/ZW), may allow for the generalization that, in vertebrates, the structural gene for Sxs antigen is contained in the genome of both sexes. Thus, the sex specificity of Sxs antigen must be a matter of genetic control. Our findings in the chicken and *Coris julis,* and similar findings obtained in the quail[10] and *Xenopus laevis*[11] support the view that in nonmammalian vertebrates this control is exerted by sex steroid hormones, i.e., estrogens in the ZW mechanism and androgens in the XY mechanism. In mammals (at least in placentalia), the *Sxs* gene appears to be controlled by the Y chromosome.

Some remarks on the possible function of Sxs antigen may be indicated on the basis of the findings reviewed briefly in this chapter. The striking association between the occurrence of Sxs antigen and the heterogametic sex has been a chief reason for the original hypothesis of Wachtel et al.[12] that this factor is essential for the development of the heterogametic gonad. As noted above, sex inversion of the homogametic sex is again associated with the occurrence of Sxs antigen. In the female chicken, Sxs antigen is first detected at the time of the differentiation of the indifferent gonad anlage, and in *Coris julis,* it is detected in connection with the sex change of females into secondary males. In XX true hermaphrodites, the development of an ovotestis is constitutive and consequently, Sxs antigen appears to be present constitutively.

This striking association is no proof of a role of Sxs antigen in the morphogenesis of the heterogametic gonad. However, such a function cannot be excluded, and there are some indications that favor this view. In the rat, Sxs antigen binds selectively to gonadal cells which have been shown to be endowed specifically with an Sxs receptor.[13] Furthermore, testicular Sertoli cells secrete Sxs antigen,[14,15] and they also bind it.[16] In contrast, male germ cells are Sxs negative up to the late diploid stages.[14] Thus, Sertoli cells should be a target for Sxs antigen, in contrast to early diploid germ cells, and therefore, a candidate function for Sxs in the developing testis could be the organization of Sertoli cells into tubular structures. It is interesting in this context that testicular morphogenesis occurs independently of germ cells.

With respect to the ovary, it has been shown that this organ also binds Sxs antigen specifically,[17] but it is not yet determined which cells of the ovary are endowed with the receptor and which are not. By analogy, one might infer that the ovarian follicular cells are at least one of the targets of Sxs antigen in the heterogametic ovary. However, as the organization of a follicle depends on the presence of germ cells (see Chapter 15), the situation in the heterogametic ovary must be different from that in the heterogametic testis. It is also not clear whether Sxs antigen is secreted at all in the heterogametic ovary, and if it is, by what cell type. Furthermore, the possibility that in the ovary, germ cells are endowed with the specific receptor for Sxs antigen cannot be excluded.

The examples discussed here may be considered another indication of a role for Sxs antigen in the primary development of the heterogametic gonad. It remains to be tested if Sxs antigen, when available as a purified protein, is itself able to bring about sex inversion, e.g., in a model system like the dissociation-reorganization experiments with chicken gonads described above, using experimental sex inversion in *Coris julis* or the in vitro reorganization of chicken gonads.

REFERENCES

1. **Wachtel, S. S.,** *H-Y Antigen and the Biology of Sex Determination,* Grune & Stratton, New York, 1983, 54, 178, 192.
2. **Wiberg, U. H.,** Facts and considerations about sex-specific antigens, *Hum. Genet.,* 76, 207, 1987.
3. **Reinboth, R., Mayerová, A., Ebensperger, C., and Wolf, U.,** The occurrence of serological H-Y antigen (Sxs antigen) in the diandric protogynous wrasse *Coris julis* (L.) (Labridae, Teleostei), *Differentiation,* 34, 13, 1987.
4. **Mittwoch, U.,** *Genetics of Sex Differentiation* Academic Press, New York, 1973, 103.
5. **Müller, U., Zenzes, M.T., Wolf, U., Engel, W., and Weniger, J.-P.,** Appearance of H-W (H-Y) antigen in the gonads of oestradiol sex-reversed male chicken embryos, *Nature (London),* 280, 142, 1979.
6. **Vergnaud, G., Page, D. C., Simmler, M. C., Brown, L., Rouyer, F., Noel, B., Botstein, D., and de la Chapelle, A.,** A deletion map of the human Y chromosome based on DNA hybridization, *Am. J. Hum. Genet.,* 38, 109, 1986.
7. **Magenis, R. E., Casanova, M., Fellous, M., Olson, S., and Sheehy, R.,** Further cytologic evidence for Xp-Yp translocation in XX males using in situ hybridization with Y-derived probe, *Hum. Genet.,* 75, 228, 1987.
8. **Schempp, W.,** personal communication, 1987.
9. **Waibel, F., Scherer, G., Fraccaro, M., Hustinx, T. W. J., Weissenbach, J., Wieland, J., Mayerová, A., Back, E., and Wolf, U.,** Absence of Y-specific DNA sequences in human 46,XX true hermaphrodites and in 45,X mixed gonadal dysgenesis, *Hum. Genet.,* 76, 332, 1987.
10. **Müller, U., Guichard, A., Reyss-Brion, M., and Scheib, D.,** Induction of H-Y antigen in the gonads of male quail embryos by diethylstilbestrol, *Differentiation,* 16, 129, 1980.
11. **Wachtel, S. S., Bresler, P. A., and Koide, S. S.,** Does H-Y antigen induce the heterogametic ovary?, *Cell,* 20, 859, 1980.
12. **Wachtel, S. S., Ohno, S., Koo, G. C., and Boyse, E. A.,** Possible role for H-Y antigen in the primary determination of sex, *Nature (London),* 257, 235, 1975.
13. **Müller, U., Wolf, U., Siebers, J.-W., and Günther, E.,** Evidence for a gonad-specific receptor for H-Y antigen; binding of exogenous H-Y antigen to gonadal cells is independent of β_2-microglobulin, *Cell,* 17, 331, 1979.
14. **Zenzes, M. T., Müller, U., Aschmoneit, I., and Wolf, U.,** Studies on H-Y antigen in different cell fractions of the testis during pubescence, *Hum. Genet.,* 45, 297, 1978.

15. **Brunner, M., Moreira-Filho, C. A., Wachtel, G., and Wachtel, S.,** On the secretion of H-Y antigen, *Cell,* 37, 615, 1984.
16. **Wolf, U. and Zenzes, M. T.,** Gonadendifferenzierung und H-Y-antigen, *Verh. Anat. Ges.,* 73, 379, 1979.
17. **Müller, U., Aschmoneit, I., Zenzes, M. T., and Wolf, U.,** Binding studies of H-Y antigen in rat tissues, *Hum. Genet.,* 43, 151, 1978.

Chapter 12

SEROLOGY OF THE MALE-SPECIFIC ANTIGEN

Vikram Jaswaney, Gwendolyn Wachtel, and Stephen Wachtel

TABLE OF CONTENTS

I. INTRODUCTION

Male-specific (H-Y) antigens are identified by the methods of transplantation,[1] cell-mediated cytotoxicity,[2] and serology.[3] Rejection of male grafts by intrastrain female mice is due to the H-Y transplantation antigen, for example, and antibodies from male-sensitized females define a male antigen in serological systems. There is some question as to whether the alternative systems recognize the same or alternative H-Y antigens,[4] and novel designations have been proposed with respect to the serological moeity (see for example Chapter 13). In this paper, we use the term H-Y antigen or serological H-Y antigen or H-Y to refer to the antigen identified by antibody, and H-Y transplantation antigen or H-Yt to refer to the antigen identified by graft rejection, without prejudice as to whether there may be one, two, or more H-Y antigens.

Serological H-Y antigen has been found in association with the heterogametic sex in more than 80 species, representing each of the major vertebrate classes.[5] This suggested an important function for H-Y, and indeed, phylogenetic conservatism was the basis for the proposition that the molecule bearing H-Y governs development of the heterogametic gonad — the testis in mouse and man.[6]

The proposition has generated research and controversy, much of the latter due to the question of supernumerary H-Y antigens, and to the formidable nature of H-Y serology.[7,8] In this paper, we review some of the techniques used in the study of H-Y antigen, and we summarize data obtained with a novel technique involving monoclonal H-Y antibodies, lymphocyte subset-specific antibodies, and flow cytometry. Our recent experience with this technique indicates differential expression of H-Y antigen among the lymphocyte subsets.

II. SEROLOGICAL METHODS

A. Cytotoxicity

Complement-mediated cytotoxicity tests for the male antigen were described as early as 1959, but 12 years elapsed before a report of methods allowing routine assay of H-Y was published.[9] Thus, in 1971, Goldberg et al.[3] described the sperm cytotoxicity test in which epididymal spermatozoa were reacted with complement and H-Y antibodies. The sperm cytotoxicity test was used for several years, especially in our laboratory, and many of the early studies relating to evolutionary persistence of H-Y were founded on that method. But the test was difficult and not always reliable. Scoring was subjective and indirect; results were based on absorption of the antibody by cells of male, female, and test subject (XYY male, for instance) and the results could be confounded by attachment of the antibody to cells of the female. (Modified sperm cytotoxicity tests have been used recently to measure the ability of antisera to detect H-Y on preimplantation embryos of the mouse.[10] The tests were designed to quantify changes in levels of ATP in solution due to cytotoxicity. For review of H-Y serological methods in X/Y sperm fractionation and embryo sex selection, see Anderson.[11])

In 1972, Scheid et al.[12] described a cytotoxicity test for H-Y on epidermal cells prepared from rat tail skin. The test offered the advantages of male and female targets, endpoints were clearcut, and cells were not likely to swim across the field, dragging debris and other cells with them, thereby nauseating the reader. But preparation of epidermal cell suspensions was a tedious affair involving multiple trypsinizations and overnight incubations, and results were complicated by lysis of female cells. So the epidermal test did not find wide applicability.

Another cytotoxicity test was described by Fellous et al.[13] The target was a cell line called "Raji", derived from a Burkitt lymphoma. The Raji test was easy to set up, and 100% cytotoxicity could be obtained with rat H-Y antibody. As in the sperm test, however, results were based on absorption; and moreover application of rat sera with the human cell line could be complicated by heteroantibody reactive with human lymphoblastoid lines.[14]

131

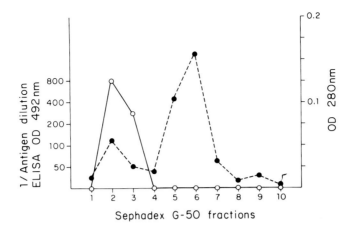

FIGURE 1. Analysis of Sephadex G-50 fractions after biotinylation of gw-16 antibody. Dotted line ●− − −● represents 1/40 dilutions of sample fractions scored for protein at OD280. Solid line ○—○ represents 1/200 dilutions of sample fractions reacted with TS (soluble H-Y) in the ELISA. The values on the vertical axis left give the highest dilutions of antigen at which OD492 values of at least 0.5 were observed. Activity of the biotinylated antibody for TS was confined to the peak at the left. (From Brunner, M., Moreira-Filho, C., Wachtel, G., and Wachtel, S. S., *J. Immunol. Methods*, 75, 203, 1984. With permission.)

B. Mixed Hemadsorption and Protein A Assays

A number of assays for H-Y were developed in the laboratory of Gloria Koo. Among those was the mixed hemadsorption hybrid antibody (MHA-HA) test[15] in which sperm cells were reacted with hybrid antibodies of anti-H-Y specificity on the one arm and anti-sheep red blood cell (SRBC) on the other. Positive reaction led to the formation of "rosettes" — sperm heads with a corona of at least three to five SRBC.

In another procedure, staphylococcal Protein-A (PA) was substituted for the hybrid antibody. PA binds to the Fc portion of IgG. When PA was coupled to SRBC, and the PA-SRBC reacted with antibody-exposed spermatozoa, the result was a corona similar to that observed in the MHA-HA test.[16]

C. Radiobinding, Fluorochrome, and Enzyme Assays

It is not our purpose to survey the entire serology of H-Y antigen, but a few other methods deserve mention here, notably those involving radiolabeling of target cells with polyclonal or monoclonal H-Y antibody and ^{125}I-conjugated protein-A or ^{125}I-conjugated rabbit anti-mouse globulin;[17-20] those involving fluorescent labeling of lymphocytes with H-Y antibody and fluorescein isothiocyanate (FITC)-conjugated goat anti-mouse Ig;[21,22] and those involving peroxidase-antiperoxidase staining for application with cultured fibroblasts.[23]

D. ELISA

The enzyme-linked immunosorbent assay has been used as a test for H-Y antigen (and H-Y antibodies) in several studies.[24,26] According to the technique outlined by Brunner et al.,[27,28] ascites fluid containing monoclonal H-Y antibody gw-16 (IgG$_{2a}$) was purified by passage through PA-Sepharose. The purified IgG was then conjugated with biotin by use of biotin-N-hydroxysuccinimide ester, and the reaction mixture fractionated by passage through a Sephadex G-50 column. When the individual fractions were analyzed for protein, two peaks were obtained (Figure 1). One of the peaks contained IgG reactive with soluble H-Y antigen in the ELISA.

For each test, the wells of a plastic microtitration plate were coated with mouse-testis

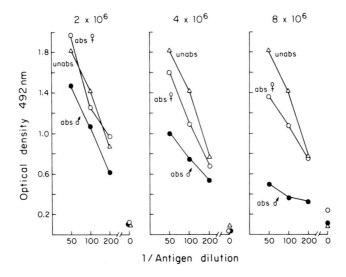

FIGURE 2.   Quantitative absorption of gw-16 antibody with spleen cells from male and female mouse. The panels, from left to right, show absorption curves for 2, 4, and 8 $\times$ 10^6 cells, respectively. Unabs denotes unabsorbed antibody; abs denotes absorption with cells of indicated sex; C denotes control, i.e., background OD in wells without antigen (TS). Biotinylated antibody was used at 1/300 dilution. (From Brunner, M., Moreira-Filho, C., Wachtel, G., and Wachtel, S. S., *J. Immunol. Methods,* 75, 203, 1984. With permission.)

supernatant (TS), Daudi-cell supernatant (DS), or conditioned medium from Sertoli cell cultures (TM-4) — each a presumptive source of H-Y antigen — and then blocked with buffer containing 2% normal mouse serum. Next, biotinylated gw-16 antibody was added, followed by an avidin-biotinylated-proxidase complex in phosphate-buffered saline (PBS) with 0.1% Tween. After incubation and washing, the substrate containing orthophenyl-enediamine and peroxide was introduced, and the reaction was scored by measurement of OD_{492} in an ELISA plate reader.

An example of the results that were obtained is given in Figure 2. For this series of tests, portions of the biotinylated antibody were diluted and absorbed with 2, 4, or 8 $\times$ 10^6 spleen cells from male or female mouse. Note the loss of reactivity after absorption with 8 $\times$ 10^6 cells of the female. In more recent tests with spleen cells of the mouse and fibroblasts of the human, considerable absorption was noted with female cells;[29] on occasion, male and female curves have been superimposed.

More recently, Bradley and Heslop[30] described an ELISA involving the use of high-titer polyclonal H-Y antibodies. The antibodies were induced in female rats by intrasplenic injection of male skin cells. In that system, aimed at biochemical characterization of H-Y, labeled Daudi proteins wer chromatographed, and the resulting fractions added to the wells of a microtitration plate. Polyclonal H-Y antibodies diluted 1/800 were added to the wells and after incubation, the plates were washed, and dilute urease-conjugated sheep anti-rat IgG was introduced. After further incubation and washing, the urea substrate was added, and the resulting color changes (yellow to purple) scored visually.

E. Particle Concentration Fluorescence Immunoassay

Particle concentration fluorescence immunoassay (PCFIA) is a method developed recently for quantification of immunoglobulin bound to the cell surface.[31] The method can be used with H-Y antibody to estimate H-Y on various target cells. In a series of preliminary tests

Table 1
PCFIA FOR H-Y ANTIGEN IN THREE CULTURED CELL LINES

Reagents	Raji	Daudi	K562
GAM-FITC[a] (control)	7,098[b] ± 1,311	4,439 ± 1,194	8,993 ± 742
Streptavidin-FITC (control)	595 ± 79	1,438 ± 193	2,371 ± 366
gw-16 (ascites); GAM-FITC	33,524 ± 2,132	24,202 ± 4,046	25,226 ± 3,866
Biotin:gw-16; streptavidin-FITC	44,176 ± 11,512	31,794 ± 5,711	25,471 ± 5,245
gw-16:FITC[c]	42,999 ± 12,530	28,798 ± 2,999	29,666 ± 4,030

[a] GAM — goat anti-mouse Ig
[b] Fluorescence 545/577 nm; arbitrary units; mean of eight readings ± SD.
[c] For this series of tests, FITC was conjugated directly to gw-16.

with the PCFIA method, we used three cultured lines as target cells: K562, a human female erythroleukemic line resembling the B cell population, and Raji and Daudi, each from a human male Burkitt lymphoma. Raji cells are H-Y$^+$; Daudi cells are negative for β2m and HLA and thus, for cell surface H-Y.[32]

For each test, 20-μℓ portions of medium containing 20,000 cells each were placed in the wells of Pandex "Epicon" plates. Polystyrene particles were placed in each well to improve filtration. Tests were performed in the Pandex FCA Fluorescence Concentration Analyzer, an automated system for the PCFIA which incorporates medium changes and washes directly in the Epicon wells. Antibody gw-16 diluted 1/20 was added to each well and the mixture incubated at room temperature. Next, a secondary reagent was added (streptavidin-FITC for use with biotinylated gw-16, or FITC-conjugated goat anti-mouse Ig for use with unmodified gw-16). After further incubation and washing, the cells were scored for bound label. The results are shown in Table 1.

F. Flow Cytometry

Laser-based flow cytometry permits direct and sensitive estimation of cell surface H-Y and rapid analysis of large numbers of cells (10,000 cells can be scored for label in a few seconds). We applied monoclonal antibody gw-16 to the study of H-Y in human white blood cells in two systems involving flow cytometry: one with FITC-conjugated goat anti-mouse Ig for single fluorochrome labeling, and the other with biotinylated gw-16.[33] The biotinylated antibody was used with streptavidin-FITC and phycoerythrin-derivative (RD1)-conjugated antibodies specific for lymphocyte antigens B1, T4, and T8. This enabled double labeling of cells, and thereby comparative evaluation of H-Y in three lymphocyte subsets: B cells (B1 antigen); T helper-inducer cells (T4 antigen), and T cytotoxic-suppressor cells (T8 antigen).

We used the EPICS 753 flow cytometer with 5 W argon laser tuned for simultaneous excitation of FITC and RD1 fluorochromes. Forward angle, 90° light scatter, and bit map gating were employed to select cell populations for evaluation. Red and green signals were measured in the gated cells and data transferred to the Coulter EASY 88 minicomputer for graphics. Data from 10,000 cells were collected for each test.

1. Single Fluorochrome Labeling

When lymphocytes were separated from whole blood and tested with gw-16 antibody and FITC-conjugated anti-mouse Ig, clearcut differences were obtained for male and female. Figure 3 depicts the results of one such test: 30% of male cells were labeled compared with 9% of female cells; intensity of fluorescence was about the same in each case. Specificity of the reaction was confirmed by absorbing the antibody with spleen cells from male or female C57BL/6 mice and then testing residual activity of the absorbed portions with human

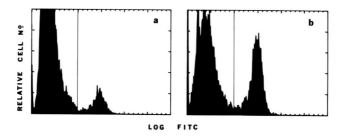

FIGURE 3.   H-Y antigen in human lymphocytes. Fluorescent staining in
EPICS 753 flow cytometer after treatment with gw-16 antibody and FITC-
labeled goat anti-mouse Ig. The vertical cursor represents the point to the
left of which fluorescence is attributable to second antibody or to autoflu-
orescence only. For each histogram, the vertical axis gives relative cell
number and the horizontal axis gives log-fluorescent intentsiy; 10,000 cells
per sample. Percent cells labeled with gw-16 in interval (a) female, 9.05;
(b) male, 30.36. (From Jaswaney, V. L. and Wachtel, S. S., *Differentiation,* 35, 115, 1988. With permission.)

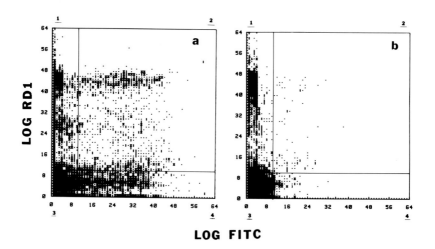

FIGURE 4.   Two-color analysis of T8 lymphocytes by flow cytometry. Scattered dot plots
for (a) male and (b) female. See text for explanation of results. (From Jaswaney, V. L. and
Wachtel, S. S., *Differentiation,* 35, 115, 1988. With permission.)

male or female lymphocytes. By that method, positive reaction was obtained only when
female-absorbed antibody was tested with male spleen cells.

2. Double Fluorochrome Labeling

Figure 4 depicts assays for H-Y antigen in T8 lymphocytes from male (a) and female (b).
This is a representative two-color analysis: lymphocytes were separated from whole blood
and incubated first with normal ascites fluid to block nonspecific Fc receptors, then with
biotinylated gw-16, and then after washing, with RD1-conjugated T8 antibody and streptavidin-FITC. After further incubation and washing, the cells were scored for red and green
label in the flow cytometer. Each square in Figure 4 is a dot plot divided into four quadrants
(marked 1, 2, 3, and 4). Quadrant 1 (upper left) shows cells scored for red (RD1) fluoro-
chrome.These are the T8 suppressor-cytotoxic cells. Quadrant 2 (upper right) shows cells
scored for red and green (FITC) fluorochromes. These are T8 cells to which H-Y antibody
has adsorbed. Quadrant 3 (lower left) depicts unstained cells, and Quadrant 4 (lower right)
depicts cells scored for FITC only.

Table 2
STAINING OF LYMPHOCYTES WITH
MONOCLONAL ANTIBODY

| Subset | Percent of circulating lymphocytes | Percent subset labeled ± SD | |
		Male (n = 8)	Female (n = 7)
B1	10—15	41.08 ± 21.68	12.44 ± 10.91
T4	40—50	5.43 ± 3.07	0.45 ± 0.41
T8	20—25	20.71 ± 12.80	1.73 ± 1.07

Two-color data for the B1, T4, and T8 subsets are summarized in Table 2. In each case, cells of the male were labeled preferentially, but the frequency of cells labeled differed from subset to subset. So there are H-Y$^+$ and H-Y$^-$ populations in each of the B1, T4, and T8 subsets, and these seem to occur in varying proportions from subset to subset and from male to male.

This is a departure from the notion that all nucleated male cells are H-Y$^+$, but the data fit well with the observations of Galbraith et al.[21] and Amice-Chambon et al.[22] regarding proportions of lymphocytes scored positive for H-Y with fluorescent antibodies. This could be due to emergence of H-Y at particular stages of differentiation, to presence of H-Y in particular stem cells and their clones, or to failure of the antibody to bind all H-Y$^+$ cells.

III. COMMENT

A. Variability of H-Y Scores

A striking feature of H-Y serology is the variability of test results, and another is the occasionally high "backgrounds" obtained for putatively H-Y negative subjects. These characteristics have generated considerable discussion,[7,34-36] yet they have endured from the days of the sperm cytotoxicity test to the current period of enzyme assays and flow cytometry.

Variability is apparent in the results summarized in Table 2. For instance, the percent of B1 cells labeled with H-Y antibody, gw-16, was 41.08 with a SD of 21.68 among eight men. But this could be expected among outbred populations, given the likelihood of alternative genetic backgrounds in the cell donors. Expression of the H-Y transplantation antigen is influenced by the major histocompatibility complex (MHC), and the same may be true for the antigen that is detected serologically. According to Ohno,[37] H-Y is anchored to the plasma membrane in association with β-2-microglobulin and cell surface components of the MHC (H-2 in the mouse, HLA in the human); thus, H-Y antibodies may recognize the perturbations induced in MHC antigens through juxtaposition with H-Y, or they may recognize the [MHC + H-Y] complex — which differs (in either case) from subject to subject depending on the specificity of particular MHC antigens.

Suppose that an antibody, such as gw-16, is MHC-restricted, having been generated in H-2^b females sensitized with cells from H-2^b males (C57BL/6). Ability of the antibody to recognize "H-Y positive" cells would depend on the extent to which the MHC antigens of those cells resembled the [H-2^b + H-Y] constellation.[38] Given the extreme diversity of MHC haplotypes in outbred populations, one might expect wide variability in the reaction of a particular MHC-restricted antibody with cells from the male or female. It would seem likely that female cells occasionally would appear to be H-Y$^+$ and male cells, H-Y$^-$.

B. H-Y in Female Cells

According to the preceding discussion, H-Y does not occur, but only *appears* to occur in female cells. But labeling of female cells in H-Y serological systems could be due to actual presence of H-Y or an H-Y-like molecule in the female. According to the scheme outlined by Wolf et al.,[39] the structural genes for H-Y are autosomal, suppressed by genes on the noninactivated portion of the X, and induced by a gene on the Y. That scheme was modified as follows.[40,41] A precursor molecule H-Yp is produced by an autosomal gene and is glycosylated by the Y and converted to H-Ya, the antigenic testis-inducer. The precursor, H-Yp, is glycosylated by the X and converted to H-Yr (H-Y related) which has no function in differentiation of the male gonad. But H-Yp and H-Ya are cross-reactive, and thus, H-Y antibodies recognize H-Yp in female cells and especially in cells from XO females which do not convert all H-Yp to H-Yr (see Chapter 10).

In fact, the antigenic specificity of H-Y may reside in a carbohydrate moiety; treatments that defeat the integrity of carbohydrate structure likewise defeat the antigenicity of the H-Y molecule.[20,30,43] So, binding of H-Y antibody in female cells could be due to fortuitous cross-reactivity of carbohydrate sequences whch are common to male and female.

Recently, Lau[43] described isolation of an H-Y gene present in male and female mice.[44] He called the relevant antigen MEA for "male-enhanced antigen" because the difference between MEA in male and female was quantitaive, and presumably under control of the Y chromosome. According to that notion, H-Y antibodies should react strongly with male cells and, to a lesser degree, with female cells. It may be worth noting that the MEA messenger was found in early embryos, and that expression of the antigen in TM-4 cells was not influenced by steroid hormones and gonadotropins. But the question remains how H-Y antibody can be generated in female mice bearing the relevant antigen (Chapter 14).

C. H-Y Antigen — Testis Inducer

McLaren et al.[45] reported absence of H-Y transplantation antigen in sex-reversed XX male mice carrying Sxr'. The mice were descended from a single female bearing the X,Sxr' chromosome and Searle's X;autosome translocation, T16H, which is preferentially activated in that combination. Because the H-Y⁻ phenotype was inherited in the XX males, the authors surmised that a rearrangement had occurred in the Sxr' segment, such that *Tdy,* the testis determinant, had been separated from the *H-Y* gene. It was accordingly proposed that H-Y antigen does not induce the mammalian testis. But consider the following.

First, the cell-mediated cytotoxicity test used by McLaren et al.[45] detects the H-Y transplantation antigen (H-Yt). In view of reports that H-Yt and H-Y serological antigen are distinct entities,[4,46] discovery of male mice negative for H-Yt should not affect the proposition that serological H-Y induces the mammalian testis.[47,48]

Second, there is evidence that antigenicity and function are separable in the molecule bearing H-Y. Thus, loss of receptor-binding capacity has been reported in a mutant Daudi-secreted H-Y that retained antigenicity.[49] It follows that Sxr' may represent the other side of the coin — loss of antigenicity in a molecule that retains function.

Finally, it is useful to define H-Y, not by immunological criteria, but by biochemical criteria. When Ohno et al.[50] reacted bovine fetal ovarian cells with ³H:Lys-labeled Daudi-secreted proteins, they obtained preferential binding of a protein of about 18,000 mol wt (proteins of similar size have been immunoprecipitated with H-Y antisera[30,51]). Presumably, the 18,000-mol-wt protein had engaged the vacant H-Y receptor of the target cells. Thus, male differentiation was observed when whole XX gonads of bovine female fetuses of 20 to 30 mm crown-rump length (30 to 45 days) were cultured for 5 days in concentrated Daudi culture medium with fetal calf serum.[50] After 3 days, transformation was represented by the sudden emergence of seminiferous cords and rete testis; after 4 days, by the loss of germ cells and somatic elements of the cortex; and after 5 days, by occurrence of mesenchymal cells and tunica albuginea. The controls — contralateral gonads from the same fetuses cultured in normal medium — remained indifferent.

ACKNOWLEDGMENT

Part of the work described here was supported by NIH grant AI-23479.

REFERENCES

1. **Eichwald, E. J. and Silmser, C. R.,** Untitled communication, *Transplant. Bull.,* 2, 148, 1955.
2. **Goldberg, E. H., Shen, F.-W., and Tokuda, S.,** Detection of H-Y (male) antigen on mouse lymph node cells by the cell to cell cytotoxicity test, *Transplantation,* 15, 334, 1973.
3. **Goldberg, E. H., Boyse, E. A., Bennett, D., Scheid, M., and Carswell, E. A.,** Serological demonstration of H-Y (male) antigen on mouse sperm, *Nature (London),* 232, 478, 1971.
4. **Simpson, E.,** Immunology of H-Y antigen and its role in sex determination, *Proc. R. Soc London Ser. B,* 220, 31, 1983.
5. **Nakamura, D., Wachtel, S. S., Lance, V., and Beçak, W.,** On the evolution of sex determination, *Proc. R. Soc. London Ser. B,* 232, 159, 1987.
6. **Ohno, S.,** Major regulatory genes for mammalian sexual development, *Cell,* 7, 315, 1976.
7. **Crichton, D. N. and Steel, C. M.,** Serologically detectable H-Y ('male') antigen: Mr or Myth?, *Immunol. Today,* 6, 202, 1985.
8. **Jaswaney, V.,** H-Y antigen and sex determination, *Immunol. Today,* 6, 350, 1985.
9. **Wachtel, S. S.,** *H-Y Antigen and the Biology of Sex Determination,* Grune & Stratton, New York, 1983.
10. **Piedrahita, J. A. and Anderson, G. B.,** Investigation of sperm cytotoxicity as an indicator of ability of antisera to detect male-specific antigen on preimplantation mouse embryos, *J. Reprod. Fertil.,* 74, 637, 1985.
11. **Anderson, G. B.,** Identification of embryonic sex by detection of H-Y antigen, *Theriogenology,* 27, 81, 1987.
12. **Scheid, M., Boyse, E. A., Carswell, E. A., and Old, L. J.,** Serologically demonstrable alloantigens of mouse epidermal cells, *J. Exp. Med.,* 135, 938, 1972.
13. **Fellous, M., Gunther, E., Kemler, R., Wiels, J., Berger, R., Guenet, J. L., Jakob, H., and Jacob, F.,** Association of the H-Y male antigen with β2-microglobulin on human lymphoid and differentiated mouse teratocarcinoma cell lines, *J. Exp. Med.,* 147, 58, 1978.
14. **Breuning, M. H., Ivanyi, P., and van Mourik, P.,** Normal mouse sera react specifically with human lymphoblastoid cell lines, *Transplant. Proc.,* 13, 1962, 1981.
15. **Koo, G. C., Boyse, E. A., and Wachtel, S. S.,** Immunogenetic techniques and approaches in the study of sperm and testicular cell surface antigens, in *Immunobiology of Gametes,* Edidin, M. and Johnson, M. H., Eds., Alden Press, Oxford, 1977, 73.
16. **Koo, G. C. and Goldberg, C. L.,** A simplified technique for H-Y typing, *J. Immunol. Methods,* 23, 197, 1978.
17. **Casanova-Bettane, M., Latron, F., Jakob, H., and Fellous, M.,** A quantitative radioimmunoassay for membranous and soluble H-Y antigen typing, *Hum. Genet.,* 58, 21, 1981.
18. **Savikurki, H., Andersson, L. C., Wachtel, S. S., and de la Chapelle, A.,** Protein A radio-assay of H-Y antigen on human leukocytes using mouse and rat antisera and monoclonal antibodies, *Hum. Genet.,* 65, 190, 1983.
19. **Nagamine, C., Reidy, J., and Koo, G. C.,** A radiobinding assay for human H-Y antigen using monoclonal antibodies, *Transplantation,* 37, 13, 1984.
20. **Shapiro, M. and Goldberg, E. H.,** Analysis of a serological determinant of H-Y antigen: evidence for carbohydrate specificity using an H-Y specific monoclonal antibody, *J. Immunogenet.,* 11, 209, 1984.
21. **Galbraith, G. M. P., Galbraith, R. M., Faulk, W. P., and Wachtel, S. S.,** Detection of H-Y antigen by fluorescence microscopy, *Transplantation,* 26, 25, 1978.
22. **Amice-Chambon, V., Amice, J., and Genetet, B.,** Evidence of H-Y antigen on human B lymphocytes, *Ann. Immunol. (Paris),* 132C, 157, 1981.
23. **Müller, U. and Bross, K.,** A highly sensitive peroxidase-antiperoxidase method for detection of H-Y antigen on cultivated human fibroblasts, *Hum. Genet.,* 52, 143, 1979.
24. **McArthur, C. P., Sengupta, S., and Smith, R. T.,** Enzyme-linked immunosorbent assay (ELISA) for the detection of antibodies against the H-Y antigen on murine sperm, *J. Immunol. Methods,* 43, 343, 1981.
25. **Farber, C. M., Liebenthal, D., Wachtel, S. S., and Cunningham-Rundles, C.,** Detection of H-Y in the ELISA, *Hum. Genet.,* 65, 278, 1984.

26. **Moreira-Filho, C. A. and Wachtel, S. S.**, Study of H-Y antigen in abnormal sex determination with monoclonal antibody and an ELISA, *Am. J. Med. Genet.*, 20, 525, 1985.

27. **Brunner, M., Moreira-Filho, C. A., Wachtel, G., and Wachtel, S.**, On the secretion of H-Y antigen, *Cell*, 37, 615, 1984.

28. **Brunner, M., Moreira-Filho, C., Wachtel, G., and Wachtel, S. S.**, H-Y typing in the ELISA, *J. Immunol. Methods*, 75, 203, 1984.

29. **Brunner, M., Jaswaney, V., and Wachtel, S.**, Reaction of monoclonal H-Y antibody in the ELISA, *J. Reprod. Immunol.*, 11, 181, 1987.

30. **Bradley, M. P. and Heslop, B. F.**, A biochemical and immunological approach to the identification of H-Y antigenic proteins secreted from Daudi cells, *Hum. Genet.*, 71, 117, 1985.

31. **Jolley, M. E., Wang, C.-H. J., Ekenberg, S. J., Zuelke, M. S., and Kelso, D. M.**, Particle concentration fluorescence immunoassay (PCFIA): a new, rapid immunoassay technique with high sensitivity, *J. Immunol. Methods*, 67, 21, 1984.

32. **Beutler, B., Nagai, Y., Ohno, S., Klein, G., and Shapiro, I. M.**, The HLA-dependent expression of testis-organizing H-Y antigen by human male cells, *Cell*, 13, 509, 1978.

33. **Jaswaney, V. L. and Wachtel, S. S.**, Differential expression of H-Y antigen on lymphocyte subsets: analysis by flow cytometry, *Differentiation*, 35, 115, 1988.

34. **Zenzes, M. T. and Reed, T. E.**, Variability in serologically detected male antigen titer and some resulting problems: a critical review, *Hum. Genet.*, 66, 103, 1984.

35. **Mayerová, A., Müller, U., Wiberg, U., Wolf, U., and Fraccaro, M.**, Comments on the paper by M. T. Zenzes and T. E. Reed, *Hum. Genet.*, 66, 110, 1984.

36. **Silvers, W. K., Gasser, D. L., and Eicher, E. M.**, H-Y antigen, serologically detectable male antigen and sex determination, *Cell*, 28, 439, 1982.

37. **Ohno, S.**, The original function of MHC antigens as the general plasma membrane anchorage site of organogenesis-directing proteins, *Immunol. Rev.*, 33, 59, 1977.

38. **Ohno, S., Epplen, J. T., and Sutou, S.**, Testis-organizing H-Y antigen as a discrete protein: its MHC restricted immune recognition and the genomic environment in which H-Y gene operates, *Hum. Genet.*, 58, 37, 1981.

39. **Wolf, U., Fraccaro, M., Mayerová, A., Hecht, T., Maraschio, P., and Hameister, H.**, A gene controlling H-Y antigen on the X-chromosome, *Hum. Genet.*, 54, 149, 1980.

40. **Adinolfi, M., Polani, P., and Zenthon, J.**, Genetic control of H-Y synthesis. A hypothesis, *Hum. Genet.*, 61, 1, 1982.

41. **Polani, P. E. and Adinolfi, M.**, The H-Y antigen and its functions: a review and a hypothesis, *J. Immunogenet.*, 10, 85, 1983.

42. **Shapiro, M. and Erickson, R. P.**, Evidence that the serological determinant of H-Y antigen is carbohydrate, *Nature (London)*, 290, 503, 1981.

43. **Lau, Y.-F.**, personal communication, 1987.

44. **Lau, Y.-F., Chan, K., Kan, Y. W., and Goldberg, E.**, Isolation of a male-specific and conserved gene using an anti-H-Y antibody, *Am. J. Hum. Genet.*, 39, (Suppl.) A142, 1987.

45. **McLaren, A., Simpson, E., Tomonari, K., Chandler, P., and Hogg, H.**, Male sexual differentiation in mice lacking H-Y antigen, *Nature (London)*, 312, 552, 1984.

46. **Andrews, P. W.**, The male-specific antigen (H-Y) and sexual differentiation, in *Genetic Analysis of the Cell Surface, Vol. XVI, Receptors and Recognition, Series B*, Goodfellow, P., Ed., Chapman and Hall, London, 1984, 159.

47. **Wachtel, G. M., Wachtel, S. S., Nakamura, D., Moreira-Filho, C. A., Brunner, M., and Koo, G. C.**, H-Y antibodies recognize the H-Y transplantation antigen, *Transplantation*, 37, 8, 1984.

48. **Wiberg, U. H. and Mayerová, A.**, Serologically H-Y antigen-negative XO mice, *J. Immunogenet.*, 12, 55, 1985.

49. **Nagai, Y., Iwata, H., Stapleton, D. D., Smith, R. C., and Ohno, S.**, Testis-organizing H-Y antigen of man may lose its receptor-binding activity while retaining antigenic determinants, in *Testicular Development, Structure and Function*, Steinberger, E. and Steinberger, A., Eds., Raven Press, New York, 1980, 41.

50. **Ohno, S., Nagai, Y., Ciccarese, S., and Iwata, H.**, Testis-organizing H-Y antigen and the primary sex-determining mechanism of mammals, *Rec. Prog. Horm. Res.*, 35, 449, 1979.

51. **Hall, J. L. and Wachtel, S. S.**, Primary sex determination: genetics and biochemistry, *Mol. Cell. Biochem.*, 33, 49, 1980.

Chapter 13

RECENT DEVELOPMENTS IN THE SEROLOGY AND BIOCHEMISTRY OF TESTICULAR SEX-SPECIFIC (H-Y) ANTIGENS

Mark P. Bradley and Barbara F. Heslop

TABLE OF CONTENTS

I. INTRODUCTION

The precise molecular signals responsible for the development of the mammalian testis from the indifferent embryonic gonad are obscure. In recent years, a sex-specific minor histocompatibility antigen (H-Y antigen) has been proposed as the primary signal initiating this event,[1] although alternative interpretations of the role of male-specific proteins in testicular development and spematogenesis have cast doubts on this original proposal.[2-4] Recent immunogenetic and biochemical data indicate the probable existence of more than one male-specific antigen with the possibility of a family of structurally similar proteins having a variety of sex-specific functions.[5-7] Molecular evidence also suggests that at least one of the genes for male-specific proteins is probably located on an autosome.[6] If this is substantiated, it is conceivable that genes on the Y chromosome are responsible for the regulation of the autosomal gene.

Before discussing male-specific antigens, it may be useful to outline the new terminology that will be used in this chapter. Because of present confusion regarding the existence of one or several male-specific antigens, Wiberg[5] has considered the various assay procedures used to detect sex-specific (Sxs) antigens and has attempted to determine what the various assay procedures are detecting at the protein level.

The male-specific antigen(s) of mammals, detected by transplantation experiments, is referred to as H-Yt. Mammalian male-specific antigen(s) detected by major histocompatibility complex (MHC) restricted H-Y specific cytotoxic T cells in vitro is called H-Yc. The H-Yt and H-Yc antigens (epitopes) may or may not be identical. Finally, sex-specific antigen(s) detected by antibodies is designated Sxs (serological sex-specific) to distinguish it from the H-Yt and H-Yc. The Sxs epitope(s) may or may not be identical to the H-Yt, but it is probably an epitope(s) different from the H-Yc. (For detailed discussions on identity or nonidentity of male-specific antigens, see Reference 5 and Chapter 10.) Discussion in this paper will be concerned only with Sxs antigens identified in gonadal tissues.

In order to elucidate the role(s) played by Sxs antigens in sex determination it is essential to identify and to characterize the molecular properties of these proteins. For this reason the research emphasis in this laboratory has centered around the purification of these proteins from a number of different tissues. It is hoped that this approach will ultimately reveal something of the unique specialization and specific function of each antigen.

In-depth biochemical studies have been hampered until recently by the lack of specific high-titer Sxs antisera. The development of techniques in this laboratory to produce specific high-titer Sxs antisera in the rat[8] has been instrumental in enabling us to begin identifying and purifying such proteins. In most of the studies described here, we have routinely used these antisera (after appropriate absorption with syngeneic female cells) at working dilutions ranging from 1:200 to 1:500. This has allowed the very specific detection of male-specific proteins using Western blots, tissue sections, and ELISAs, and has also provided antibody for the preparation of the immunoaffinity columns used for the separation of cells and proteins.

II. SEROLOGY OF TESTIS Sxs ANTIGENS

Early serological studies employing cytotoxic assays established that certain cells within the rat testis express Sxs antigens at defined stages of testicular development. Using assays designed to test the absorptive capacity of testicular cells, Müller et al.[9,10] demonstrated that the testis was a secretory organ for Sxs antigen and that this secretion increased with age. Maximal antigen expression was detected in the adult. This work was extended to define the various cell populations within the testis that might be involved in this secretion.[11] Sertoli cells were found to be Sxs positive at all stages of development (18 to 90 days post partum) and to secrete Sxs antigen in culture. Leydig cells were also found to be Sxs positive, but they did not secrete Sxs antigen. Germ cells were Sxs positive only in animals over 24 days old; they neither secreted nor bound exogenous antigen at any stage of development. Such findings raise the question of the origin of the Sxs antigen on germ cells.

The secretion of Sxs antigens by Sertoli cells has been questioned by other investigators[12] who failed to find any serological evidence for such proteins in Sertoli cell-conditioned culture medium. These studies, however, used low-titer Sxs antisera combined with relatively insensitive assay procedures and may not have been capable of identifying small amounts of Sxs antigens in the secretions. It was subsequently demonstrated by Brunner et al.[13] using a more sensitive ELISA technique that mouse testicular and presumptive Sertoli cell lines both secreted Sxs antigens. Our own studies on testicular extracts and Sertoli cell secretions using high-titers Sxs antiserum in conjunction with a more sensitive urease ELISA[14] further support the original results of Brunner et al.[13]

Indications of the distribution of Sxs antigens on developing spermatids were presented by Koo et al.[15] after studying testicular cells from mice at various stages of development. The expression of Sxs antigen increased after meiosis; the greatest amount was present on late spermatids. This post-meiotic increase suggested the possibility of haploid gene expression in spermatids. In a later study, Hoppe and Koo[16] noted that epididymal sperm appear to lose some Sxs antigenic activity as they proceed through the epididymis during the post-testicular maturation phase. Thus, mature sperm probably express less Sxs antigen than is expressed by testicular spermatids.

Results from immunohistochemical studies in our laboratory — on the expression of Sxs antigen by cell populations within rat, mouse, and sheep testis — have enabled visualization of the distribution of Sxs antigens on testicular cells. The results of these studies agree with the findings in many of the earlier serological studies. In testis from 20-day-old rats, intense immunoperoxidase staining of cellular material was evident within the seminiferous tubules; the distribution of peroxidase-positive material was consistent with its presence in Sertoli cell cytoplasm. It is difficult to be certain whether some of the seminiferous epithelial cells closely associated with Sertoli cell cytoplasm were Sxs positive or negative. However, the cell surface and cytoplasm of the spermatogonia and at least some of the spermatocytes appeared to lack significant immunoperoxidase staining at this stage of development. However, a population of large spermatocytes did express Sxs antigen.[17] In testis from sexually mature adult rats, there was little staining of Sertoli cell cytoplasm but significantly more staining on the various germ cell populations — especially on the spermatogonia.[17]

Within seminiferous tubules of the mature rat testis, it is possible to observe spermatozoa which still contain residual droplets.[17] After reaction with Sxs antiserum, only some of these residual droplets showed intense immunoperoxidase staining while others were virtually unstained. The possibility cannot be discounted that these appearances reflect haploid gene expression of a Sxs antigen on a defined population of sperm. Evidence already exists that some proteins in sperm arise as a result of haploid gene expression;[18] observations that make it reasonable to postulate that the same mechanism accounts for a subpopulation of sperm expressing Sxs antigens by this mechanism.

To address this possibility further, we looked at the expression of Sxs antigen on the surface of ram and human spermatozoa and attempted to correlate this expression with sperm genotype.

III. EXPRESSION OF Sxs ANTIGEN ON MAMMALIAN SPERMATOZOA

The expression of Sxs antigen on the plasma membrane of spermatozoa is now well documented.[19-24] The basis of this expression is, however, still unknown. Using immuno-peroxidase and immunofluorescence staining techniques we attempted to map the plasma membrane distribution of Sxs antigen expression on ram sperm at two stages of post-testicular development.[24]

In caput sperm, staining was observed on both the midpiece of the flagellum and on the post-acrosomal region of the head. Because of the refractile nature of sperm under the light microscope, however, such staining was not always clearly seen. A notable feature of these sperm was the intense staining of the residual droplets present on many of them, while in others very little immunoperoxidase staining could be seen. This observation has been consistent in all of these experiments. Not more than 45% of the sperm in any given sample of ram sperm ever stained Sxs positive in this manner. This finding accords with previous observations on the apparent differential expression of Sxs antigen on testicular rat spermatids.[17]

Caudal and ejaculated ram spermatozoa (which lack residual droplets) were also examined using immunofluorescence. Antigen labeling was confined to the posterior head region and to the midpiece of the flagellum. In these experiments also, it was observed that fewer than 50% of spermatozoa label positively with Sxs antiserum. It could not be ruled out that such a differential labeling pattern might be due to some technical problem inherent in experimental protocols. However, the consistency with which these results were obtained suggests that Sxs positive sperm do indeed represent a minority.

In a further attempt to examine the basis of the differential expression of Sxs antigen by mammalian spermatozoa, we constructed a cell affinity column to which Sxs antiserum was immobilized. This column was used to separate the sperm into two distinct populations on the basis of their Sxs phenotypes.

Human spermatozoa are particularly suitable for ascertaining whether Sxs antigen expression is related to the presence of a Y chromosome. It is possible to distinguish between X chromosome- and Y chromosome-bearing human spermatozoa by staining with the fluorescent dye, quinacrine, which binds to DNA in the Y chromosome.[25] This procedure allows for the quantitative assessment of any sperm sexing separation procedure using Sxs antisera.

The results of ten separate experiments are presented in Figure 1. In each of the experiments, it was observed that human semen appears to contain slightly more X chromosome-bearing sperm ($54 \pm 11\%$) (mean $\pm$ SD) than Y chromosome-bearing sperm ($44 \pm 10\%$). Although this is not a statistically significant difference ($p > 0.1$), the slight deviation from the expected ratio of 50 X:50 Y spermatozoa in semen has been noted previously by other investigators[25] and certainly accords with observations on the immunoperoxidase labeling of ram spermatozoa with Sxs antiserum.[24]

When not more than 10×10^6 cells were applied to the affinity column, a significantly enriched population of X chromosome-bearing spermatozoa (quinacrine negative) was found in the samples of spermatozoa which did not adhere to the column (flow-through fraction) ($74 \pm 16\%$), compared with the number of X chromosome-bearing sperm present in the loading sample ($54 \pm 11\%$; $p < 0.02$). Although the flow-through fractions often contained Y chromosome-bearing spermatozoa, the percentage ($24 \pm 14\%$) was significantly reduced compared with that found in the original sample ($44 \pm 10\%$; $p < 0.02$). This contamination was probably due to overloading of the column beyond its total binding capacity. Generally, close to 60% of the original number of cells loaded onto the column were subsequently recovered in the flow-through fraction. Sperm, specifically eluted from the column, were greatly enriched for quinacrine positive (Y chromosome-bearing) spermatozoa compared with the original sample ($88 \pm 12\%$; $p < 0.001$). Only a small percentage of X chromosome-bearing spermatozoa ($12 \pm 14\%$) was detected in the eluted fraction. The yield of cells in the eluted fraction was consistently less than the expected 40% which should remain in this sample. This was possibly due to incomplete elution of spermatozoa from the affinity support, as well as rupture during the experiment.

This study demonstrates that it is possible to distinguish between human Y chromosome-bearing and X chromosome-bearing sperm on the basis of a cell surface (Sxs) antigen. These

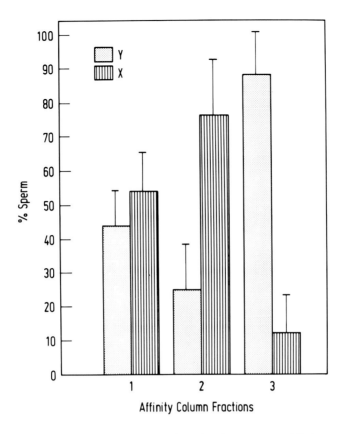

FIGURE 1. Immunoaffinity separation of X chromosome- and Y chromosome-bearing human sperm genotypes. The percentage of quinacrine positive (Y) sperm, and quinacrine negative (X) sperm, in each column fraction: (1) sperm sample prior to loading onto the column, (2) sperm which did not adhere to the column, and (3) sperm eluted from the column with excess Sxs antiserum. The results are the means ± SD of ten separate experiments.

findings now pose important questions relating to the origin of this antigen on sperm, i.e., whether it results from haploid gene expression during spermatogenesis, or whether the protein is produced elsewhere (possibly Sertoli cells) and adsorbed or bound to specific receptors on the sperm membrane. Such receptors may conceivably be products of genes on the Y chromosome. The identification of testicular proteins which bind to sperm in a sex-specific manner would be an important first step in elucidating their function. Recent suggestions that male-specific proteins in the testis may have an important regulatory role in spermatogenesis are consistent with this general concept.[26]

IV. BIOCHEMISTRY OF TESTICULAR SEX-SPECIFIC ANTIGENS

In 1986, in collaboration with our colleagues at the Institute of Human Genetics and Anthropology in Freiburg, FRG, we began a biochemical study of sex-specific proteins of the testis. The soluble fraction from a testicular homogenate was identified as containing sex-specific proteins when assayed in an ELISA with female absorbed Sxs antiserum[14] and some of these proteins have now been isolated.

The primary tissue source of these proteins has been ovine testes because it provides large amounts of protein for the selective isolation of potentially very small quantities of Sxs

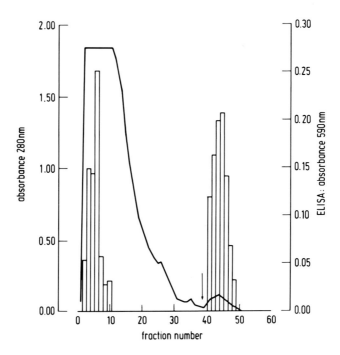

FIGURE 2.   Elution profile of ovine testicular extract from a lentil lectin affinity column. Protein was applied to the column and the column, which was then washed with 50 mM Tris-HCl, 150 mM NaCl, pH 8.4. Bound glycoproteins were eluted with 0.2M α-methyl-mannoside in 50 mM Tris-HCl, 150 mM NaCl, pH 8.4 (arrow). Protein eluting from the column was monitored at 280 nm. 2 mℓ fractions were collected and 30-μℓ samples were assayed by ureas ELISA[14] using Sxs antiserum to identify those fractions which contained Sxs-antigenic proteins. The ELISA was read spectrophotometrically at 590 nm. Sxs-positive fractions are indicated by the histograms. Sxs-positive proteins which did not bind to the column initially were passaged over the column for as many times as necessary to recover all of the protein.

proteins. In order to remove the bulk of nonspecific protein from tissue homogenates, salt precipitation was used as a preliminary fractionation step. All sex-specific antigenic activity was precipitated from solution with 60% ammonium sulfate. The protein fraction was then resuspended in buffer and dialyzed extensively before passage over a lentil lectin affinity column. Lentil lectin specifically binds glycoproteins containing glucose and mannose residues. Column fractions were collected and assayed by ELISA[24] to identify those fractions which contained Sxs antigenic activity.

Figure 2 shows the results of a typical elution profile. Antigenic activity was always detected in the fractions specifically eluted with α-methyl-mannoside from the lentil lectin column. However, substantial amounts of activity were also detected in the material which did not bind to the column. Passage of this unbound material back through the lentil lectin column resulted in the binding of this original unbound material to the column, with a subsequent total recovery of all of the sex-specific protein activity.

These results, which demonstrate the interaction of sex-specific proteins with a lectin, accord with other studies which indicate that Sxs antigens are glycolsylated.[27,28] The earlier studies indicated that galactose moieties were important determinants in the binding of Sxs antibodies to their respective antigens. The results of our work also demonstrate that Sxs proteins have glucose and mannose moieties in their carbohydrate structure. It is possible that these sugars are important for antibody recognition and binding because it has consistently

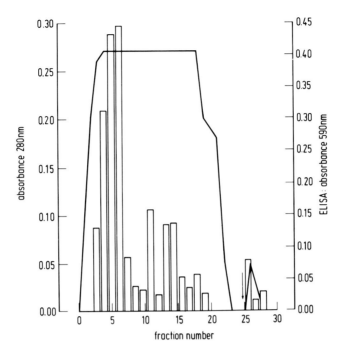

FIGURE 3.   Chromatography of pooled Sxs-positive glycoproteins from
the lentil lectin column on a rat IgG affinity column (nonimmune female
serum). The column was washed with 5 mM Tris-HCl, 150 mM NaCl,
0.025% deoxycholate, pH 8.4. Proteins bound to the column were eluted
with DEA/DOC, pH 11. Protein eluting from the column was monitored
at 280 nm. 2-mℓ fractions were collected and 30 $\mu\ell$ of each fraction were
assayed in a urease ELISA to determine those which contain Sxs proteins.
The ELISA was read spectrophotometrically at 590 nm. Sxs-positive fractions are represented by the histograms. Most of the Sxs antigenic activity
was found in the material which did not adhere to the column.

been observed that failure to remove α-methyl-mannoside (the sugar used to elute bound
proteins from the column) from the proteins eluted from the lentil lectin affinity column,
prevents subsequent reactivity of these proteins with immobilized Sxs antibodies. The in-
hibitory effect is abolished only after complete dialysis of the protein fraction, a step which
presumably removes the α-methyl-mannoside from the protein environment.

Our other experiments with ^{125}I-Daudi or ^{125}I-Sertoli cell secretions have also shown that
the sex-specific antigens in these preparations are retained on lentil lectin columns. It is
evident from these investigations and from others[27,28] that future studies of Sxs-specific
proteins should involve detailed analyses of the carbohydrate composition of Sxs proteins
from somatic and gonadal sources. It is conceivable that serological differences between
Sxs antigens present on cells from various tissues may result from differences in the glycosylation patterns of a core protein, analogous to those observed for blood group antigens.[7]

Sex-specific testicular glycoproteins eluted from the lentil lectin column were purified
further by use of two immunoaffinity steps. The first involved the passage of the glycoproteins
over an affinity column to which IgG from a nonimmune HS female rat had been coupled.
This step was undertaken to remove from the glycoprotein fraction those proteins which
nonspecifically bound to female rat IgG molecules. It was assumed that the same nonspecific
background reactivity would also be associated with the Sxs immune serum used in the final
immunoaffinity step. This column prepared from nonimmume serum thus served as a pre-
clearing step.

Figure 3 shows the elution profile from this column. Each fraction was assayed by ELISA

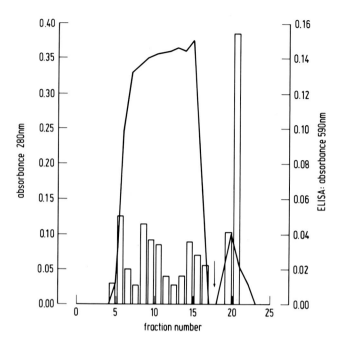

FIGURE 4.   Elution profile of Sxs-positive glycoproteins (recovered from
the flow-through off the nonimmune affinity column) after application to
the Sxs immunoaffinity column. The column was washed with 5 m*M* Tris-
HCl, 150 m*M* NaCl, 0.025% deoxycholate, pH 8.4. Bound protein was
eluted with DEA/DOC, pH 11. Proteins eluting from the column were
monitored at 280 nm. 2-m*ℓ* fractions were collected and assayed for Sxs
proteins as described previously. The ELISA was read spectrophotometrically at 590 nm. Sxs-positive fractions are represented by the histograms.
Significant amounts of Sxs antigens were found in the eluted protein from
this column. Sxs-positive material that did not bind to the column initially
was subsequently repassaged through the column until all of the Sxs proteins were recovered.

with Sxs antiserum in order to identify fractions that contained the Sxs antigenic proteins.
It was observed that most of the Sxs antigenic activity was recovered in the flow-through
fraction, and that only a small amount of Sxs activity was present in the eluted fraction.

The picture was effectively reversed when the flow-through fraction from the nonimmume
IgG column was passed over the Sxs immune affinity column (Figure 4). Here, a substantial
amount of the protein was bound to the column.

Significantly more Sxs antigenic activity was detected in the material eluted from the
column than was detected in the material which did not bind to the column. Repassage of
the Sxs positive fractions present in this unbound protein always resulted in specific binding
of this antigenic material to the column. Therefore, initial failure to bind this protein on the
column can probably be attributed to loading of the column beyond its total antigen-binding
capacity.

Examination of proteins eluted from the Sxs immune column on SDS polyacrylamide gels
showed several proteins ranging in molecular weights from 20,000 to 75,000 (Figure 5).
The most prominent protein species had a molecular weight of approximately 20,000. Other
less distinct bands were present characterizing proteins with molecular weights of 29,000;
43,000; 50,000; 60,000; and 75,000. The 29,000 mol wt protein (which might be a doublet)
was poorly stained with Coomassie blue, a property possibly related to its carbohydrate
composition. This protein had a molecular weight similar to that of a sex-specific protein

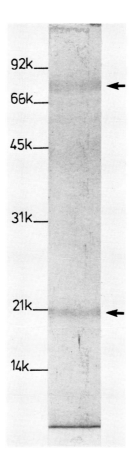

FIGURE 5. 15% SDS polyacrylamide gel of Sxs proteins eluted from the Sxs immunoaffinity column. The position of relative standard molecular weight markers (BioRad) are indicated.

previously identified in the plasma membranes of ram spermatozoa.[24] It is unlikely that the protein of molecular weight 29,000 identified in the testicular preparations described in this study originated from the membranes of testicular sperm because the procedure used to prepare the testicular extract selects for soluble proteins rather than for those proteins which are membrane bound. It is possible that the molecular weight 29,000 membrane-bound protein identified in epididymal sperm may have been acquired from cellular secretions during a testicular phase of sperm development.

Recently, a glycoprotein of molecular weight 29,000 was identified in rat Sertoli cell secretions as well as on the membranes of mature spermatozoa.[29-31] Rabbit antisera produced to this protein were used to map the cellular distribution of the protein within the rat testis. Labeling was evident in the Sertoli cell cytoplasm, the heads and tails of spermatids, and the residual droplets. The distribution pattern was virtually identical to that which we observed for Sxs proteins within the rat testis as well as on rat and ram spermatozoa. No comment was made, however, as to the sex specificity of the protein. It is conceivable that the proteins identified by Sylvester and Griswold[29-31] are the same as some of the Sxs proteins recognized by our Sxs antiserum. Confirmation awaits further biochemical characterization of these proteins from both sources.

The question that now needs to be considered is whether the sex-specific proteins which have been identified in the testis are unique to this tissue or whether they are also found in other cells and tissues. The possibility exists that some of these proteins may be shared between somatic and gonadal tissues, while others may be specific for individual tissues.

ACKNOWLEDGMENTS

We thank Pam Salmon for her expert assistance with the immunohistochemical work described in this paper, and Peter Swarbrick for help with the affinity chromatography. Discussions with Dr. Ulf Wiberg, Institute for Human Genetics and Anthropology, Freiburg, FRG, relating to many aspects of this work are greatly appreciated. This work was supported by the Medical Research Council of New Zealand and the Anderson & Telford Trust.

REFERENCES

1. **Ohno, S.**, *Major Sex-Determining Genes*, Springer-Verlag, Berlin, 1979.
2. **McLaren, A., Simpson, E., Tomonari, K., Chandler, P., and Hogg, H.**, Male sexual differentiation in mice lacking H-Y antigen, *Nature (London)*, 312, 552, 1984.
3. **Simpson, E., McLaren, A., Chandler, P., and Tomonari, K.**, Expression of H-Y antigen by female mice carrying Sxr, *Transplantation*, 37, 17, 1984.
4. **Simpson, E., Chandler, P., Hunt, R., Hogg, H., Tomonari, K., and McLaren, A.**, H-Y status of X/X Sxr' male mice: in vivo tests, *Immunology*, 57, 348, 1986.
5. **Wiberg, U.**, Facts and considerations about sex-specific antigens, *Hum. Genet.*, 76, 207, 1987.
6. **Lau, Y.-F., Chan, K., Kau, Y. W., and Goldberg, E.**, Isolation of a male-specific and conserved gene using an anti H-Y antibody, *Am. J. Hum. Genet.*, Suppl. 39, A142, 1986.
7. **Polani, P. E. and Adinolfi, M.**, The H-Y antigen and its functions: a review and a hypothesis, *J. Immunogenet.*, 10, 85, 1983.
8. **Bradley, M. P. and Heslop, B. F.**, Elicitation of a rapid and transient antibody response to H-Y antigen by intrasplenic immunisation, *Transplantation*, 39, 643, 1985.
9. **Müller, U., Aschmoneit, I., Zenzes, M. T., and Wolf, U.**, Binding studies of H-Y antigen in rat tissues, *Hum. Genet.*, 43, 151, 1978.
10. **Müller, U., Siebers, J. W., Zenzes, M. T., and Wolf, U.**, The testis as a secretory organ for H-Y antigen, *Hum. Genet.*, 45, 209, 1978.
11. **Zenzes, M. T., Muller, U., Aschmoneit, I., and Wolf, U.**, Studies on H-Y antigen in different cell fractions of the testis during pubescence, *Hum. Genet.*, 45, 297, 1978.
12. **Gore-Langton, R. E., Tung, P. S., and Fritz, I. B.**, The absence of specific interactions of Sertoli-cell-secreted proteins with antibodies against H-Y antigen, *Cell*, 32, 289, 1983.
13. **Brunner, M., Moreira-Filho, C. A., Wachtel, G., and Wachtel, S.**, On the secretion of H-Y antigen, *Cell*, 37, 615, 1984.
14. **Bradley, M. P., Ebensperger, C., and Wiberg, U.**, Determination of the serological sex-specific (Sxs) antigen (H-Y antigen) in birds and mammals using high titer antisera and a sensitive urease ELISA, *Hum. Genet.*, 76, 352, 1987.
15. **Koo, G., Mittl, L. R., and Goldberg, G. L.**, Expression of H-Y antigen during spermatogenesis, *Immunogenetics*, 9, 2293, 1979.
16. **Hoppe, P. C. and Koo, G. C.**, Reacting mouse sperm with monoclonal H-Y antbodies does not influence sex ratio of eggs fertilized in vitro, *J. Reprod. Immunol.*, 6, 1, 1985.
17. **Bradley, M. P. and Heslop, B. F.**, The distribution of sex-specific antigens within the testis: an immunohistochemical study, *Hum. Genet.*, in press, 1987.
18. **Distell, R. J., Kleene, K. C., and Hecht, N. B.**, Haploid gene expression of a mouse-tubulin gene, *Science*, 224, 68, 1984.
19. **Goldberg, E. H., Boyse, E. H., Bennett, D., Scheid, M., and Carswell, E. E.**, Serological demonstration of H-Y (male) antigen on mouse sperm, *Nature (London)*, 232, 478, 1971.
20. **Hausman, S. J. and Palm, J.**, Serologic detection of a male-specific membrane antigen in the rat, *Transplant. Proc.*, 5, 307, 1973.
21. **Koo, G., Stackpole, C., Boyse, E., Hammerling, U., and Lardis, M. P.**, Topographical location of H-Y antigen on mouse spermatozoa by immunoelectronmicroscopy, *Proc. Natl. Acad. Sci. U.S.A.*, 70, 1502, 1973.
22. **Zaborski, P.**, Detection of H-Y antigen on mouse sperm by the use of *Staphylococcus aureus*, *Transplantation*, 27, 348, 1979.

23. **Tung, P. S., Gore-Langton, R. E., and Fritz, I. B.,** An objective sperm cytotoxicity assay for male-specific antisera based on ATP levels of unlysed cells: application to assay of H-Y antigen, *J. Reprod. Immunol.,* 4, 315, 1982.

24. **Bradley, M. P., Forrester, I. T., and Heslop, B. F.,** The identification of a male specific antigen on the flagellar plasma membrane of epididymal spermatozoa, *Hum. Genet.,* 75, 362, 1987.

25. **Sumner, A. T., Robinson, J., and Evans, H. J.,** Distinguishing between X, Y and YY bearing human spermatozoa by fluorescence and DNA content, *Nature (London),* New Biol. 229, 131, 1971.

26. **Burgoyne, P. S., Levy, E. R., and McLaren, A.,** Spermatogenic failure in male mice lacking H-Y antigen, *Nature (London),* 320, 170, 1986.

27. **Shapiro, M. and Erickson, R. P.,** Evidence that the serological determinant of H-Y antigen is carbohydrate, *Nature (London),* 290, 503, 1981.

28. **Shapiro, M. and Goldberg, E.,** Analysis of a serological determinant of H-Y antigen: evidence for carbohydrate specificity using an H-Y specific monoclonal antibody, *J. Immunogenet.,* 11, 209, 1984.

29. **Sylvester, S. R. and Griswold, M. D.,** Rat Sertoli cells and epididymal epithelium secrete a protein found in mature sperm, *Ann. N.Y. Acad. Sci.,* 438, 561, 1984.

30. **Sylvester, S. R., Skinner, M. K., and Griswold, M. D.,** A sulfated glycoprotein synthesized by Sertoli cells and by epididymal cells is a component of the sperm membrane, *Biol. Reprod.,* 31, 1087, 1984.

31. **Griswold, M. D., Roberts, K., and Bishop, P.,** Purification and characterisation of a sulfated glycoprotein secreted by Sertoli cells, *Biochemistry,* 25, 7265, 1986.

Chapter 14

ARE MALE-ENHANCED ANTIGEN AND SEROLOGICAL H-Y ANTIGEN THE SAME?

Yun-Fai Lau

TABLE OF CONTENTS

I. MALE SEX DETERMINATION IN MAMMALS

The determination of male sex is a complex process in mammals. The primary gene responsible for the activation of the testicular differentiation is located on the Y chromosome.[1-4] This gene has been termed the testis-determining factor (*TDF*) in man and *Tdy* in the mouse. During the early stages of mammalian gonadogenesis, the *TDF* gene is expressed; products of *TDF* interact with the products of other autosomal and/or X-linked genes to complete the process of differentiation of the embryonic testes. At present, *TDF* and other genes involved in this differentiative cascade have not been completely isolated and characterized in molecular terms.

Recent studies of DNA derived from sex-reversed patients with Y chromosome-specific DNA probes have demonstrated that a major proportion of XX males has inherited Y-specific sequences — presumably the *TDF* gene(s) — in the paternal X chromosome as a result of aberrant X-Y interchange during paternal meiosis.[5,6] Similarly, most XY female patients have lost the TDF segment of the Y chromosome, and thereby develop as females.[2] These studies demonstrate clearly the importance of *TDF* in the initiation of the male-determining process. However, in a minor population of sex-reversed patients, gain or loss of Y chromosome-specific DNA sequences has not been confirmed.[7]

Genetic studies have clearly shown that autosomal genes are also involved in primary sex determination in the mouse. Eicher and co-workers[1,8] found that when the Y chromosome from *Mus domesticus* was introduced onto the C57BL/6 *(M. musculus)* genomic background, all XY mice developed either two ovaries or various combinations of ovotestes and ovaries. Their studies indicated the possibility that this sex-reversed condition was a result of incompatibility of the *Tdy* locus on the *Mus domesticus* Y chromosome and a testis-determining autosomal (*Tda-1*) locus in the C57BL/6 strain. Eicher and Washburn[1] also identified other autosomal loci involved in the primary testis-determining events in mice. Among these *Tda* loci, a dominantly inherited trait, called T-associated sex reversal or *Tas*, is closely linked to the hairpin-tail mutation of the T/t complex on chromosome 17 of the mouse.[9] The chromosomal locations of other *Tda* loci are uncertain. Recent studies by Nagamine et al.[10,11] indicated that the various combinations of ovotestes and ovaries found in this condition may result from the actions of a series of loci involved in a common testis-determining pathway.

II. FUNCTION(S) OF THE H-Y ANTIGEN

One of the major hypotheses on the mechanism of testis differentiation in mammals invokes the histocompability-Y, or H-Y antigen. Based on the male-specific expression and genetic conservation of this antigen, Wachtel et al.[12] proposed that H-Y is the molecule that induces the indifferent gonads to form embryonic testes. A simple model would locate the H-Y structural gene on the Y chromosome. But alternative models have been proposed by various investigators to account for results obtained in different laboratories.[13-16] According to a recent version,[17] the structural gene for H-Y antigen is not on the Y chromosome, but on an autosome. During gonadogenesis, a regulatory gene present on the Y chromosome is expressed; its products, either directly or indirectly, stimulate the transcription of the H-Y structural gene and this leads to the biosynthesis of H-Y antigen. This antigen then induces the indifferent gonads to differentiate into testes. The relationship between the H-Y regulatory gene and *TDF* of the human or *Tdy* of the mouse, and the relationship between the H-Y structural gene and the *Tda* and *Tas* loci of the mouse are not clear at the present time.

In recent years, the H-Y hypothesis has been challenged critically,[18-10] and data separating the *TDF* and *Tdy* loci from the *H-Y* loci of human and mouse have been reported.[21-23] Burgoyne et al.[24] have postulated that the H-Y antigen may be a spermatogenic factor in mammals.

Much of the controversy regarding the function of H-Y antigen has arisen from the operational definitions of the molecule.[16,25,26] Although the H-Y antigen was first demonstrated by Eichwald and Silmser[27] in transplantation experiments, male ("H-Y") antigens can also be detected by other immunologic means — notably antibody-mediated serological tests[27] and cell-mediated cytotoxicity assays.[29,30] Evidence for the sex-determining role of H-Y antigen has been obtained mostly from studies using the serological methods, whereas evidence for the spermatogenic function of H-Y antigen has come from transplantation and cell-mediated cytotoxicity tests. At present, there are no conclusive data to support either of the hypothetical functions of the H-Y antigen.

Given the various immunological methods for detection of the H-Y antigen, the question remains as to whether the alternative assays identify the same or different antigens.[18,30,31] The purification of H-Y antigen has been unsuccessful in some laboratories[32] (but see Reference 33). Presumably, specific polyclonal antibodies are synthesized in sensitized female hosts in the serological assays, but the titers of most antisera are usually low and unstable.[31,34] Monoclonal antibodies are difficult to obtain, and most of them are specific only for the carbohydrate moiety of the molecule.[35,36] The problem of identification of the function(s) of H-Y antigen cannot be adequately addressed until the gene(s) for this antigen (or group of antigens) is isolated by some of the recently developed expression cloning methods and is characterized in molecular terms.

III. MOLECULAR ISOLATION OF A CANDIDATE GENE FOR THE SEROLOGICAL H-Y ANTIGEN

Advances in recombinant DNA technology have enabled isolation of specific genes by use of either prokaryotic or eukaryotic expression systems. The prokaryotic expression system, λ gt11, offers an opportunity to isolate cDNA genes based on antibody reaction to hybrid proteins made in bacterial hosts.[37] In this type of immunoscreening, small amounts of antisera can be used in scaled-down experiments. In collaboration with Ellen Goldberg of the University of New Mexico, we recently isolated, with specific antisera, the cDNA of a candidate gene for the serological H-Y antigen from a mouse testicular cDNA library constructed in λ gt11.[38] Since the amino acid sequence of the serological H-Y antigen is not yet available for comparison with that deduced from our cDNA clone, we called this candidate gene the "male-enhanced antigen" or *MEA* gene.

IV. IS MEA THE SEROLOGICAL H-Y ANTIGEN?

This is a key question the answer to which certainly will lead to other questions regarding the functions of both antigens. Presently, there is no definitive answer to this question. As was the case above, much of the problem lies within the operational definitions of the H-Y antigen. Evidence has been presented suggesting that transplantation and cell-mediated cytotoxicity (cytotoxic T cell) methods detect the same antigen.[30] Although these assays can be used to demonstrate the existence of the antigen, they are not suitable for the purification of the molecule. On the other hand, the serological methods which involve the biosynthesis of specific antibodies by the B lymphocytes against this antigen can be used in immunochemical analysis. As noted above, however, the serological assays may detect a different male-specific antigen than that detected by transplantation and cytotoxic T cell tests. Accordingly, the terms, "serologically detected male antigen" (SDM antigen)[18] and "serologically detected male predominant antigen" (SDMP antigen),[31] have been proposed for the H-Y antigen demonstrated by serological methods. Whether or not these immunological methods detect the same or different antigens, the availability of specific antisera enables investigators to study the molecular properties of the antigen.[15] This is especially significant as high-titer sera have recently been generated.[39] Such improvement increases the technical feasibility of purification and eventual molecular characterization of the proteins reactive with anti-H-Y sera.

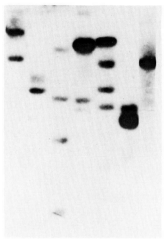

FIGURE 1. Southern analysis of DNA from different mammals. DNA samples were isolated from testes of each species and human placenta, digested with the restriction endonuclease EcoR 1, size-fractionated in 0.8% agarose gel, and transferred to nitrocellulose filter paper. The filter was then hybridized with ^{32}P-labeled mouse MEA cDNA probe in a hybridization mixture containing 50% formamide at 42°C for 20 hr. After extensive washing, the filter was air-dried and exposed to X-ray film. Positive hybridization of the mouse MEA cDNA to other mammalian DNA suggests a phylogenetic conservation of this gene sequences in mammals.

In the absence of purified H-Y antigen and its corresponding amino acid sequence, we cannot, at present, confirm the identity of MEA and serological H-Y antigen. However, our preliminary molecular characterization of the *MEA* gene has indicated that MEA shares a number of properties attributed to the serological H-Y antigen on the basis of immuno-chemical and biochemical studies. The detailed studies of the *MEA* gene will be reported elsewhere.[40] However, it may be useful to outline here some of the characteristics of the *MEA* gene in order to illustrate this point. First, the anti-H-Y sera that we used in our immunoscreening of the mouse testicular cDNA library, consistently identified three independent recombinant bacteriophages harboring cDNA inserts derived from the same (MEA) mRNA, suggesting that the sera were specific for the MEA transcripts. Second, DNA-sequencing studies of both human and mouse MEA cDNA indicated that their corresponding mRNAs encoded polypeptides of 185 and 174 amino acids, respectively. The calculated molecular weights for the human and mouse MEA were 20,000 and 18,600 respectively. Earlier studies of the serological H-Y antigen sggested a molecular weight range of 18,000 to 20,000. Third, the cDNA of both species showed considerable homology in nucleic acid and protein levels. Using the mouse cDNA as a probe in Southern blots, genetic conservation could be extended to other mammalian species, such as guinea pig, rabbit, bull, rat, and dog (Figure 1). Significantly, the phylogenetic conservation of the *MEA* gene was associated

Table 1
SIMILAR PROPERTIES OF SEROLOGICAL H-Y ANTIGEN AND MEA

Serological H-Y antigen[a]	Male-enhanced antigen[b]
Male-specific glycoprotein	Male-enhanced transcription
Molecular weight about 18,000—20,000	Molecular weight about 18,600—20,000
Phylogenetically conserved, immunologic	Phylogenetically conserved, genomic
Detected in mammalian testes	Expressed in mammalian testes
Expressed in Sertoli cells	Transcribed in Sertoli cells
Sperms contain concentrated amounts	Spermatids transcribe most abundantly
Located on autosome or X chromosome (?)	Located on human chromosome 6 and mouse chromosome 17

[a] Results from immunological and biochemical studies.[16,25,26,33]
[b] Results from molecular analyses.[17,38,40,42]

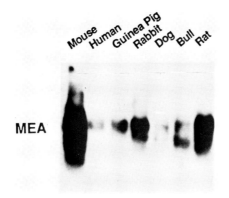

FIGURE 2. Northern analysis of poly A + RNA isolated from different mammalian testes. Two µg of poly A + RNA from each species were denatured by glyoxalation, electrophoresed in a 1.2% agarose gel, and transferred to Biodyne membrane filter. The filter was hybridized similarly with a ^{32}P-labeled mouse MEA cDNA probe as described in the legend of Figure 1. Positive hybridization of the mouse MEA cDNA to all testicular RNA samples indicates that *MEA* gene is conservatively transcribed in this mammalian organ. The MEA transcripts are about 1 kb in size.

with testis-specific expression in the mammals, as indicated by Northern RNA analysis (Figure 2). Table 1 summarizes some of the similarities between the serological H-Y antigen and the MEA antigen. Preliminary studies showed that other cell types, such as embryonic carcinoma cells, 14-day mouse fetuses, and adult Leydig cells also expressed the *MEA* gene. These results strongly suggest that the *MEA* gene is a candidate gene for the serological H-Y antigen.

V. THE FUNCTIONS OF THE *MEA* GENE

Preliminary chromosome mapping with a somatic cell hybrid harboring mouse chromosome 17 in the Chinese hamster background[41] has assigned the mouse *MEA* gene to this chromosome.[42] Even though we have not further localized the *MEA* gene, occurrence of the *MEA* gene in this chromosome is interesting. First, the mouse major histocompatibility complex, H-2, is also located on chromosome 17, establishing the likely linkage of MEA to this complex, as in human chromosome 6. Second, a locus, *Hye*, that has been postulated to regulate the expression of H-Y antigen is mapped between the T/t and H-2 systems on chromosome 17.[43] Third, Washburn and Eicher[9] have demonstrated a dominantly inherited trait, *Tas*, that is closely linked to the *T/t* locus and is involved in the testis-determining process in the mouse. At present, the exact relationship of *MEA* to *Hye*, *Tas*, and *T/t* and *H-2* loci is unknown. Lastly, studies on the genomic sequences harboring the structural gene for *MEA* have identified two other testis-specific genes, designated as gene *A* and *B*, that are linked within 40 to 60 kb of DNA to the *MEA* gene. This finding signifies the possibility that the *MEA* gene is a member of a gene cluster that specifically expresses abundantly in the testis. The location of the *MEA* gene cluster on chromosome 17 places these genes on the portion of the mouse genome that is important for embryogenesis, spermatogenesis, and sex determination.[44] Further molecular and genetic studies of the *MEA* gene cluster should provide information regarding the precise functions of the *MEA* genes in these developmental processes.

REFERENCES

1. **Eicher, E. M. and Washburn, L. L.,** Genetic control of primary sex determination in mice, *Annu. Rev. Genet.,* 20, 327, 1986.
2. **Page, D. C.,** Sex reversal: deletion mapping the male-determining function of the human Y chromosome, *Cold Spring Harbor Symp. Quant. Biol.,* 51, 229, 1986.
3. **Seboun, E., Leroy, P., Casanova, M., Magenis, E., Boucekkine, C., Disteche, C., Bishop, C., and Fellous, M.,** A molecular approach to the study of the human Y chromosome ad anomalies of sex determination in man, *Cold Spring Harbor Symp. Quant. Biol.,* 51, 237, 1986.
4. **de la Chapelle, A.,** Genetic and molecular studies on 46,XX and 45,X males, *Cold Spring Harbor Symp. Quant. Biol.,* 51, 249, 1986.
5. **Ferguson-Smith, M. A.,** X-Y chromosomal interchange in the aetiology of true hermaphroditism and of XX Klinefelter's syndrome, *Lancet,* 2, 475, 1966.
6. **Petit, C., de la Chapelle, A., Levilliers, J., Castillo, S., Noel, B., and Weissenbach, J.,** Abnormal terminal X-Y interchange accounts for most but not all cases of human XX maleness, *Cell,* 49, 595, 1987.
7. **Vergnaud, G., Page, D. C., Simmler, M.-C., Brown, L., Rouyer, F., Noel, B., Botstein, D., de la Chapelle, A., and Weissenbach, J.,** A deletion map of the human Y chromosome based on DNA hybridization, *Am. J. Hum. Genet.,* 38, 109, 1986.
8. **Eicher, E. M., Washburn, L. L., Whitney, J. B., III, and Morrow, K. E.,** *Mus poschiavinus* Y chromosome in C57BL/6J murine genome causes sex reversal, *Science,* 217, 535, 1982.
9. **Washburn, L. L. and Eicher, E. M.,** Sex reversal in XY mice caused by dominant mutation on chromosome 17, *Nature (London),* 303, 338, 1983.
10. **Nagamine, C. M., Taketo, T., and Koo, G. C.,** Morphological development of the mouse gonad in *tda-1* XY sex reversal, *Differentiation,* 33, 214, 1987.
11. **Nagamine, C. M., Taketo, T., and Koo, G. C.,** Studies on the genetics of *tda-1* XY sex reversal in the mouse, *Differentiation,* 33, 223, 1987.
12. **Wachtel, S. S., Ohno, S., Koo, G. C., and Boyse, E. A.,** Possible role for H-Y antigen in the primary determination of sex, *Nature (London),* 257, 235, 1975.
13. **Haseltine, F. P. and Ohno, S.,** Mechanisms of sex differentiation, *Science,* 211, 1272, 1981.
14. **Wolf, U., Fraccaro, M., Mayerová, A., Hecht, T., Maraschio, P., and Hameister, H.,** A gene controlling H-Y antigen on the X chromosome, *Hum. Genet.,* 54, 149, 1980.
15. **Polani, P.E. and Adinolfi, M.,** The H-Y antigen and its functions: a review and a hypothesis, *J. Immunogenet.,* 10, 85, 1983.
16. **Wachtel, S. S.,** *H-Y Antigen and the Biology of Sex Determination,* Grune & Stratton, New York, 1983.
17. **Lau, Y.-F., Chan, K., Kan, Y. W., and Goldberg, E.,** Male-enhanced expression and genetic conservation of a gene isolated with an anti-H-Y antibody, *Trans. Assoc. Am. Physicians,* 100, 45, 1987.
18. **Silvers, W. K., Gasser, D. L., and Eicher, E. M.,** H-Y antigen, serologically detectable male antigen and sex determination, *Cell,* 28, 439, 1982.

19. **Goodfellow, P. N.,** The case of the missing H-Y antigen, *Trends Genet.,* 2, 87, 1986.
20. **Simpson, E.,** The H-Y antigen and sex reversal, *Cell,* 44, 813, 1986.
21. **McLaren, A., Simpson, E., Tomonari, K., Chandler, P., and Hogg, H.,** Male sexual differentiation in mice lacking H-Y antigen, *Nature (London),* 312, 552, 1984.
22. **Simpson, E., Chandler, P., Hunt, R., Hogg, H., Tomonari, K., and McLaren, A.,** H-Y status of X/X *Sxr'* male mice: *in vivo* tests, *Immunology,* 57, 345, 1986.
23. **Simpson, E., Chandler, P., Goulmy, E., Disteche, C. M., Ferguson-Smith, M. A., and Page, D. C.,** Separation of the genetic loci for the H-Y antigen and for testis determination on human Y chromosome, *Nature (London),* 326, 876, 1987.
24. **Burgoyne, P. S., Levy, E. R., and McLaren, A.,** Spermatogenic failure in male mice lacking H-Y antigen, *Nature (London),* 320, 170, 1986.
25. **Müller, U.,** The H-Y antigen: identification, functions, and role in sexuality, in *The Y Chromosome,* Part A, Sanberg, A. A., Ed., Alan R. Liss, New York, 1985, 63.
26. **Andrews, P.W.,** The male-specific antigen (H-Y) and sexual differentiation, in *Genetic Analysis of the Cell Surface,* Goodfellow, P., Ed., Chapman and Hall, London, 1984, 159.
27. **Eichwald, E. J. and Silmser, C. R.,** Untitled communication, *Transplant. Bull.,* 2, 148, 1955.
28. **Goldberg, E. H., Boyse, E. A., Bennett, D., Scheid, M., Carswell, E. A.,** Serological demonstration of H-Y (male) antigen on mouse sperm, *Nature (London),* 232, 478, 1971.
29. **Goldberg, E. H., Shen, F., and Tokuda, S.,** Detection of H-Y (male) antigen on mouse lymph node cells by the cell to cell cytotoxicity test, *Transplantation,* 15, 334, 1973.
30. **Simpson, E., McLaren, A., and Chandler, P.,** Evidence for two male antigens in mice, *J. Immunogenet.,* 15, 609, 1982.
31. **Zenzes, M. T. and Reed, T. E.,** Variability in serologically detected male antigen titer and some resulting problems: a critical review, *Hum. Genet.,* 66, 103, 1984.
32. **Gore-Langton, R. E., Tung, P. S., and Fritz, I. B.,** The absence of specific interactions of Sertoli-cell-secreted proteins with antibodies directed against H-Y antigen, *Cell,* 32, 289, 1983.
33. **Bradley, M. P. and Heslop, B. F.,** A biochemical and immunological approach to the identification of H-Y antigenic proteins secreted from Daudi cells, *Hum. Genet.,* 71, 117, 1985.
34. **Shapiro, M. and Erickson, R. P.,** Genetic effects on quantitative variation in serologically detected H-Y antigen, *J. Reprod. Immunol.,* 6, 197, 1984.
35. **Shapiro, M. and Erickson, R. P.,** Evidence that the serological determinant of H-Y antigen is carbohydrate, *Nature (London),* 290, 1, 1981.
36. **Shapiro, M. and Goldberg, E. H.,** Analysis of a serological determinant of H-Y antigen: evidence for carbohydrate specificity using an H-Y specific monoclonal antibody, *J. Immunogenet.,* 11, 209, 1984.
37. **Young, R. A. and Davis, R. W.,** Efficient isolation of genes by using antibody probes, *Proc. Natl. Acad. Sci. U.S.A.,* 80, 1194, 1983.
38. **Lau, Y.-F., Chan, K., Kan, Y. W., and Goldberg, E.,** Isolation of a male-specific and conserved gene using an anti-H-Y antibody, *Am. J. Hum. Genet.,* 39 (Suppl.), 142A, 1986.
39. **Bradley, M. P. and Heslop, B. F.,** Elicitation of a rapid and transient antibody response to H-Y antigen by intrasplenic immunization, *Transplantation,* 39, 634, 1985.
40. **Lau, Y.-F., Chan, K., Kan, Y. W., and Goldberg, E.,** manuscript in preparation, 1988.
41. **D'Eustachio, P. and Ruddle, F. H.,** Somatic cell genetics and gene families, *Science,* 220, 919, 1983.
42. **Lau, Y.-F.,** Localization of the gene for the male-enhanced antigen on human and mouse chromosomes, in *Genetic Markers of Sex Differentiation,* Haseltine, F. P., McClure, M. E., and Goldberg, E. H., Eds., Plenum Press, New York, 1987, 161.
43. **Kralova, J. and Lengerova, A.,** H-Y antigen: genetic control of the expression as detected by host-versus-graft popliteal lymph node enlargement assay maps between the T and H-2 complexes, *J. Immunogenet.,* 6, 429, 1979.
44. **Erickson, R. P.,** Genetics of the cell surface of the preimplantation embryo: studies on antigens determined by chromosome 17 in the mouse, in *Genetics Analysis of the Cell Surface,* Goodfellow, P., Ed., Chapman and Hall, London, 1984, 143.

Part IV: Sex Reversal

Chapter 15

GENETICS OF XX AND XO SEX REVERSAL IN THE MOUSE

Paul S. Burgoyne

TABLE OF CONTENTS

I. INTRODUCTION

It has been accepted for decades that the mammalian Y chromosome carries genetic information which is responsible for diverting the indifferent gonad toward the testicular pathway. In the simplest view, the Y is envisaged as carrying a single testis-determining gene that triggers a cascade of gene activity required for testis differentiation. Even if this simple view is correct, it is entirely possible that some of the genes in this cascade are also located on the Y. In this review we will show how analysis of inherited sex reversal brought about by the Y-derived factor Sxr and its derivative Sxr', has contributed to our knowledge of genes on the mouse Y chromosome.

II. TESTIS DIFFERENTIATION

Before considering XX and XO sex reversal, it is important to have a basic understanding of how the Y chromosome acts in normal testis differentiation. In essence, the Y chromosome acts by *preempting* ovarian differentiation; the cells of the indifferent gonad are diverted to form testicular cell types *before* the progam of ovarian differentiation is initiated. Three gonad-specific cell lineages are involved:

1. The germ cell lineage, which in the fetal ovary proceeds through the stages of oogenesis, but in the testis is diverted to prespermatogenesis.
2. A "supporting cell" lineage which forms follicle cells in the fetal ovary, but is diverted to form Sertoli cells in the fetal testis. These Sertoli cells secrete the factor responsible for the regression of the female duct system in fetal life.
3. A "steroid cell" lineage which at puberty forms estrogen-secreting theca cells in the ovary, but in the fetal testis is diverted to form testosterone-secreting Leydig cells. In conjunction with the differentiation of Leydig cells there is the development of a complex vascular connective tissue framework enabling testosterone export from the testis, and hence the systemic effects of testosterone on the genital tract.

Although we do not yet fully understand how the Y chromosome acts in altering the fate of these cell lineages, some important points have emerged. For the germ cell lineage we know that the initial commitment to the male pathway does not require a Y in the germ line; the Y, through the activity of the testis-determining gene, brings about this commitment by creating the necessary testicular environment.[1,2] However, the Y is necessary in the male germ line at a later stage (see Sections III and VIII). From recent studies of the fate of XX cells in XX ↔ XY chimeric testes,[3] it has become clear that cells lacking a Y can form functional Leydig cells, and it is our belief that Leydig cell differentiation is triggered by the Sertoli cells. It is in the differentiation of the Sertoli cells that the Y appears to act directly; the Sertoli cells of XX ↔ XY chimeras are exclusively XY.[3]

III. THE BASIC GENETICS OF *Sxr*

In 1971, Cattanach et al.[4] described an inherited form of sex reversal in mice. The factor responsible, dubbed *sex-reversed (Sxr),* caused XX and XO mice to develop as males. Sxr exhibited an autosomal pattern of inheritance in that carrier males transmitted the factor to half their XX and half their XY progeny, but attempts over the next two decades to find the autosomal location of *Sxr* were unsuccessful.[5] The solution to this puzzle was an intriguing one. In 1982, it was shown that the Sxr factor consisted of a small segment of Y-chromosomal material[6] (encompassing the testis-determining gene). This segment is located at the centromeric end of the normal mouse Y,[6] but in XY*Sxr* carrier males there is an extra distally located copy which can be transferred to the X during male meiosis;[6,7] the autosomal pattern

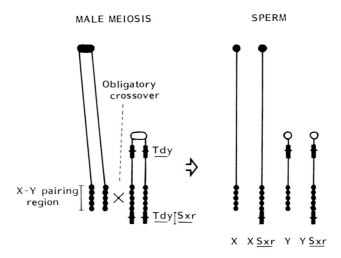

FIGURE 1. The transmission of *Sxr*. Because *Sxr* in XY*Sxr* carrier males is distal to the obligatory X-Y crossover, four types of sperm are produced in equal proportions; X and Y*Sxr* sperm, derived from the unrecombined chromatids, and X*Sxr* and Y sperm, derived from the recombinant chromatids. The four types of offspring — XX female, XX*Sxr* male, XY male, and XY*Sxr* carrier male — are consequently produced in equal proportions giving a "pseudoautosomal" pattern of inheritance.[8]

of inheritance is thus explained by a single obligatory crossover in the X-Y pairing region (Figure 1).[8-10] This breakthrough in our understanding of the inheritance of *Sxr* in many ways simplified interpretation of the sex reversal phenomenon, so those seeking an introduction to this subject may wish to begin with reviews published since 1982.[11-13]

IV. X*Sxr*O MALES

X*Sxr*O males can be produced by mating XO females with XY*Sxr* carrier males. These X*Sxr*O males carry *Sxr* attached distally to their single X chromosome. By comparing X*Sxr*O males with normal XY males we can determine which Y chromosome functions are provided by *Sxr* and we can identify which are not.

X*Sxr*O males look and behave like normal males, but they are completely sterile.[4,5] They are positive for the male-specific transplantation antigen H-Y.[14] Gross examination of the reproductive tract reveals no abnormalities except that the testes are on average about half the normal size. Histologically, these testes appear to have normal somatic components. Deficiencies in the germ cell component account for the reduced testis size.[4,5] Although all stages of spermatogenesis are present, there is a marked shortage of the condensing spermatid stages, and the few sperm produced have abnormal heads.[4] The majority of spermatids in these mice are in fact diploid, but the head abnormalities affect haploid and diploid alike.[15,16]

What Y chromosome functions does *Sxr* provide? Clearly, *Sxr* carries the Y-chromosomal information need for testis determination and for H-Y expression. We shall refer to the genes controlling these two functions as *Tdy* and *Hya*, respectively. We do not yet know whether *Hya* is the structural gene for H-Y, or whether it is regulatory. Recently, Levy and Burgoyne[17] demonstrated that XO germ cells in XO/XY mosaics rarely attain and never progress beyond the meiotic prophase; the authors attributed this failure to lack of a Y-chromosomal "spermatogenesis" gene (*Spy*). Since X*Sxr*O germ cells are not subject to this block, Sxr must also include *Spy*. The Y-chromosomal functions not fulfilled by *Sxr* relate to the later stages of spermatogenesis and have recently been discussed at length by Burgoyne.[18] First, the Y has an important role as a pairing partner for the X during meiotic prophase;[15] sex chromosome univalence leads (for unknown reasons) to severe spermatogenic losses, hence the rarity of mature sperm in X*Sxr*O males. Second, Eicher and co-workers[13,19] have suggested that there is a sperm morphogenesis gene on the Y which is lacking from Sxr, thus accounting for the sperm abnormalities in X*Sxr*O males. We are currently trying to test that suggestion by introducing a pairing partner for the X of X*Sxr*O males without providing any part of the Y chromosome.

V. XX*Sxr* MALES

XX*Sxr* males can be produced by mating XY*Sxr* males to normal XX females (Figure 1). They have *Sxr* attached distally to one of their X chromosomes. By comparing XX*Sxr* males with X*Sxr*O males we can see what additional abnormalities arise from the presence of a second X chromosome.

The only major difference between adult XX*Sxr* and X*Sxr*O males is that XX*Sxr* testes are much smaller than X*Sxr*O testes and typically lack spermatogenic cells.[4,5] Other minor differences (perhaps secondary effects of the germinal failure) include the presence of immature Sertoli cells[20] and abnormalities of the epididymides.[21] Although germ cells are absent in the adult, they do initially colonize XX*Sxr* fetal gonads.[4] The majority of these germ cells differentiate into T_1-prospermatogonia, but in most fetal XX*Sxr* testes, a small proportion of the germ cells fails to make this transition to the male pathway and instead form fetal oocytes (i.e., they enter meiosis during fetal life).[22] Oocytes are not found in fetal X*Sxr*O testes.[22] Whichever pathway is taken, the germ cells degenerate, most having disappeared by the time of birth.[4] Sometimes a few oocytes survive beyond birth and begin to grow inside the testis cords,[23] but oocytes have never been seen in adult XX*Sxr* testes.

There are thus two separate abnormalities of germ cell development in XX*Sxr* testes which must in some way be due to the presence of the second X chromosome. First, some of the germ cells fail to divert to the male pathway. Second, the germ cells that do enter the male pathway as T_1-prospermatogonia die before or soon after birth. The diversion of germ cells to the male pathway is known to be brought about by the testicular environment (Section II), so the finding of groups of oocytes in most fetal XX*Sxr* testes implies a local inadequacy of the testicular environment. We believe this inadequacy must be a manifestation of the X inactivation which occurs in XX*Sxr* (but not X*Sxr*O) males. When *Sxr* is located on an inactive X, spreading of X-inactivation could prevent the action of *Tdy* in some cells, and thus perturb testicular differentiation. We will discuss this more fully in Section VI. The death of XX*Sxr* germ cells that do enter the male pathway is also a consequence of the presence of two X chromosomes, but the mechanism is very different. This germinal failure is a consequence of excess X chromosome activity in the germ line. Proof of this comes from observations on those rare XX*Sxr* testes which have some "pockets" of spermatogenesis. This spermatogenesis is in fact supported by X*Sxr*O germ cells[24] (which presumably arose by nondisjunction), so loss of the second X is sufficient to allow an XX*Sxr* germ cell to survive in an XX*Sxr* testis. Consistent with this X-dosage-effect in the male germ line, has been the finding that X inactivation is repealed in fetal XX*Sxr* T_1-prospermatogonia,[25] just as it is in fetal XX oogonia.[26]

VI. T16/X*Sxr* MALES, FEMALES, AND HERMAPHRODITES

In adult female mice heterozygous for the reciprocal X-autosome translocation T(X;16)16H (hereafter abbreviated as T16), the normal X is always the inactive one (Figure 2), probably as a result of the death at the early egg cylinder stage of cells in which the 16^X chromosome was inactivated.[27] T16/X*Sxr* mice can be produced by mating T16/X females to XY*Sxr* males, thus creating mice in which *Sxr* is always attached to the inactive X. T16/X*Sxr* mice, as compared to XX*Sxr* mice, should therefore show a heightening of the consequences of

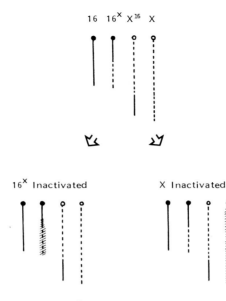

FIGURE 2. X inactivation in T16/X females. Initially random X inactivation in the epiblast (from which the embryo proper is derived) leads to inactivation of either the 16^X or the normal X. The X^{16} cannot be inactivated because it lacks the X inactivation center. Cells which inactivate the 16^X die because of inappropriate X activity from the X^{16} and perhaps also because of spreading of X inactivation into the autosomal part of the 16^X. Thus the only cells which survive have the normal X inactivated. At this early stage of development death of half the cells in the epiblast can be compensated for easily, and subsequent development is normal.

X inactivation spreading into *Sxr*, although we need to be alert for possible deleterious effects of the translocation itself.

T16/X*Sxr* mice were first described in 1982 by Cattanach and colleagues[28] and by McLaren and Monk.[29] Both groups reported that these mice could develop as males, females, or hermaphrodites, but there was a marked difference between the two studies in the proportion developing as males (85% and 50%, respectively). Subsequently, McLaren[30] demonstrated an intriguing association between the frequency of males and the "strength" of the *Xce* allele carried by the normal X. The *Xce* (*X chromosome-controlling element*) locus is involved in X inactivation, and a "strong" allele reduces the frequency of inactivation of an X carrying it, when in competition with an X carrying a "weak" *Xce* allele.[31] In T16/X*Sxr* mice, a "weak" *Xce* allele on the normal X appears to favor males, and a "strong" allele appears to favor hermaphrodites or females, implying that a "strong" allele favors the spread of X inactivation into *Sxr*.[30] It is not surprising (in view of the double X dosage) that T16/X*Sxr* males lack germ cells and are sterile; in fact, they have smaller testes than their XX*Sxr* sibs have.[28,30] T16/X*Sxr* females, on the other hand, are fertile and transmit *Sxr* to half their progeny.[29,29] Irrespective of sexual phenotype, all T16/X*Sxr* mice are H-Y antigen positive according to the cytotoxic T cell assay.[32]

The features of T16/X*Sxr* mice which distinguish them from XX*Sxr* mice is their range of sexual phenotype and the additional reduction in testis size seen in T16/X*Sxr* males. That these features are due to a heightening of the X inactivation effects on *Sxr*, rather than to a direct effect of the translocation, can be demonstrated by comparing T16/X*Sxr* with T16*Sxr*/X mice. Both have the translocation, but in the former, *Sxr* is on the inactive X, whereas in the latter, *Sxr* is on the 16^X chromosome (Figure 2) which is active. T16*Sxr*/X mice are always male, and they have the same size testes as their XX*Sxr* sibs,[33] so a direct effect of the translocation cannot explain the difference between XX*Sxr* mice and T16/X*Sxr* mice.

In what follows, I will attempt to outline how I envisage the range of gonadal phenotype, from XX*Sxr* and T16/X*Sxr* testes to T16/X*Sxr* ovaries, can be generated simply by varying the proportion of cells expressing *Tdy*. The first important point follows from the finding that T16/X*Sxr* mice are not always female. This tells us that when *Sxr* is attached to an inactive X, *Tdy* is not always inactivated. T16/X*Sxr* mice therefore can be mosaics for *Tdy* expression,[28,29] just as XX*Sxr* mice can; it is simply that the bias has been altered in favor of *Tdy* inactivity. If *Tdy* exerts its testis-determining effect by triggering cells of the supporting cell lineage to form Sertoli cells (see Section II) then XX*Sxr* and T16/X*Sxr* gonads should initially retain some uncommitted supporting cells in which *Tdy* has been inactivated. We shall refer to these uncommitted cells as "prefollicular" cells since they have the potential to form follicle cells. However, realization of this potential requires the presence of oocytes, since they act as "templates" during folliculogenesis, and when Sertoli cells are present the odds are stacked against the formation and the survival of oocytes. To begin with, the Sertoli cells actively try to sequester all the germ cells within testis cords and foster the conditions needed to "persuade" them to form T_1-prospermatogonia rather than oocytes. In XX*Sxr* testes a few of the germ cells inside cords do develop as oocytes, presumably due to some influence of nearby prefollicular cells, but they cannot survive long in this location. If there are too few Sertoli cells to sequester all the germ cells, then some will remain outside the cords and enter meiosis in the prefollicular cell environment (I would classify such a gonad as a fetal ovotestis). As a general rule, these escapees are also doomed because some diffusible substance produced by testicular tissue (not testosterone,[34] but possibly the Müllerian regression factor) happens to be toxic to oocytes in early meiotic prophase.[35-37] Consequently, most fetal ovotestes lose their oocytes and appear postnatally as testes. Only when the testicular component is very minor will ovarian tissue survive into the postnatal period. It is in such cases that incomplete masculinization of the genital tract and external genitalia are also frequently encountered. Although fetal T16/X*Sxr* gonads have not been studied we would expect most to be ovotestes or ovaries. The small size of the testes of T16/X*Sxr* males, as compared to XX*Sxr* testes, may reflect the fact that T16/X*Sxr* testes are derived from fetal ovotestes. In XO/XY mouse mosaics there is compelling evidence that fetal ovotestes evolve into abnormally small testes.[17,38]

VII. AN H-Y NEGATIVE VARIANT OF *Sxr*

In 1975, Wachtel et al. put forward the stimulating hypothesis that the male-specific transplantation antigen H-Y was responsible for testis determination.[39] The H-Y positivity of XX*Sxr* mice[40] was consistent with this, and in the ensuing years the hypothesis became widely accepted. However, in 1984, McLaren et al. described a variant of *Sxr* (designated *Sxr'*) which although male determining, no longer conferred H-Y antigenicity as determined by the cytotoxic T cell assay[41] and, subsequently, by skin grafting.[42] The first mouse carrying *Sxr'* was in fact a female and was picked out as an exceptional H-Y negative T16/X*Sxr* female; all previous females of this genotype had been H-Y positive. Initially, it was considered that the H-Y negativity could have resulted from an unusually efficient spread of X inactivation in this particular mouse, but the H-Y negativity proved to be heritable. This led Ohno[43] to suggest a heritable change which favored the spread of X inactivation, but this is now ruled out by the finding that X*Sxr'*O mice are H-Y negative.[14] We therefore must accept that testis determination and H-Y expression in mice are controlled by separate loci. This conclusion has recently been extended to the human in which the testis-determining function of the Y maps distally on the short arm, whereas the control of H-Y expression maps either to the long arm or possibly very proximally on the short arm.[44,45]

It has recently been suggested that a male-specific antigen detected serologically (originally also called H-Y) could be the testis-determining molecule.[46] In this context it is important to know whether XX*Sxr'* mice carry this serologically determined antigen, but attempts to type these mice have so far been unsuccessful. I believe that if this antigen is involved in testis differentiation, its involvement is subsequent to Sertoli cell differentiation.[35]

VIII. H-Y NEGATIVE X*Sxr'*O MALES

The realization that H-Y transplantation antigen was not involved in testis determination prompted a new search for a male-specific function for this antigen. Our approach was to compare H-Y positive X*Sxr*O males with H-Y negative X*Sxr'*O males.[14] As described earlier (Section III), X*Sxr*O germ cells, unlike XO germ cells, can take part in spermatogenesis because Sxr includes the Y-chromosomal spermatogenesis gene *Spy*.[17,18] X*Sxr*O testes, consequently, have all stages of spermatogenesis, although the later stages are abnormal. X*Sxr'*O mice, on the other hand, proved to have testes which appeared at first to be devoid of germ cells. However, a more detailed study revealed occasional groups of spermatogonia or early meiotic prophase stages, although no cells reached the first meiotic metaphase.[14] This behavior of X*Sxr'*O germ cells is indistinguishable from the behavior of XO germ cells in XO/XY mosaics.[17] This led to the conclusion that the change from *Sxr* to *Sxr'* involved the loss of *Spy* function. Since it also involved the loss of H-Y expression, this raised the possibility that H-Y antigen was the mediator of *Spy* activity.[14]

IX. SUMMARY

The sex-reversal factor, Sxr, causes XO and XX mice to develop as males because Sxr includes the Y-chromosomal testis-determining gene, *Tdy*. Some years ago it was proposed that the male-specific antigen, H-Y, was involved in testis determination with H-Y expression controlled by *Tdy*, but the discovery of H-Y negative XX*Sxr'* males ruled out this possibility. In searching for an alternative function for H-Y, we have found that absence of H-Y antigen correlates with an early block in spermatogenesis, raising the possibility that H-Y antigen is involved in spermatogenesis.

NOTE ADDED IN PROOF

It is now known that *Tdy* is located on the minute short arm of the normal mouse Y chromosome. Figure 1 is thus incorrect in showing the non-Sxr copy of *Tdy* located proximally on the long arm (the minute short arm is not shown).[47,48]

REFERENCES

1. **McLaren, A.,** Does the chromosomal sex of a mouse germ cell affect its development?, in *Current Problems in Germ Cell Differentiation,* McLaren, A. and Wylie, C. C., Eds., Cambridge University Press, London, 1983, 225.
2. **McLaren, A.,** Meiosis and differentiation in mouse germ cells, in *Controlling Events in Meiosis,* Evans, C. W. and Dickinson, H. G., Eds., Company of Biologists, Cambridge, England, 1984, 7.
3. **Burgoyne, P. S., Buehr, M., Koopman, P. A., Rossant, J., and McLaren, A.,** Cell-autonomous action of the testis-determining gene: Sertoli cells are exclusively XY in XX ↔ XY chimric mouse testes, *Development,* 102, 443, 1988.
4. **Cattanach, B. M., Pollard, C. E., and Hawkes, S. G.,** Sex reversed mice: XX and XO males, *Cytogenetics,* 10, 318, 1971.

5. **Lyon, M. F., Cattanach, B. M., and Charlton, H. M.,** Genes affecting sex differentiation in mammals, in *Mechanisms of Sex Differentiation in Animals and Man,* Austin, C. R. and Edwards, R. G., Eds., Academic Press, New York, 1981, 329.

6. **Singh, L. and Jones, K. W.,** Sex reversal in the mouse (*Mus musculus*) is caused by a recurrent nonreciprocal crossover involving the X and an aberrant Y chromosome, *Cell,* 28, 205, 1982.

7. **Evans, E. P., Burtenshaw, M. D., and Cattanach, B. M.,** Meiotic crossing over between the X and Y chromosomes of male mice carrying the sex-reversing (*Sxr*) factor, *Nature (London),* 300, 443, 1982.

8. **Burgoyne, P. S.,** Genetic homology and crossing over in the X and Y chromosomes of mammals, *Hum. Genet.,* 61, 85, 1982.

9. **Eicher, E. M.,** Primary sex determining genes in mice, in *Prospects for Sexing Mammalian Sperm,* Amann, R. P. and Seidel, G. E., Eds., Colorado Associated University Press, Boulder, 1982, 121.

10. **Hansmann, I.,** Sex reversal in the mouse, *Cell,* 30, 331, 1982.

11. **Cattanach, B. M.,** Sex reversed mice and sex determination, *Ann. N.Y. Acad. Sci.,* 513, 27, 1988.

12. **McLaren, A.,** Sex reversal in the mouse, *Differentiation,* 23 (Suppl.), S93, 1983.

13. **Eicher, E. M. and Washburn, L. L.,** Genetic control of primary sex determination in mice, *Annu. Rev. Genet.,* 20, 327, 1986.

14. **Burgoyne, P. S., Levy, E. R., and McLaren, A.,** Spermatogenic failure in male mice lacking H-Y antigen, *Nature (London),* 320, 170, 1986.

15. **Burgoyne, P. S. and Baker, T. G.,** Meiotic pairing and gametogenic failure, in *Controlling Events in Meiosis,* Evans, C. W. and Dickinson, H. G., Eds., Company of Biologists, Cambridge, England, 1984, 349.

16. **Levy, E. R. and Burgoyne, P. S.,** Diploid spermatids: a manifestation of spermatogenic impairment in XO *Sxr* and T31H/+ male mice, *Cytogenet. Cell Genet.,* 42, 159, 1986.

17. **Levy, E. R. and Burgoyne, P. S.,** The fate of XO germ cells in the testes of XO/XY and XO/XY/XYY mouse mosaics: evidence for a spermatogenesis gene on the mouse Y chromosome, *Cytogenet. Cell Genet.,* 42, 108, 1986.

18. **Burgoyne, P. S.,** The role of the mammalian Y chromsome in spermatogenesis, *Development,* 101 (Suppl.), 133, 1987.

19. **Eicher, E. M., Phillips, S. J., and Washburn, L. L.,** The use of molecular probes and chromosomal rearrangements to partition the mouse Y chromosome into functional regions, in *Recombinant DNA and Medical Genetics,* Messer, A. and Porter, I. H., Eds., Academic Press, New York, 1983, 57.

20. **Chung, K. W.,** A morphological and histochemical study of Sertoli cells in normal and XX sex-reversed mice, *Am. J. Anat.,* 139, 369, 1974.

21. **LeBarr, D. K. and Blecher, S. R.,** Epididymides of sex-reversed XX mice lack the initial segment, *Dev. Genet.,* 7, 109, 1986.

22. **McLaren, A.,** The fate of germ cells in the testis of fetal sex-reversed mice, *J. Reprod. Fertil.,* 61, 461, 1981.

23. **McLaren, A.,** Oocytes in the testis, *Nature (London),* 283, 688, 1980.

24. **Lyon, M. F.,** Sex chromosome activity in germ cells, in *Physiology and Genetics of Reproduction,* Fuchs, F. and Continho, E. M., Eds., Plenum Press, New York, 63, 1974.

25. **McLaren, A. and Monk, M.,** X-chromosome activity in the germ cells of *sex-reversed* mouse embryos, *J. Reprod. Fertil.,* 63, 533, 1981.

26. **West, J.,** X chromosome expression during mouse embryogenesis, in *Genetic Control of Gamete Production and Function,* Crosignani, P. G., Rubin, B. L., and Fraccaro, M., Eds., Academic Press, New York, 1982, 49.

27. **McMahon, A. and Monk, M.,** X-chromosome activity in female mouse embryos heterozygous for Pgk-1 and Searle's translocation, T(X;16)16H, *Genet. Res.,* 41, 69, 1983.

28. **Cattanach, B. M., Evans, E. P., Burtenshaw, M., and Barlow, J.,** Male, female and intersex development in mice of identical chromosome constitution, *Nature (London),* 300, 445, 1982.

29. **McLaren, A. and Monk, M.,** Fertile females produced by inactivation of an X chromosome of 'sex-reversed' mice, *Nature (London),* 300, 446, 1982.

30. **McLaren, A.,** Sex ratio and testis size in mice carrying *Sxr* and T(X;16)16H, *Dev. Genet.,* 7, 177, 1986.

31. **Cattanach, B. M.,** Control of X chromosome inactivation, *Annu. Rev. Genet.,* 9, 1, 1975.

32. **Simpson, E., McLaren, A., Chandler, P., and Tomonari, K.,** Expression of H-Y antigen by female mice carrying *Sxr, Transplantation,* 37, 17, 1984.

33. **McLaren, A.,** personal communication, 1987.

34. **Arrau, J., Roblero, L., Cury, M., and Gonzalez, R.,** Effect of exogenous sex steroids upon the number of germ cells and the growth of foetal ovaries grafted under the kidney capsule of adult ovariectomised hamsters, *J. Embryol. Exp. Morphol.,* 78, 33, 1983.

35. **Burgoyne, P. S., Ansell, J. D., and Tournay, R.,** Can the indifferent XX gonad be sex-reversed by interaction with testicular tissue?, in *Serono Symp. Rev. 11, Development and Function of the Reproductive Organs,* Eshkol, A., Eckstein, N., Dekel, N., Peters, H., and Tsafriri, A., Eds., Ares-Serono Symposia, Rome, 1986, 23.

36. **Byskov, A. G. and Saxen, L.,** Induction of meiosis in fetal mouse testis, *in vitro, Dev. Biol.,* 52, 193, 1976.
37. **Macintyre, M. N., Hunter, J. E., and Morgan, A. H.,** Spatial limits of activity of fetal gonadal inductors in the rat, *Anat. Rec.,* 138, 137, 1960.
38. **Eicher, E. M., Beamer, W. G., Washburn, L. L., and Whitten, W. K.,** A cytogenetic investigation of inherited true hermaphoroditism in BALB/cWt mice, *Cytogenet. Cell Genet.,* 28, 104, 1980.
39. **Wachtel, S. S., Ohno, S., Koo, G. C., and Boyse, E. A.,** Possible role for H-Y antigen in the primary determination of sex, *Nature (London),* 257, 235, 1975.
40. **Bennett, D., Mathieson, B. J., Scheid, M., Yanagisawa, K., Boyse, E. A., Wachtel, S., and Cattanach, B. M.,** Serological evidence for H-Y antigen in *Sxr,* XX sex-reversed phenotypic males, *Nature (London),* 265, 255, 1977.
41. **McLaren, A., Simpson, E., Tomonari, K., Chandler, P., and Hogg, H.,** Male sexual differentiation in mice lacking H-Y antigen, *Nature (London),* 312, 552, 1984.
42. **Simpson, E., Chandler, P., Hunt, R., Hogg, H., Tomonari, K., and McLaren, A.,** H-Y status of X/X *Sxr'* male mice: *in vivo* tests, *Immunology,* 57, 345, 1986.
43. **Ohno, S.,** The Y-linked testis determining gene and H-Y plasma membrane antigen gene: are they one and the same?, *Endocr. Rev.,* 6, 421, 1985.
44. **Simpson, E., Chandler, P., Goulmy, E., Disteche, C. M., Ferguson-Smith, M. A., and Page, D. C.,** Separation of the genetic loci for the H-Y antigen and for testis determination on human Y chromosome, *Nature (London),* 326, 876, 1987.
45. **Simpson, E., Chandler, P., McLaren, A., Goulmy, E., Page, D. C., and Ferguson-Smith, M. A.,** Mapping the H-Y gene, *Development,* 101(Suppl.), 157, 1987.
46. **Wiberg, U. H.,** Facts and considerations about sex-specific antigens, *Hum. Genet.,* 76, 207, 1987.
47. **McLaren, A., Simpson, E., Epplen, J. T., Studer, R., Koopman, P., Evans, E. P., and Burgoyne, P. S.,** Location of the genes controlling H-Y antigen expression and testis determination on the mouse Y chromosome, *Proc. Natl. Acad. Sci. U.S.A.,* 85, in press, 1988.
48. **Roberts, C., Weith, A., Michot, J. L., Mattei, M. G., and Bishop, C. E.,** Molecular and cytogenetic evidence for the location of *Tdy* and H-Y on the mouse Y chromosome short arm, *Proc. Natl. Acad. Sci. U.S.A.,* 85, in press, 1988.

Chapter 16

GATA-GACA SIMPLE REPEATED DNA AS A MEANS TO DISSECT THE *SEX-REVERSED (Sxr)* MUTATION OF THE MOUSE

Jörg T. Epplen and Anne McLaren

TABLE OF CONTENTS

I. INTRODUCTION

Male sex determination in mammals remains a most interesting, yet in molecular terms still poorly understood, developmental system.[1] The Y chromosome, or better, a small part thereof, plays a crucial role in the differentiation process. In general, mammals with XY sex chromosome constitution are male[2] while individuals with two X chromosomes are female.[3] The presence of a Y chromosome (XY, XXY, XYY) causes the undifferentiated gonad to develop as testis and causes the organism to develop the male phenotype. Absence of the Y chromosome (XX, XO, XXX) results in the development of ovaries[4] and consequently, the female phenotype. Deeper insight into the sex determination mechanisms has been provided by exceptions to the rule. For example, sex-reversed individuals are defined as either XX males with pure testicular tissue or as XY individuals with only ovaries. The sex-reversal (designated *Sxr*) mutation in the mouse[5] causes the development of XX*Sxr* males in a regular manner among the progeny of carrier males (XY*Sxr*). In carrier mice, the Sxr factor is located at the distal tip of the Y chromosome[6,7] and is transferred to the X chromosome by crossing-over during male meiosis.[8] Probably the Sxr DNA arose by duplication of a pericentrometric part of the Y chromosome. This duplicated piece of the Y chromosome containing the testis-determining gene carries a high concentration of so-called Bkm (banded krait minor satellite) DNA.[6] A major part of Bkm DNA has been shown to consist of simple repetitive tandem sequences of the GATA-GACA family.[9] The involvement of simple repetitive Bkm sequences in testis determination has been a matter of speculation.[6,10]

Simple repetitive DNA has been known since the early 1970s, but it still retains nearly all its mysteries.[11] Even today, eminent authorities in the biological sciences elaborate, e.g., on the "*meaninglessness* of simple repeats".[12] We summarize here very briefly what is known about a specific family of simple repeats, the GATA-GACA sequences. So far, we have used these GATA-GACA simple repeats as tools in experiments to tackle various seemingly unrelated problems where other approaches are not suitable or completely lacking. We describe some new experimental results pertaining to the mouse Y chromosome and the *Sxr* and *Sxr'* mutations, mainly obtained with chemically synthesized GATA-GACA oligonucleotides as probes. Since with oligonucleotides virtually single-base-specific hybridizations can be performed,[13] the specificity of the probe and the fidelity of the hybridization results are not in question.

II. GATA-GACA SEQUENCES ARE ORGANIZED SPECIES-SPECIFICALLY AND THEIR EXPRESSION IS DIFFERENTIALLY REGULATED

Bkm DNA (with its main component the GATA-GACA simple repeats[9]) has been identified in snake species of the family Colubridae.[14] Though interspersed throughout the snake genome, GATA-GACA simple repeats are primarily concentrated on the heteromorphic sex chromosome of these species. The GATA-GACA sequences could be preparatively isolated from Ag-CsCl gradients due to their sex-specific organization.[6,9] In contrast to other satellite DNAs, simple GATA-GACA tandem repeats occur in every eukaryotic species examined so far by hybridization analysis.[15,16] In several insect species, e.g., *Drosophila melanogaster*, GATA-GACA repeats are intensively interspersed with many other sequences on the sex chromosomes as well as on the autosomes. Yet in *Ephestia kuehniella*, a moth, GATA-GACA repeats are limited to blocks in only a few hypervariable autosomal restriction fragments.[17,18] Studies on the organization of simple GATA-GACA sequences in vertebrate genomes have revealed manifold patterns which deviate from the bulk of other simple repeated DNA. GATA-GACA sequences are often intensively interspersed with all other major abundance classes of DNA sequences in reptile, avian, and mammalian genomes. Ubiquitous interspersion of GATA-GACA, simple repeats has been best documented in mouse[13] and man.[19,20] These properties make GATA-GACA sequences ideal tools for the simultaneous analysis of multiple-restriction fragment-length polymorphisms thus enabling, e.g., individual-specific DNA fingerprint analysis in humans.[19]

GATA-GACA simple repeats have been rapidly reorganized during rodent evolution. GATA-GACA sequences are again ubiquitously interspersed in the chromosomes of *Ellobius lutescens*, but they are significantly reduced on the X chromosome.[21] The opposite phenomenon has been shown in *Microtus agrestis*.[22] In the large heterochromatic block on the X chromosome of this species, there is an increased concentration of GATA sequences demonstrable by hybridization *in situ* of oligonucleotide and cloned, nick-translated probes to metaphase chromosomes. GACA sequences are comparatively rare. Further data indicate that the GATA repeats are organized also in higher-order structures in this specific heterochromatin, giving rise to "hidden satellite" bands with certain restriction enzymes. In the hamster, we found no sex differences. In the mouse, roughly one quarter of all the GATA-GACA sequences are concentrated in the pericentromeric region of the Y chromosome,[6,9] besides their ubiquitous interspersion with other sequences of the X and autosomes. The accumulation of GATA-GACA sequences in this chromosomal region is responsible for the sex difference observed in hybridization with appropriate probes. This region is at least partially duplicated in the XY*Sxr* carrier mouse.

Before data on other rodents were available, the apparent conservation of Bkm DNA or GATA-GACA simple repeated sequences in the heteromorphic sex chromosome over the evolutionary distance from snakes to mice triggered extensive speculation on their direct or indirect role in gonadal differentiation.[6,23] A functional role for GATA-GACA repeats should be reflected in their appearance as mRNAs. The expression of the GATA-GACA and the complementary TATC-TGTC sequences was therefore investigated in many mouse tissues, using a variety of different probes:[24] oligonucleotides and "single-stranded" and conventional nick-translated DNA. Organ-specific transcripts were detected, and the transcription of ubiquitous GATA-GACA-RNAs was found to be differentially regulated in various organs. In contrast to an unconfirmed report,[23] no sex differences were ever observed with simple repeat probes. Nevertheless, GATA-GACA-containing cDNA clones could be isolated from several libraries of different cell types. These cDNAs were sequenced and found to contain inverted repeats reminiscent of the inverted repeats of transposons.[25] The transposon character of some members of the GATA-GACA repeat family would explain elegantly the rapid spreading of the simple repeats throughout eukaryote genomes.[17,25] Rapid amplification and contraction of the simple repeats would be especially facilitated in chromosomal regions with reduced crossover frequencies like the sex chromosomes. Similarly, in the early evolution of sex chromosomes, sequences with the ability of rapid amplification could have facilitated the separation of a homologous pair of incipient sex chromsomes by extensive, instantaneous sequence diversification on the original homologs. As for the functional relevance of GATA-GACA repeats, no internally repetitive, highly hydrophilic translation products of the sense-strand (TATC-TGTC) have yet been identified.[24]

III. GATA-GACA SEQUENCES DISTINGUISH BETWEEN MALE AND FEMALE MOUSE DNA AS WELL AS Sxr AND Sxr′ DNA

When XY*Sxr* carrier males are mated to normal females, four types of offspring are produced: normal XX females, normal XY males, XY*Sxr* carrier males, and sex-reversed XX*Sxr* males. Like normal XY males and XY*Sxr* carrier males, XX*Sxr* males type positive for the male-specific histocompatibility antigen H-Y.[26] Recently, a variant Sxr factor, designated *Sxr′,* has been described. XX*Sxr′* individuals are also sex-reversed, sterile, phenotypic males, but type H-Y negative.[27] In view of the male-specific organization of GATA-GACA sequences in mice, we decided to investigate the hybridization behavior of these sequences with DNA from XX*Sxr*, XX*Sxr′*, XY*Sxr,* and XY*Sxr′* males.

It has been shown that hybridization of uncloned Bkm sequences can be used to distinguish

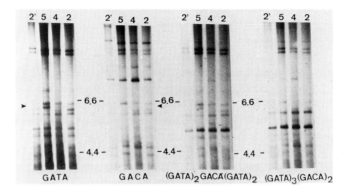

FIGURE 1. GATA-GACA oligonucleotide probes and their derivatives differentiate normal XY, Sxr and Sxr' DNA. Liver DNA of XX*Sxr'*, XX*Sxr*, XY*Sxr*, and XY male mice was digested with the restriction enzyme Hae III to completion; DNA was electrophoresed in a 0.7% agarose gel; the gel was dried and sequentially hybridized with the ^{32}P-labeled probes indicated [(GATA)$_4$: hybridization and stringent wash temperature 35°C; (GACA)$_4$: 43°C; (GATA)$_2$ GACA (GATA)$_2$: 47°C; (GATA)$_3$ (GACA)$_2$: 49°C]. Probe labeling as well as hybridization and washing of the gels has been described previously.[13] Dried gels were exposed to X-ray film for up to 1 day without intensifying screen. Before any subsequent hybridization, efficient removal of the oligonucleotide probe was monitored by control exposures. The numbers on top of the lanes identify the patterns observed: 2 and 2', XX*Sxr* and XX*Sxr'*; 4, XX*Sxr* variant; and 5, XY*Sxr* carrier. The arrows point to GATA-positive but GACA-negative bands distinguish the patterns of XY and Sxr mice. Fragment lengths are indicated in kilobases.

male and female mouse DNA digested with certain restriction enzymes.[28] An equivalent result has been obtained with cloned satellite DNA from the snake species *Elaphe radiata*.[9] The DNA of XX*Sxr* mice showed a similar pattern to that of normal males. The simple repeated constituents of Bkm DNA, the GATA-GACA sequences, have recently given improved hybridization results when oligonucleotide probes were used.[15] Given the resolution power of in-gel hybridization with oligonucleotide probes, we attempted to distinguish normal male and XX*Sxr* DNA. Since this consistently proved possible, we extended our investigations into two backcross pedigrees.[38] XY*Sxr* and XY*Sxr'* males have been backcrossed onto C57BL/6 (B6) females for approximately ten generations. Mice from these pedigrees were coded, and their DNA isolated and hybridized with the GATA-GACA repeat probes and several derivatives thereof.

Using the two most informative restriction enzymes (Hae III and Hinf I), DNAs from about 100 different mice were digested and restriction fragments separated by agarose gel electrophoresis. In-gel hybridization with GATA-GACA oligonucleotide probes revealed some 15 different hybridization patterns, some of which were repeatedly observed (Figures 1 and 2). At first sight, some of the patterns seemed closely related. The explanation became obvious after decoding the DNAs of the mice. The Sxr and Sxr' patterns were regularly inherited with only two exceptions (see next paragraph and Figures 3 and 4). In order to generate the carrier patterns of XY*Sxr* and XY*Sxr'* mice, the patterns of the normal Y chromosome and that of Sxr or Sxr' DNA were superimposed (see Figures 1 to 4: patterns 1 + 2 = 3; 1 + 4 = 5; 1' + 2' = 3'). The normal Y chromosomes gave a consistent hybridization pattern, different for the two pedigrees (patterns 1 and 1'). This polymorphism is not surprising since in the CB pedigree the Y chromosome is descended from the original Y chromosome in which the *Sxr* mutation was first detected.[5] In contrast in the CB' pedigree, a new Y chromosome from a stock of mice carrying *Tabby* was introduced before the

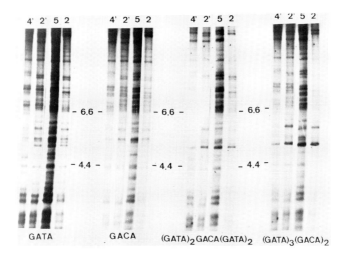

FIGURE 2. Hinf I-digested mouse DNA from XY, XXSxr, XXSxr', and XYSxr males can be distinguished by GATA-GACA oligonucleotide probes. For methodological details see legend to Figure 1. Numbers on top of the lanes identify the patterns observed; 2 and 2', XXSxr and XXSxr'; 4', XXSxr' variant; and 5, XYSxr carrier. Fragment lengths are indicated in kilobases.

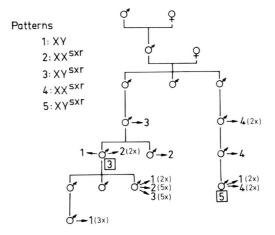

FIGURE 3. Pedigree of the CB strain in which XYSxr males are continuously backcrossed onto C57BL/6 female mice. Only the male XYSxr carriers are indicated after the first two backcrosses. Boxed numbers mean that the XYSxr carrier status (pattern no. 3) was confirmed by hybridization analysis; numbers at the tip of arrows identify hybridization patterns observed in the progeny of the respective carrier male (numbers in brackets indicate in how many progeny the patterns were observed). Note that XXSxr pattern no. 2 was regularly inherited in the left-hand section of the pedigree while the variant pattern no. 4 is seen only in the right-hand section.

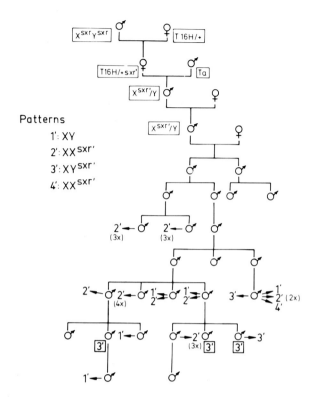

Patterns

1': XY

2': XX$^{Sxr'}$

3': XY$^{Sxr'}$

4': XX$^{Sxr'}$

FIGURE 4. Pedigree of the CB' strain where the *Sxr'* mutation is backcrossed onto C57BL/6 female mice. The *Sxr'* variant was first demonstrated in the daughter of an X*Sxr* Y*Sxr* fertile male which had been mated to a female carrying the T16H translocation. This fertile *Sxr*-carrying female[37] typed unexpectedly negative for the H-Y antigen and so did her XX*Sxr'* male progeny. Mating the T16H/ + *Sxr'* female to a male with the *Tabby* (*Ta*) coat color mutation resulted in a fertile male carrying the Sxr' factor. Successive backcrossing onto C57BL/6 female mice resulted in the pedigree shown (in the lower part only the XY*Sxr'* carrier males are depicted). 3' means that the XY*Sxr'* carrier status has been confirmed by hybridization pattern analysis. Numbers indicated by arrows identify hybridization patterns observed in the progeny of the respective carrier male (numbers in brackets indicate in how many offspring the patterns were actually observed). Note that the XX*Sxr'* pattern no. 4 was only observed once and no further generations of the respective carrier male could be examined.

backcross program began (see Figure 4). We observed one variant Sxr and one variant Sxr' pattern arising within the backcross pedigrees. One of the variant patterns was again regularly inherited by the offspring of the respective carrier mice; the other was apparently transmitted only to an XX male and hence was lost. Thus, at least 1 unequal meiotic recombination took place among 28 meiotic events in the CB pedigree, and 1 among 35 in the CB' pedigree (~3%). We have no positive indication that any of the normal Y chromosomes of the two pedigrees were involved in unequal exchange: all 24 and 22 Y-chromosomal patterns in the CB and CB' pedigrees, respectively, were invariant. In independently inbred B6 mice we also found no variable patterns, though the number of meiotic events covered cannot be quantified exactly. Studies to clarify the rate of unequal recombination in the Y chromosome of inbred strains are currently under way.

IV. ARE GATA-GACA SEQUENCES ACTIVELY INVOLVED IN UNEQUAL RECOMBINATION IN THE Sxr AND Sxr′ REGIONS?

Gene order is a characteristic of all members of a certain species. But eukaryotic genomes are dynamic structures that are reorganized during evolution and in special cases in ontogeny.[29] Homologous recombination events break and rejoin the DNA at exactly the same positions. Consequently, so-called unequal recombination between homologous sequences leads to deletion and/or duplication, and usually to deleterious effects for the individual. Homologous recombination is thought to occur anywhere along the chromosome. On the other hand, programmed site-specific unequal recombination is limited to a small number of loci, e.g., to immunoglobulin and to T lymphocyte antigen receptor genes.[30] A few DNA recognition sequences have been invoked in hotspots of recombination, e.g., in prokaryotes but also in vertebrates including man. The eight base pair Chi (crossing-over hotspot instigator) sequence of Lambda phage[31] shares some sequence homology with the 16 base pair core of a human hypervariable "minisatellite":[32]

5′-GCTGGTGG   (Lambda Chi sequence)

  | | | | | |

5′-GGAGGTGGGCAGGAXG   (Human minisatellite)

Hotspots for the initiation of gene conversion in the mammalian major histocompatibility complex (MHC) have been assumed by several investigators, e.g., Brégégère.[33] Steinmetz[34] and Kobori et al.[35] went on to search for the actual sequences involved in the event. A so-called E_β hotspot in the mouse is characterized by the simple quadruplet repeat $(CAGG)_{7-9}$; an $A\beta 3/A\beta 2$ hotspot harbors the $(CAGA)_6$ sequence in certain inbred mouse strains [note the identity with the aforementioned $(GACA)_n$ sequences]. When there were only four CAGA copies present, the hotspot was reported to be inactive.[34]

What kind of bearing, if any, do these findings have on the interpretation of unequal recombination in mouse Y chromosome DNA? A frequency of unequal recombination of approximately 3% in both the CB and CB′ pedigrees seems at first sight a very high rate. An even higher rate (7%) of unequal recombination has, however, been found in the pseudoautosomal Y-chromosomal region of the mouse.[36] Though formal proof is still lacking, direct involvement of repetitive sequences in unequal recombination has also been suggested in this case.[36] Hence GATA-GACA sequences, if not instrumental in the recombination mechanism itself, can at least be used as effective tools for the analysis of unequal recombination.

ACKNOWLEDGMENTS

We thank Andrea Becker, Indrajit Nanda, and Roland Studer for their experimental contributions and Mrs. G. Eichhorn for typing this manuscript, an early version of which was discussed with Roland Studer and Indrajit Nanda.

REFERENCES

1. **Eicher, E. M. and Washburn, L. L.,** Genetic control of primary sex determination in mice, *Annu. Rev. Genet.,* 20, 327, 1986.
2. **Jacobs, P. A. and Strong, J. A.,** A case of human intersexuality having a possible XXY sex determining mechanisms, *Nature (London),* 183, 302, 1959.

3. **Ford, C. E., Jones, K. W., Polani, P. E., de Almeida, J. C., and Briggs, J. H.,** A sex chromosome anomaly in a case of gonadal dysgenesis (Turner's syndrome), *Lancet,* 1, 711, 1959.

4. **Welshons, W. J. and Russell, L. B.,** The Y chromosome as the bearer of male determining factors in the mouse, *Proc. Natl. Acad. Sci. U.S.A.,* 45, 560, 1959.

5. **Cattanach, B. M., Pollard, C. E., and Hawkes, S. G.,** Sex-reversed mice: XX and XO males, *Cytogenetics,* 10, 318, 1971.

6. **Singh, L. and Jones, K. W.,** Sex reversal in the mouse (*Mus musculus*) is caused by a recurrent nonreciprocal crossover involving the X and an aberrant Y chromosome, *Cell,* 28, 205, 1982.

7. **Evans, E. P., Burtenshaw, M., and Cattanach, B. M.,** Cytological evidence for meiotic crossing-over between the X and Y chromosome of male mice carrying the sex reversing (Sxr) factor, *Nature (London),* 300, 443, 1982.

8. **Burgoyne, P. S.,** Genetic homology and crossing-over in the X and Y mammals, *Hum. Genet.,* 61, 85, 1982.

9. **Epplen, J. T., McCarrey, J. R., Sutou, S., and Ohno, S.,** Base sequence of a cloned snake W-chromosome DNA fragment and identification of a male-specific putative mRNA in the mouse, *Proc. Natl. Acad. Sci. U.S.A.,* 79, 3798, 1982.

10. **Chandra, H. S.,** Sex determination: a hypothesis based on noncoding DNA, *Proc. Natl. Acad. Sci. U.S.A.,* 82, 1165, 1985.

11. **Skinner, D. M., Beattie, W. G., Blattner, F. R., Stark, B. P., and Dahlberg, J. E.,** The repeat sequence of a hermit crab satellite deoxyribonucleic acid is $(-T-A-G-G-)_n$, $(-A-T-C-C-)_n$, *Biochemistry,* 13, 3930, 1974.

12. **Doolittle, W. F.,** Genome evolution in review, *Trends Genet.,* 3, 82, 1987.

13. **Hochgeschwender, U., Weltzien, H. U., Eichmann, K., Wallace, R. B., and Epplen, J. T.,** Preferential expression of a defined T-cell receptor β-chain gene in hapten-specific cytotoxic T-cell clones, *Nature (London),* 322, 376, 1986.

14. **Singh, L., Purdom, I. F., and Jones, K. W.,** Satellite DNA and evolution of sex chromosomes, *Chromosoma,* 59, 43, 1976.

15. **Schäfer, R., Ali, S., and Epplen, J. T.,** The organization of the evolutionarily conserved GATA/GACA repeats in the mouse genome, *Chromosoma,* 93, 502, 1986.

16. **Epplen, J. T. and Ohno, S.,** On DNA, RNA and sex determination, in *Selected Topics in Molecular Endocrinology,* Lau, Y.-F., Ed., Oxford University Press, New York, 1988.

17. **Traut, W.,** Hypervariable Bkm DNA loci in a moth, Ephestia kuehniella: does transposition cause restriction fragment length polymorphism (RFLP)?, *Genetics,* 115, 493, 1987.

18. **Traut, W. and Epplen, J. T.,** unpublished observations, 1987.

19. **Ali, S., Müller, C. R., and Epplen, J. T.,** DNA fingerprinting by oligonucleotide probes specific for simple repeats, *Hum. Genet.,* 74, 239, 1986.

20. **Arnemann, J., Jakubiczka, S., Schmidtke, J., Schäfer, R., and Epplen, J. T.,** Clustered GATA repeats (Bkm sequences) on the human Y chromosome, *Hum. Genet.,* 73, 301, 198.

21. **Vogel, W., Steinbach, P., Djalali, M., Mehnert, K., Ali, S., and Epplen, J. T.,** Chromosome no. 9 of *Ellobins lutesceus* is the X-chromosome, *Chromosoma,* 96, 112, 1988.

22. **Nanda, I., Neitzel, H., Sperling, K., Studer, R., and Epplen, J. T.,** Simple GA$_C^T$A repeats characterize the X chromosomal heterochromatin of *Microtus agrestis,* European field role (Rodentia, Cricetidae), *Chromosoma,* 96, 213, 1988.

23. **Singh, L., Phillips, C., and Jones, K. W.,** The conserved nucleotide sequences of Bkm, which define Sxr in the mouse, are transcribed, *Cell,* 36, 111, 1984.

24. **Schäfer, R., Bölz, E., Becker, A., Bartels, F., and Epplen, J. T.,** The expression of the evolutionarily conserved GATA/GACA repeats in mouse tissues, *Chromosoma,* 93, 496, 1986.

25. **Epplen, J. T., Cellini, A., Romero, S., and Ohno, S.,** An attempt to approach the molecular mechanisms of primary sex determination: W- and Y-chromosomal conserved simple repetitive DNA sequences and their differential expression in mRNA, *J. Exp. Zool.,* 228, 305, 1983.

26. **Bennett, D., Mathieson, B. J., Scheid, M., Yanagisawa, K., Boyse, E. A., Wachtel, S., and Cattanach, B. M.,** Serological evidence for H-Y antigen in Sxr, XX sex-reversed phenotypic males, *Nature (London),* 265, 255, 1977.

27. **McLaren, A., Simpson, E., Tomonari, K., Chandler, P., and Hogg, H.,** Male sexual differentiation in mice lacking H-Y antigen, *Nature (London),* 312, 552, 1984.

28. **Singh, L., Purdom, I. F., and Jones, K. W.,** Conserved sex-chromosome-associated nucleotide sequences in eukaryotes, *Cold Spring Harbor Symp. Quant. Biol.,* 45, 805, 1981.

29. **Borst, P. and Greaves, D. R.,** Programmed gene rearrangements altering gene expression, *Science,* 235, 658, 1987.

30. **Epplen, J. T., Chluba, J., Hardt, C., Hinkkanen, A., Steimle, V., and Stockinger, H.,** Mammalian T-lymphocyte antigen receptor genes: genetic and nongenetic potential to generate variability, *Hum. Genet.,* 75, 300, 1987.

31. **Smith, G. R., Kunes, S. M., Schultz, D. W., Taylor, A., and Trinan, K. L.,** Structure of Chi hotspots of generalized recombination, *Cell,* 24, 429, 1981.

32. **Jeffreys, A. J., Wilson, V., and Thein, S. L.,** Hypervariable minisatellite regions in human DNA, *Nature (London),* 314, 67, 1985.

33. **Brégégère, F.,** A directional process of gene conversion is expected to yield dynamic polymorphism associated with stability of alternative alleles in class I histocompatibility antigens gene family, *Biochimie,* 65, 229, 1983.

34. **Steinmetz, M.,** Hotspots of homologous recombination in mammalian genomes, *Trends Genet.,* 3, 7, 1987.

35. **Kobori, J. A., Strauss, E., Minard, K., and Hood, L.,** Molecular analysis of the hotspot of recombination in the murine major histocompatibility complex, *Science,* 234, 173, 1986.

36. **Harbers, K., Soriano, P., Müller, U., and Jaenisch, R.,** High frequency of unequal recombination in pseudoautosomal region shown by proviral insertion in transgenic mouse, *Nature (London),* 324, 682, 1986.

37. **McLaren, A. and Monk, M.,** Fertile females produced by inactivation of an X chromosome of *sex-reversed* mice, *Nature (London),* 300, 446, 1982.

38. **McLaren, A.,** unpublished results, 1987.

Chapter 17

tda-1 XY SEX REVERSAL IN THE MOUSE

Claude M. Nagamine and Gloria C. Koo

TABLE OF CONTENTS

I. INTRODUCTION

Several years ago, Eicher and colleagues[1-3] reported that XY hermaphrodites and XY females were produced when the Y chromosome of wild *Mus musculus domesticus* mice was placed on the C57BL/6 (B6) genetic background. The XY sex reversal occurred only in the first and subsequent backcross generations; the F_1 generation was phenotypically normal. This finding was confirmed later by Gropp[4] and more recently by our laboratory.[5-7] The data demonstrated that the mammalian Y chromosome itself is not completely male determining and that other genes are probably involved in testis determination.

On the basis of these data, Eicher and colleagues[1-3] proposed that the testis-determining gene on the Y^{Dom} chromosome (Tdy^{Dom}) interacts with a dominant testis-determining autosomal gene, $Tda\text{-}1^{Dom}$, to produce normal XY males in the (*M. m. domesticus* × B6) F_1 generation. However, Tdy^{Dom} interacts improperly with the putative recessive B6 gene called $tda\text{-}1^{B6}$. All mice that possess Tdy^{Dom} and are homozygous for $tda\text{-}1^{B6}$ develop as hermaphrodites or as XY females.

Since the original reports, little information was published to substantiate the hypothesis, and it was not known if these genetic interactions occurred in other inbred strain combinations. Recently, we developed a consomic strain of B6 mice (B6.Y^{Dom}) possessing the Y chromosome of wild *M. m. domesticus* (Tirano, Italy), and extended the breeding of the B6.Y^{Dom} strain to other inbred strains.[5-7] Using this animal model, we gained valuable insight into *tda-1* XY sex reversal and normal sex determination. Our ultimate goal is the eventual mapping of the mammalian sex-determining genes.

II. GENERATION OF THE B6.Y^{DOM} STRAIN

The B6.Y^{Dom} strain was obtained by mating *M. m. domesticus* males to B6 females. The F_1 males were backcrossed to B6 females.[5] In subsequent generations, "male" phenotypes, which in fact are cryptic hermaphrodites[6] or overt hermaphrodites, were backcrossed to B6 females. The backcrossing to B6 females is being continued; we are currently in the N13 backcross generation.

Table 1 shows the accumulated breeding data of eight B6.Y^{Dom} backcross generations. All phenotypes are determined postnatally according to the gross anatomy of the gonads and accessory ducts as seen under the dissecting microscope. "Male" phenotype animals possessed bilateral testicular gonads with associated epididymides and vasa deferentia. Female phenotypes possessed bilateral ovarian gonads with associated oviducts and uteri. Overt hermaphrodites possessed either an ovarian gonad and contralateral testicular gonad with respective accessory structures, or an ovarian or testicular gonad with a contralateral ovotestis associated with both Müllerian and Wolffian duct derivatives.[5]

The breeding data confirmed the original observations on the generation of XY hermaphrodites and XY females.[1-4] Furthermore, in each backcross generation, fertile "male" phenotypes and hermaphrodites, although uncommon, allowed us to maintain the breeding of the B6.Y^{Dom} strain. We concluded that the B6.Y^{Dom} strain is stable and is in no danger of extinction. Two points of interest arise from the B6.Y^{Dom} breeding data:

1. The same Y^{Dom} chromosome could give rise to XY "male" phenotypes, XY hermaphrodites, and XY females, all on the same B6 genetic background.
2. The percentages of "male", hermaphroditic and female phenotypes from the N4 generation onward were relatively constant, suggesting that the penetrance of the Y^{Dom} chromosome could be modulated by other genes. Indeed, preliminary data suggested that a gene or group of genes in the pseudoautosomal region might be involved.[7]

Table 1
POSTNATAL PROGENY OF B6.Y^{Dom}
STRAIN OF THE N2 TO N9 GENERATIONS
CATEGORIZED BY MALE,
HERMAPHRODITIC, OR FEMALE
PHENOTYPES[5]

Generation	Male (%)	Hermaphroditic (%)	Female (%)
N2	10 (34)[a]	2 (7)	17 (59)
N3	33 (40)	5 (6)	44 (54)
N4	49 (25)	26 (13)	120 (62)
N5	53 (21)	38 (15)	163 (65)
N6	47 (26)	22 (12)	111 (62)
N7	10 (14)	13 (19)	46 (67)
N8	9 (10)	14 (15)	68 (75)
N9	8 (13)	8 (13)	47 (74)

[a]   Number of offspring (percent of total).

III. MATINGS OF *MUS MUSCULUS DOMESTICUS* OR B6.Y^{Dom} WITH OTHER INBRED STRAINS

We extended the breeding of *M. m. domesticus* to four other inbred strains: SJL, BALB/c, BALB.B10, and C3H/An (BALB.B10 is a congenic with BALB/c, but carries the *H-2b* major histocompatibility haplotype). Although crosses with B6 and SJL females produced fertile F$_1$ males, crosses with the other three strains resulted in complete sterility (BALB/c, BALB.B10) or in very low numbers of sperm of inferior quality (C3H/An). The sterility/low fertility problem is probably not due to the Y^{Dom} chromosome per se, since B6.Y^{Dom} males produced fertile F$_1$ males when mated to C3H/An females[5] (see below). Therefore, sterility is probably due to incompatible interaction of autosomal genes of *M. m. domesticus* with those of BALB/c and C3H/An.

Crossing *M. m. domesticus* males to SJL females resulted in fertile F$_1$ males. When backcrosses to SJL females were carried out, no overt hermaphrodites were identified in a total of 29 N2-N4 postnatal progeny, suggesting that this strain possessed the dominant *Tda-1* allele previously known only in wild *M. m. domesticus*.[5]

To further determine the *tda-1* allele of other inbred strains, and to confirm the status of SJL, AKR, BALB/c, C3H/An, and C3H/He females with respect to the dominant *Tda-1* allele, days 14 to 15 fetuses were examined for aberrant testicular differentiation. If the strains possessed the *tda-1^{B6}* allele, XY sex reversal identical to that seen in B6.Y^{Dom} would be expected in F$_1$ fetal progeny. However, if the strains possessed the dominant *Tda-1* allele, all gonads should be normal. If testicular differentiation were abnormal but not identical to B6.Y^{Dom}, then either the female strain possessed different *tda-1* alleles or modifying genes were present. As before, all (SJL × B6.Y^{Dom}) F$_1$ fetuses were normal, confirming presence of the dominant *Tda-1* allele in SJL, as discussed above.

The *tda-1* allele of the other strains was less clear. Although crosses with AKR, BALB/c, C3H/An, and C3H/He all resulted in aberrant testes, the effect was considerably less than that seen in the B6.Y^{Dom} strain, suggesting that either different *tda-1* alleles exist in these strains or other genes were modifying *tda-1* XY sex reversal. These results are summarized in Table 2.

The presence of a recessive *tda-1* allele in AKR is noteworthy since this strain, along with SJL, possesses a *M. m. domesticus* Y chromosome.[8-11] No evidence of aberrant testicular differentiation has been reported in AKR or SJL strains. Whereas absence of sex reversal

Table 2
SUMMARY OF *tda-1* XY SEX-REVERSAL BREEDING DATA

Mating

Mus musculus domesticus ♂	×	C57BL/6 ♀	Fertile F_1 males
		BALB/c ♀	Sterile F_1 males
		BALB.B10 ♀	Sterile F_1 males
		C3H/An ♀	Sterile F_1 males
B6.Y^{Dom}	×	BALB/c ♀	Aberrant testes
		C3H/An ♀	Aberrant testes
		C3H/He ♀	Aberrant testes
		AKR ♀	Aberrant testes
		SJL ♀	Normal testes
(*M. m. domesticus* ♂ × B6 ♀) ♂ × B6 ♀			Aberrant testes
(*M. m. domesticus* ♂ × SJL ♀) ♂ × SJL ♀			Normal testes

Table 3
NUMBER OF MALE, FEMALE, AND ABERRANT MALE PROGENY SIRED BY N3 MALES OF B6.Y^{SJL} AND B6.Y^{AKR}, AS DETERMINED EITHER (a) ON DAY 14-15 OF GESTATION OR (b) POSTNATALLY[5]

Sire	N*	Males (%)	Females (%)	Aberrant males (%)
(a) B6.Y^{SJL}	14	43 (49.5)	43 (49.5)	1 (1)
B6.Y^{AKR}	6	1 (2)	19 (42)	25 (56)
(b) B6.Y^{SJL}	8	22 (49)	23 (51)	0
B6.Y^{AKR}	8	18 (50)	18 (50)	0

*   Number of litters.

in SJL could be due to the presence of the dominant *Tda-1* allele, a similar explanation cannot be given for AKR. As outlined below, the system turns out to be rather more complex than expected.

IV. ANALYSIS OF SIMILAR Y^{DOM} CHROMOSOMES FROM AKR AND SJL MALES

It has been hypothesized that *tda-1* XY sex reversal is due to the improper interaction between the testis-determining gene on the *M. m. domesticus* Y chromosome (Tdy^{Dom}) and the *tda-1^{B6}*. Eicher[1] has proposed that at least two *Tdy* alleles are present (Tdy^{Dom}, and Tdy^{B6}). Recombinant DNA data have demonstrated that SJL and AKR strains both possess the *M. m. domesticus* Y chromosome.[8-11] Whether the *Tdy* allele in all strains of *M. m. domesticus* origin are identical is not known. In order to answer this question, we placed the Y chromosomes of SJL and AKR on the B6 background and examined the progeny for *tda-1* XY sex reversal. B6.Y^{AKR} and B6.Y^{SJL} males of the N3 backcross generation were backcrossed to B6 females and the fetal progeny examined at days 14 to 15 of gestation and postnatally.

Table 3 summarizes the results of these experiments: 1/87 (1%) of SJN3 (SJLY on B6 background) fetuses had aberrant testes. In contrast, 25/45 (56%) of AN3 (AKRY on B6 background) fetuses had aberrant testes.[5] It appears that the Y^{AKR} chromosome produced considerably more aberrant testes on the B6 background. Morphologically and histologically, the aberrant gonads were similar to fetal ovotestes seen in crosses between B6.Y^{Dom} and inbred strains; i.e., the ovarian-type regions were localized at the cranial and caudal poles of the gonad, the tunica albuginea over the affected regions were either thin or absent, and meiotic germ cells were less numerous than were those in B6.Y^{Dom} ovotestes.

Surprisingly, when B6.Y^{AKR} or B6.Y^{SJL} progeny were examined postnatally, no evidence of hermaphroditism was found either at the morphological or histological level and the sex ratio was 1:1.[5] Thus, in the B6.Y^{AKR} strain, although testicular development could be severely affected at days 14 to 15 of fetal development, the effect was impermanent and the testes appeared capable of recuperating. It was possible that the less pronounced effect on testicular differentiation in B6.Y^{AKR}, as compared to B6.Y^{Dom} in the original report,[5] was due to residual AKR genes since only the N4 generation was examined. However, ongoing backcrossing of the B6.Y^{AKR} strain to B6 (present backcross generation = N7) resulted in neither an amelioration nor worsening of the original reported "fetal hermaphroditism", and overt hermaphrodites still have not been found postnatally.

To date, the data therefore suggest that the differences seen originally at the N4 generation between B6.Y^{Dom}, B6.Y^{AKR}, and B6.Y^{SJL} are due to differences in the Y chromosomes themselves and not to residual AKR and SJL genes. The capabilities of these three Y chromosomes in inducing aberrant testes in B6 can be placed in a hierarchy of Y^{Dom} > Y^{AKR} > Y^{SJL}. We conclude that either different *Tdy* alleles exist in strains derived from *M. m. domesticus* or other genes on the Y chromosome, located centromeric from the pseudoautosomal region, are present that modify the effect of *Tdy*.

V. MOLECULAR ANALYSIS OF THE Y CHROMOSOMES OF *DOMESTICUS* ORIGIN

To test for differences in the Y chromosomes, we used the recombinant DNA probe specific for the murine Y chromosome, pY353/B,[9] to analyze the restriction fragment-length polymorphism (RFLP) patterns of the genomic DNA of various mouse strains possessing the Y chromosome of *M. m. domesticus* (AKR, B6.Y^{AKR}, B6.Y^{Dom}, BFM/2, NCS, SJL, SWR, WLA/76, WMP). Our results showed that the Y chromosome of *M. m. domesticus* can be divided into at least five different groups:

1. B6.Y^{Dom}, *M. m. domesticus* (Tirano, Italy)
2. AKR, B6.Y^{AKR}, NCS
3. SJL, SWR
4. BFM/2 (Montpellier, France), WLA/76 (Toulouse, France)
5. WMP (Monastir, Tunisia)

The RFLP data confirmed our genetic data indicating that the Y chromosomes of B6.Y^{Dom}, B6.Y^{AKR}, and B6.Y^{SJL} are different since they differ in their ability to induce aberrant testicular development (see above). The RFLP data also predict that when the Y chromosomes of NCS and SWR are backcrossed to B6, the degree of aberrant testicular development should be identical to that obtained with B6.Y^{AKR} and B6.Y^{SJL}, respectively.

To further study the cause of *tda-1* XY sex reversal, genomic DNA of males, hermaphrodites, and XY females in the B6.Y^{Dom} strain were prepared and examined for RFLP with DNA probes specific for the murine Y chromosome (both positive and negative for the *Sxr* mutation) and with probes common to X and Y chromosomes but absent on autosomes. To date we have found no evidence of any differences in the Y chromosomes that give rise to the different XY phenotypes in the B6.Y^{Dom} strain. These findings confirm previous karyotyping and genetic data suggesting that the Y chromosome in the wild *M. m. domesticus* is not abnormal.[1-3,5]

VI. HISTOLOGICAL ANALYSIS OF THE GONADS OF B6.Y^{Dom} MICE

Detailed description of the gonads of XY hermaphrodites and females were given by Nagamine et al.[6] The main findings are summarized here:

1. Although testicular phenotypes were identified by gross examination of fetal XY gonads (days 14 to 18), histological studies of 34 gonads revealed that they were all ovotestes.
2. Postnatally, however, true testes were found and they appeared normal. The apparent discrepancy could be due to the combination of (a) incorporation of ovarian cells into testicular tissues; (b) degeneration of ovarian cells; and (c) difficulty in identifying ovarian cells.
3. XY ovaries exhibited hyperplasia of the lutea cells, as though the premature loss of XY oocytes had led to persistence of corporae lutea.
4. XY ovotestes always retained the ovarian tissue at the cranial and caudal ends and the testicular tissue in the center. This pattern of distribution appeared to be common among various etiologies of murine ovotestes,[6] suggesting a testis-inducing factor in the center of the gonad.

VII. CONCLUSIONS

Our genetic analyses of the *tda-1* XY sex reversal confirmed earlier findings. In addition, we have made the following observations:

1. That the B6.Y^{Dom} strain is genetically stable (N13 backcross generation), and that the various XY phenotypes (''males'', overt hermaphrodites, XY females) appear to be constant after the N4 generation.
2. That *tda-1* or *tda-1*-related XY sex reversal is not unique to B6, although the effect seen in other strains tested is considerably less marked.
3. That the SJL strain possesses the dominant *Tda-1* allele previously known only in wild *M. m. domesticus.*
4. That all Y chromosomes of *M. m. domesticus* origin do not necessarily induce *tda-1* XY sex reversal when placed into B6, and that a hierarchy in the induction of XY sex reversal can be established: Y^{Dom} (Tirano) > Y^{AKR} > Y^{SJL}.
5. That the Y chromosomes of B6.Y^{Dom}, AKR, and SJL also possess different RFLP patterns with the murine Y-chromosomal probe, pY353/B, thus confirming our genetic data.
6. That histological and morphological similarities between the ovotestes in B6.Y^{Dom} and those found in other murine hermaphroditic conditions of various etiology suggest that all ovotestes may be formed through a common pathway.

At present little is known about the causes of XY females in man. The *tda-1* XY sex-reversal data suggest that perhaps incompatible Y chromosomal products could account for the etiology of at least some human XY females. The B6.Y^{Dom} and B6.Y^{AKR} strains will continue to be powerful tools for molecular and genetic studies to further our understanding of the genes and gene products involved in sex determination and sexual differentiation.

ACKNOWLEDGMENTS

We thank Dr. Michael Potter for providing the stock *M. m. domesticus* under NCI contract no. 1-CB-25584 and Dr. Colin Bishop for use of recombinant DNA probes. This research was funded in part by National Institutes of Child Health and Human Development grants HD-18067 and HD-18669, NIH grant CA-08748 and a grant from the Association pour le Developpement de l'Institut Pasteur.

REFERENCES

1. **Eicher, E.,** Primary sex determining genes in mice, in *Prospects for Sexing Mammalian Sperm,* Amann, R. P. and Seidel, G. E., Jr., Eds., Colorado Associated University Press, Boulder, 1982, 121.

2. **Eicher, E., Washburn, L. L., Whitney, J. B., III, and Morrow, K. E.,** *Mus poschiavinus* Y chromosome in the C57BL/6J murine genome causes sex reversal, *Science,* 217, 535, 1982.

3. **Eicher, E. and Washburn, L. L.,** Inherited sex reversal in mice: identification of a new primary sex-determining gene, *J. Exp. Zool.,* 228, 297, 1983.

4. **Gropp, A.,** Sex reversal in XY mice by a foreign Y-chromosome, in *Prospects for Sexing Mammalian Sperm,* Amann, R. P. and Seidel, G. E., Jr., Eds., Colorado Associated University Press, Boulder, 1982, 136.

5. **Nagamine, C. M., Taketo, T., and Koo, G. C.,** Studies on the genetics of *tda-1* XY sex reversal in the mouse, *Differentiation,* 33, 223, 1987.

6. **Nagamine, C. M., Taketo, T., and Koo, G. C.,** Morphological development of the mouse gonad in *tda-1* XY sex reversal, *Differentiation,* 33, 214, 1987.

7. **Nagamine, C. M. and Koo, G. C.,** Evidence that the X-Y pairing/recombination region may be involved in *tda-1* inherited sex reversal, in *Genetic Markers of Sex Differentiation,* Plenum Press, New York, in press, 1988.

8. **Lamar, E. E. and Palmer, E.,** Y encoded, species-specific DNA in mice: evidence that the Y chromosome exists in two polymorphic forms in inbred strains, *Cell,* 37, 171, 1984.

9. **Bishop, C. E., Boursot, P., Baron, B., Bonhomme, F., and Hatat, D.,** Most classical *Mus musculus domesticus* laboratory mouse strains carry a *Mus musculus musculus* Y chromosome, *Nature (London),* 315, 70, 1985.

10. **Nishioka, Y. and Lamothe, E.,** Isolation and characterization of a mouse Y chromosomal repetitive sequence, *Genetics,* 113, 417, 1986.

11. **Nagamine, C. M.,** unpublished data, 1987.

Chapter 18

GENETICS OF THE XY FEMALE: A COMPARATIVE STUDY

Marijo Kent and Stephen Wachtel

TABLE OF CONTENTS

I. INTRODUCTION

Differentiation of the gonad is a critical event in sex determination. In the presence of testes, the embryo is masculinized, and in the presence of ovaries, the embryo is feminized — regardless of karyotype. Errors of secondary sex differentiation occur when the gonad develops abnormally, or when the gonad develops in association with the sex chromosomes of the opposite sex. True hermaphroditism — development of testes and ovaries in the same individual — is a classic example of abnormal sex differentiation. In that case, the steroid hormones of both sexes are secreted, and the result is a mingling of the male and female secondary traits.

Among the mammals, differentiation of the gonad may be compromised by sex chromosome mosaicism, by loss of a sex chromosome, or by deletion or translocation of a critical sex-determining sequence. In the human, for example, XX/XY mosaics may become true hermaphrodites, XO embryos develop the stigmata of Turner's syndrome, and XYp⁻ embryos may become females with dysgenetic gonads.[1]

In this review, we shall address the syndrome of XY gonadal dysgenesis with particular attention to development of that condition in the horse. The "XY sex reversal syndrome" of the horse is widespread among several breeds. Affected mares have the karyotype of a stallion, 64,XY, but develop as females with varying degrees of masculinization. The syndrome may serve as a model for XY gonadal dysgenesis and related conditions in the mammals generally.[2]

II. GONADAL DYSGENESIS IN XO FEMALES

Evidently, two X chromosomes are required for normal development of the ovary.[3] In humans, development of the 45,X ovary resembles that of the normal 46,XX ovary for about the first trimester of gestation.[4] Then, there is a progressive loss of germ cells and a breakdown of ovarian architecture. The follicles become atretic and there is an increase in the amount of connective tissue. The gonad is represented at around the time of birth by a "streak" that is endocrinologically inert. As a result, 45,X women are sterile and develop the stigmata of Turner's syndrome, including Müllerian hypoplasia, sparse pubic and axillary hair, and minimal breast development.

A similar situation develops in other species. In the horse, for instance, embryos with XO become small infertile females with degenerative ovaries and eunuchoid musculature.[5,6] As in the human, the Müllerian derivatives tend to be underdeveloped. In rodent species such as the mouse, the XO female is fertile, but the reproductive life of the ovary is foreshortened; evidently that is due to premature loss of germ cells.[7]

III. GONADAL DYSGENESIS IN XY FEMALES IN THE HUMAN

Development of the XY ovary may parallel that of the XO ovary. In the human, early development of the XY ovary is normal, but there is a progressive loss of germ cells as in the XO ovary, and a breakdown of ovarian architecture. The follicles become atretic, and the XY ovary too is represented at around the time of birth by streaks of connective tissue devoid of endocrinologic activity,[8] although follicular structures may persist in certain cases into young adulthood.[9]

There are certain differences between 45,X gonadal dysgenesis and 46,XY gonadal dysgenesis. Women with the latter condition rarely exhibit the Turner stigmata; testicular tubules are found occasionally in the 46,XY dysgenetic gonad; and there is a high incidence of gonadoblastoma and dysgerminoma in females with a partial or intact Y.[10,11]

The condition is sometimes referred to as Swyer's syndrome.[12] Clinical signs include delayed puberty and amenorrhea. The phenotype may vary from feminized to masculinized although most affected individuals are eunuchoid. The gonads in 46,XY gonadal dysgenesis may be dysgenetic ovaries or dysgenetic testes.[13,14] Gonadotropin secretion is elevated. There is a depressed level of serum testosterone and estradiol. Development of mammary tissue is variable. The external genitalia are feminine, but cliteromegaly is not uncommon.[15]

A. Genealogy and Genetics

Human cases of XY gonadal dysgenesis may seem to occur spontaneously, but the fact that multiple cases are seen infrequently within human kindreds may reflect the difficulty of clinical pedigree analysis. Simpson et al.[16] reviewed 18 cases of familial XY gonadal dysgenesis. In three of the pedigrees, X-linked recessive or male-limited autosomal dominant genes were indicated. In the other 15 pedigrees, the only affected members were siblings, and the mode of inheritance could not be ascertained. Müller[17] reported the condition in three siblings and a maternal second cousin; this indicated X-linked inheritance. Simpson[18] analyzed 24 pedigrees by segregation analysis; in 12 of the pedigrees only one patient had been diagnosed; results in the other 12 pedigrees were not different from what might be expected in cases of male-limited (XY limited) autosomal recessive inheritance, although X-linked recessive and male-limited autosomal dominant genes could not be excluded.

Recently, Warner et al.[19] surveyed 55 cases of 46,XY gonadal dysgenesis in which affected individuals were evaluated for H-Y antigen: 41 of the patients typed H-Y$^+$ and 14 typed H-Y$^-$. Given the occurrence of H-Y$^+$ and H-Y$^-$ classes of 46,XY gonadal dysgenesis, and given the alternative modes of inheritance indicated above (X-linked recessive and autosomal dominant or autosomal recessive), Simpson et al.[16] proposed that the condition could have alternative etiologies. The proposition has been supported by recent reports that Y-specific DNA sequences were reduced in copy number in some cases of 46,XY gonadal dysgenesis, but not in others.[20,21] (There is an account of XY gonadal dysgenesis in a chacma baboon.[22] The animal was a large female, eunuchoid in appearance, and amennorheic. The external genitalia were feminine but hypoplastic; mammary glands were underdeveloped. Laparotomy revealed a hypoplastic uterus, normal tubes and streak gonads with ovarian stroma, and fibrous tissue with rete ovarii; there were no testicular structures. The karyotype was 42,XY — that of a normal male.)

IV. XY FEMALES IN CATTLE

There are reports of XY gonadal dysgenesis in the bovine. Chapman et al.[23] described an anestrus 60,XY Charlais heifer with small, underdeveloped ovaries, and Gustavsson et al.[24] described the 60,XY karyotype in five heifers in Swedish Red and White cattle. Four of the XY females were related and were diagnosed as representing cases of "testicular feminization", although male genital tracts were present internally and an enlarged clitoris was present externally. Pedigree analysis suggested X-linked or autosomal modes of inheritance. The fifth female, unrelated to the others, had feminine external genitalia with a slightly enlarged clitoris. The gonads were hypoplastic ovaries with anovular cords and rete ovarii.

V. FERTILE XY FEMALES OF THE WOOD LEMMING

In the Scandinavian wood lemming, *Myopus schisticolor*, there is a 4:1 sex ratio favoring the female. Study of the sex chromosomes reveals two classes of female: one has a female karyotype (32,XX), and the other has a male karyotype (32,XY). In this species, XY females are fertile and anatomically indistinguishable from XX females. But XY females produce only daughters, whereas XX females produce male and female offspring.[25]

The Y chromosome of the XY female is evidently intact and no different from the Y of the normal male. But G-banding analysis reveals patterns in the short arm of the X in the XY female that are different from those found in the short arm of the X in the XY male. So there are two kinds of X chromosomes in the wood lemming; one is denoted X and the other is denoted X*.[26] Embryos with X*Y become fertile females and embryos with XY become males.

Wiberg et al.[27] evaluated several wood lemmings of H-Y antigen. The X*Y females were typed H-Y[+]. On that basis, it was suggested that a gene for the H-Y gonadal receptor was located in the short arm (p) of the wood lemming X chromosome, and that the normal function of that gene had been compromised by a rearrangement in X*p.

VI. XY[Pos] FEMALES OF THE MOUSE

When the Y chromosome of the subspecies *Mus poschiavinus* was introduced into the C57BL/6 inbred background, XY[Pos] mice become phenotypic females or hermaphrodites; none developed normal testes.[28] Although XY[Pos] females rarely bred, hermaphrodites with male genitalia could breed. With increased backcrossing to the C57BL/6 strain, increased numbers of females were generated.

On the basis of these and related studies,[29] it was proposed that the Y chromosomes of the subspecies *M. poschiavinus* and *M. domesticus* carry testis-determining genes that are different from those carried on the Y of the C57BL/6 (B6) (*M. musculus*). This gene (*tdy*) functions in concert with another autosomal (or X-linked) gene called *tda-1*. The *tdy*[B6] gene "recognizes" the *tda-1*[B6] but not *tda-1*[Dom]. Thus, mice become hermaphrodites or females when they are homozygous for the B6 allele *tda-1*[B6] and bear the Y[Pos] or Y[Dom], and they become normal males when they are homozygous for *tda-1*[Dom]/*tda-1*[Dom] (or heterozygous for *tda-1*[Dom]/*tda-1*[B6]) and carry a Y[B6] or Y[Dom]. Accordingly, it was proposed that polymorphism of *tdy* could be a factor in the evolution of mammalian species.[28]

VII. THE XY SEX REVERSAL SYNDROME OF THE HORSE

Hughes and Trommerhausen-Smith[30] and Trommerhausen-Smith et al.[31] described gonadal dysgenesis in three XY mares and one 63,X/64,XY mosaic mare. In each case, the reproductive tract was intact and feminine, yet infantile. The gonads were small and firm, and consisted of undifferentiated ovarian stroma. Germinal cells were scarce, and estrus cycles were erratic. Sharp et al.[32] reported a similar case in 5-year-old mare. Despite minute ovaries (2 cm × 2 cm) and an underdeveloped uterus, the XY mare produced a normal 64,XX filly; she has produced five additional progeny, including an XY female, and is pregnant at the time of writing.

A. Pedigree Studies

More recently, we surveyed 286 horses from six unrelated pedigrees.[2] Among the mares, 38 had the male karyotype (64,XY). The study stemmed from eight original cases, referred for cytogenetic evaluation. The original eight XY mares were called Propositi I to VIII. Propositi I, IV, V, and VII were not related. They were found in Pedigrees 1, 2, 3, 4, and 5, respectively (Figure 1). Propositi II, III, and VIII were paternal half siblings and were found in Pedigree 6 (Figure 2).

The XY sex reversal syndrome is heritable. In Pedigrees 1 to 4, 8, and 9, transmission occurred through carrier mares, whereas in Pedigrees 6 and 7, transmission occurred through carrier stallions CS-I and CS-II. Only one XY mare was diagnosed in Pedigree 5; this pedigree is not included in the present analysis (Figure 1). In Pedigrees 1 to 4, 8, and 9, infertile XY mares occurred in 2, 3, 2, 3, 2, and 2 successive generations. These pedigrees

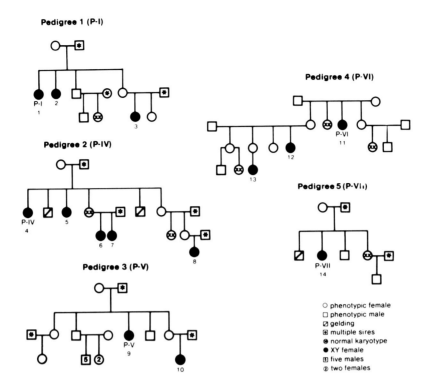

FIGURE 1. XY females of the horse, Pedigrees 1 to 5. In Pedigrees 1 to 4 transmission is through carrier females. The mode of transmission cannot be ascertained in Pedigree 5. (From Kent, M., Shoffner, R., Buoen, L., and Weber, A. F., *Cytogenet. Cell Genet.*, 42, 8, 1986. With permission.)

resemble human pedigrees with cases of XY gonadal dysgenesis in which X-linked recessive or autosomal sex-limited dominant genes were implicated.[16]

In Pedigree 6, all but two of the XY females were paternal half siblings. Transmission through a common stallion and resulting sex ratios provide further evidence that the revelant gene is autosomal, male (XY) limited and dominant (Figure 2).

B. Clinical Studies

On the basis of our clinical evaluation, we characterized four grades or classes of sex reversal as follows.[33]

Class I — XY females in this class resemble normal females. Testicular development is completely suppressed. The reproductive tract develops as in normal XX mares; ovaries may be normal in size and can be functional.

Class II — Mares in this category exhibit the classical signs of gonadal dysgenesis: the gonads are small and firm and there is little follicular development; germ cells are scarce. Müllerian development is normal, and the external genitalia are feminine without evidence of virilization.

Class III — Horses in this class are intersexes with gonadal dysgenesis. Gonads are streaks with rete testis in some cases and rete ovarii in others. The Müllerian derivatives are present but infantile on palpation and abnormal histologically; only a few nonfunctional glands are present in the laminar propria. A key feature of Class III sex reversal is cliteromegaly. Rectovaginal distance is increased by about 3 cm. Some of the mares in this class have measureable titers of serum testosterone, and all are tall and virilized when compared with XX mares. Behavior is generally agressive.

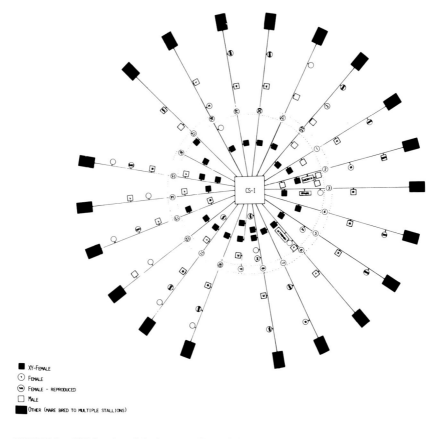

FIGURE 2.   XY females of the horse. Pedigree 6 showing transmission through carrier stallion (CS-1). This stallion produced 23 XY females when bred to 21 different unrelated mares. (From Kent, M., Shoffner, R., Buoen, L., and Weber, A. F., *Cytogenet. Cell Genet.*, 42, 8, 1986. With permission.)

Class IV — Horses in this class are virilized intersexes. The gonads are ovotesticular or testicular. The uterus and cervix are hypoplastic and may be absent in some cases. Cliteromegaly is likely to occur. In some cases, the clitoris resembles a glans penis. Rectovaginal distance is increased by more than 3 cm. These mares are tall and masculinized in appearance and behavior.

C. H-Y Antigen in XY Mares

H-Y ("male") antigen is identified by antibodies from male-grafted female mice. On account of widespread phylogenetic occurrence, it was proposed that H-Y antigen is involved in some critical aspect of gonadal differentiation. That proposal is challenged by the recent observation that certain XX*Sxr'* male mice are negative for H-Y as detected by the cell-mediated cytotoxicity (CMC) test;[34] but there is some question whether the H-Y antigen detected by CMC is the same as that detected by antibody,[35] and moreover Burgoyne et al.[36] have now shown that X*Sxr'*O male mice (H-Y$^-$) are remarkable for absence of spermatogenesis (and see Chapter 15).

We studied H-Y in white blood cells (WBC) from 18 sex-reversed XY mares, 8 normal mares, and 7 normal stallions. The WBC were reacted with monoclonal H-Y antibody gw-16 (IgG), and then after washing, with fluorochrome-conjugated goat-anti-mouse-Ig. After washing again, the cells were scored for label in a flow cytometer. By this technique, male cells were labeled preferentially in comparison with female cells, whereas the values for

Table 1
H-Y ANTIGEN IN XY MARES

Sex phenotype	Sex chromosomes	N	H-Y phenotype
Normal female	XX	8	8($-$)
Normal male	XY	7	7($+$)
Class I	XY	3	1($\pm$)[a]
Class II	XY	7	8($-$), 1($\pm$)
Class III	XY	5	1($\pm$), 4($+$)
Class IV	XY	3	3($+$)

[a] This is a retrospective assignment; fluorescent values were mid-way between the values obtained for normal male and female controls. This horse, a fertile XY mare, was typed H-Y$^\pm$ in the sperm cytotoxicity test by Sharp et al.[32]

Table 2
Bkm IN XY MARES

Sex phenotype	Sex chromosomes	N	In N cases Bkm hybridized to Autosome		Y
Normal female	XX	5	—		—
Normal male	XY	5	—		5
Class I	XY	1	1		—
Class II	XY	6	6[a]	>	6
Class III	XY	9	5	<	9[b]
Class IV	XY	3	—		3

[a] In four cases, the signal was strongest in the autosome, and in one case, in the Y.
[b] In five cases in which Bkm hybridized with DNA on the Y and on the autosome, the signal was stronger on the Y.

WBC from XY females ranged between those for normal males and normal females. Among 18 XY females, 8 were typed H-Y$^-$ and 7 were typed H-Y$^+$. In three cases, H-Y phenotype was intermediate between the phenotypes for normal male and female. It is perhaps worth noting that H-Y phenotypes in this study were generally correlated with the degree of virilization and testicular differentiation (Table 1).

D. Molecular Genetics of XY Sex Reversal in the Horse

We evaluated 19 sex-reversed mares with the Bkm probe which hybridizes preferentially with W-specific DNA of the snake and Y-DNA of the mouse.[37] In our preliminary *in situ* study, Bkm hybridized with sequences in the distal end of the Y in prometaphase spreads from normal stallions. No hybridization signal was detected in XX mares. The probe consists of a highly repetitive GATA-GATA sequence.

Our preliminary results with *in situ* hybridization suggest that perturbation (deletion ?) of the Y sequence has occurred in some of the XY mares (Table 2). In the more viril Class IV mares, hybridization of the Bkm probe was similar to that in normal stallions (distal end of the Y short arm). In Class III mares, some hybridization occurred at the telomeric region of a large acrocentric autosome — possibly chromosome 17 — but most signal occurred at the distal end of the Y short arm. In Class II mares, more hybridization occurred on the autosome and less on the Y, whereas in Class I mares, hybridization was specific for the autosome.

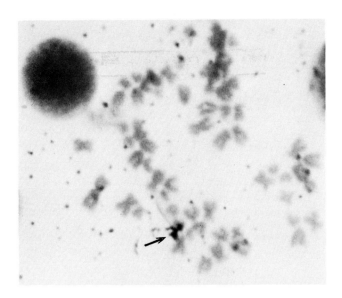

FIGURE 3.   Preferential hybridization of Bkm probe in XY stallion.
Arrow shows area of dense radiolabeling in Y chromosome.

These data are preliminary, but they indicate that a repetitive sequence related to Bkm is present on the Y chromosome in the normal stallion (Figure 3). The sequence is hetero-chromatic, and thus, may not be transcribed. Within this sequence, there is a small critical region of euchromatin. Even though the metaphase-banded horse Y chromosome appears to be entirely heterochromatic, a small euchromatic region can be seen in the same region of the proposed sequence in prophase or prometaphase spreads. It remains to be determined whether this region may code for a transcribable regulatory factor involved in sex determination.

VIII. CONCLUSIONS

Four classes of XY gonadal dysgenesis have been identified in the horse, corresponding to the degree of gonadal transformation and general sex phenotype. Retrospective analysis has revealed a correlation of sex phenotype, Bkm hybridization pattern, and H-Y phenotype. Females in Classes I and II are feminine (and Class I females are occasionally fertile). Most are H-Y$^-$, and all carry deleted Y chromosomes. Females in Classes III and IV are more masculine and infertile. Most are H-Y$^+$ and carry comparatively intact Y chromosomes.

The mechanisms leading to sex reversal of the XY embryo would seem without benefit in the horse because most XY females are infertile in that species. Yet we derive some insight in our study of XY sex reversal: it may be inferred that the condition, in one form at least, is due to deletion of Y sequences related to Bkm and associated with H-Y, and that the condition subsumes a gamut of intersex classes that can be categorized according to the amount of Y-specific DNA deleted or otherwise compromised. It follows that differentiation of the normal XY testis depends on the integrity of a Bkm-like sequence, which resides on the equine Y in association with genes that determine expression of H-Y antigen and dif-ferentiation of the testis.

Similar patterns of genetic and phenotypic heterogeneity have been described in human cases of XY gonadal dysgenesis. It remains to be determined whether the human condition too is due to deletion of a critical Y sequence (see Chapter 25), and whether the degree of feminization can be correlated with appearance of the sequence on an autosome and with H-Y phenotype.[19] In any event, the horse may serve as a useful model for the study of familial XY gonadal dysgenesis in the human and other outbred species.

It has been proposed that the alternative H-Y$^-$ and H-Y$^+$ classes in human XY gonadal dysgenesis represent loss of testis-determining sequences and mutation of the H-Y "receptor", respectively.[15] In light of the present discussion, it may be asked whether those classes represent, instead, alternative magnitudes of Y deletion.

An X-linked variety of XY gonadal dysgenesis has been demonstrated unambiguously in the wood lemming[26] and human,[8,16] but there is, so far, no clear-cut evidence for a similar condition in the horse. It is sufficient to say that X-linked gonadal dysgenesis is not ruled out by genealogic studies in the horse, and that X-linked genes in one species are likely to be X-linked in another ("Ohno's law").[38] So, X-linked forms of XY sex reversal are likely to be discovered in the horse; and it may be inferred that XY sex-reversed syndromes have multiple genetic bases among the mammals generally.

ACKNOWLEDGMENT

Supported in part by NIH Grant AI-23479.

REFERENCES

1. **Wachtel, S. S.,** *H-Y Antigen and the Biology of Sex Determination,* Grune & Stratton, New York, 1983.
2. **Kent, M., Shoffner, R., Buoen, L., and Weber, A. F.,** XY sex reversal syndrome in the domestic horse, *Cytogenet. Cell Genet.,* 42, 8, 1986.
3. **Short, R. V.,** Sex determination and differentiation of the mammalian gonad, *Int. J. Androl.,* Suppl. 2, 21, 1978.
4. **Singh, R. P. and Carr, D. H.,** The anatomy and histology of XO human embryos and fetuses, *Anat. Rev.,* 155, 369, 1966.
5. **Chandley, A. C., Fletcher, J., Rossdale, P. D., Peace, C. K., Thorne, J. P., Short, R. V., and Allen, W. R.,** Chromosome abnormalities as a cause of infertility in mares, *J. Reprod. Fertil.,* Suppl. 23, 377, 1975.
6. **Hughes, I. P., Benirschke, K., Kennedy, P. C., and Trommerhausen-Smith, A.,** Gonadal dysgenesis in the mare (case report of 5 cases), *J. Reprod. Fertil.,* Suppl. 23, 385, 1975.
7. **Lyon, M. F. and Hawker, S. G.,** Reproductive lifespan in irradiated and unirradiated chromosomally XO mice, *Genet. Res.,* 21, 185, 1973.
8. **Bernstein, R., Koo, G. C., and Wachtel, S. S.,** Abnormality of the X chromosome in human 46,XY female siblings with dysgenetic ovaries, *Science,* 207, 768, 1980.
9. **Russell, M. H., Wachtel, S. S., Davis, B. W., Cahill, L. T., Groos, E., Niblack, G. D., and Burr, I. M.,** Ovarian development in 46,XY gonadal dysgenesis, *Hum. Genet.,* 60, 196, 1982.
10. **Schellhas, H. F.,** Malignant potential of the dysgenetic gonad I, *Obstet. Gynecol.,* 44, 298, 1974.
11. **Simpson, J. L.,** *Disorders of Sexual Differentiation: Etiology and Clinical Delineation,* Academic Press, New York, 1976.
12. **Swyer, G. I. M.,** Male pseudohermaphroditism: a hitherto undescribed form, *Br. Med. J.,* 2, 709, 1955.
13. **Moltz, L., Schwartz, U., Pickartz, H., Hammerstein, J., and Wolf, U.,** XY gonadal dysgenesis: aberrant testicular differentiation in the presence of H-Y antigen, *Obstet. Gynecol.,* 58, 17, 1981.
14. **Wolman, S. R., McMorrow, L. E., Roy, S., Koo, G. C., Wachtel, S. S., and David, R.,** Aberrant testicular differentiation in 46,XY gonadal dysgenesis: morphology, endocrinology, serology, *Hum. Genet.,* 55, 1, 1980.
15. **Wolf, U.,** XY gonadal dysgenesis and the H-Y antigen, *Hum. Genet.,* 47, 269, 1979.
16. **Simpson, J. L., Blagowidow, N., and Martin, A. O.,** XY gonadal dysgenesis: genetic heterogeneity based upon clinical observations, H-Y antigen status and segregation analysis, *Hum. Genet.,* 58, 91, 1981.
17. **Müller, U.,** Genetics of H-Y antigen and its gonad-specific receptor, *Bibl. Anat.,* 24, 111, 1983.
18. **Simpson, J. L.,** Genetic forms of gonadal dysgenesis in 46,XX and 46,XY individuals, *Semin. Reprod. Endocrinol.,* 1, 93, 1983.
19. **Warner, B. A., Monsaert, R. P., Stumpf, P. G., Kulin, H. E., and Wachtel, S. S.,** 46,XY gonadal dysgenesis: is oncogenesis related to H-Y phenotype or breast development?, *Hum. Genet.,* 69, 79, 1985.

20. **Müller, U., Lalande, M., Donlon, T., and Latt, S. A.,** Moderately repeated DNA sequences specific for the short arm of the human Y chromosome are present in XX males and reduced in copy number in an XY female, *Nucleic Acids Res.,* 14, 1325, 1986.
21. **Müller, U., Donlon, T., Schmid, M., Fitch, N., Richer, C.-L., Lalande, M., and Latt, S. A.,** Deletion mapping of the testis determining locus with DNA probes in 46,XX males and in 46,XY and 46,X,dic(Y) females, *Nucleic Acids Res.,* 14, 6489, 1986.
22. **Bielert, C., Bernstein, R., Simon, G. B., and van der Walt, L. A.,** XY gonadal dysgenesis in a chacma baboon (*Papio ursinus*), *Int. J. Primatol.,* 1, 3, 1980.
23. **Chapman, H. M., Bruere, A. N., and Jaine, P. M.,** XY gonadal dysgenesis in a Charlais heifer, *Am. Reprod. Sci.,* 1, 9, 1978.
24. **Gustavsson, I., Settergren, I., Gustafsson, H., and Larsson, K.,** Testicular feminization and XY gonadal dysgenesis in Swedish Red and White cattle, in *Abstracts: 73rd Annual Meeting of the American Society of Animal Science,* 1981, 160.
25. **Fredga, K., Gropp, A., Winking, H., and Frank, F.,** A hypothesis explaining the exceptional sex ratio in the wood lemming (*Myopus schisticolor*), *Hereditas,* 85, 101, 1977.
26. **Herbst, E. W., Fredga, K., Frank, F., Winking, H., and Gropp, A.,** Cytological identification of two X-chromosome types in the wood lemming (*Myopus schisticolor*), *Chromosoma,* 69, 185, 1978.
27. **Wiberg, U., Mayerová, A., Müller, U., Fredga, K., and Wolf, U.,** X-linked genes of the H-Y antigen system in the wood lemming (*Myopus schisticolor*), *Hum. Genet.,* 60, 163, 1982.
28. **Eicher, E. M., Washburn, L. L., Whitney, J. B., III, and Morrow, K. E.,** *Mus poschiavinus* Y chromosome in C57BL/6J murine genome causes sex reversal, *Science,* 217, 535, 1982.
29. **Eicher, E. and Washburn, L. L.,** Inherited sex reversal in mice: identification of a new primary sex-determining gene, *J. Exp. Zool.,* 228, 297, 1983.
30. **Hughes, J. P. and Trommerhausen-Smith, A.,** Infertility in the horse associated with chromosomal abnormalities, *Aust. Vet. J.,* 53, 253, 1977.
31. **Trommerhausen-Smith, A., Hughes, J. P., and Neely, D. P.,** Cytogenetic and clinical findings in mares with gonadal dysgenesis, *J. Reprod. Fertil.,* 27, 271, 1979.
32. **Sharp, A. J., Wachtel, S. S., and Benirschke, K.,** H-Y antigen in a fertile XY female horse, *J. Reprod. Fertil.,* 58, 157, 1980.
33. **Kent, M. G., Shoffner, R. N., Hunter, A., Elliston, K. O., Schroder, W., Tolley, E., and Wachtel, S. S.,** XY sex reversal syndrome in the mare: clinical and behavioral studies, H-Y phenotype, *Hum. Gent.,* in press, 1988.
34. **McLaren, A., Simpson, E., Tomonari, K., Chandler, P., and Hogg, H.,** Male sexual differentiation in mice lacking H-Y antigen, *Nature (London),* 312, 552, 1984.
35. **Simpson, E., McLaren, A., and Chandler, P.,** Evidence for two male antigens in mice, *Immunogenetics,* 15, 609, 1982.
36. **Burgoyne, P. S., Levy, E. R., and McLaren, A.,** Spermatogenic failure in male mice lacking H-Y antigen, *Nature (London),* 320, 170, 1986.
37. **Singh, L. and Jones, K. W.,** Sex reversal in the mouse (*Mus musculus*) is caused by a recurrent nonreciprocal crossover involving the X and an aberrant Y chromosome, *Cell,* 28, 205, 1982.
38. **Ohno, S.,** *Sex Chromosomes and Sex-Linkes Genes,* Springer-Verlag, New York, 1967.

Part V: Biology of Development

Chapter 19

AN EXPERIMENTAL TEST OF THE PRODUCTION LINE HYPOTHESIS OF OOGENESIS IN THE FEMALE MOUSE

Paul E. Polani and John A. Crolla

TABLE OF CONTENTS

I. INTRODUCTION

In 1968, Henderson and Edwards,[1] in an important paper in *Nature (London)*, proposed a production line hypothesis (PLH) of mammalian oogenesis. According to this, oogonia committed to meiosis earlier on in fetal life are released soon after puberty as mature oocytes, while those committed later in fetal life are released at correspondingly later postnatal ovulations. The idea was based on experimental findings in two strains of mice in the oocytes of which they detected, as the animal grew older, a decrease in chiasma (Xma) frequency and a parallel increase in chromosomal univalents, a well-known inverse relationship.[2] As Xma, related to crossover recombination, are formed during pachynema and so during fetal or perinatal life of the female, and are accepted to remain unchanged during the prolonged (dictyate) resting period before resumption of meiosis prior to ovulation, Henderson and Edwards[1] postulated that oocytes committed to meiosis early have more Xma than those formed later. The decrease of Xma in the latter, and the consequent loss of the mechanical function of Xma that normally keep the homologous chromosomes together as bivalents to the first meiotic division would lead to their random segregation. This, in turn, would result in frequent nondisjunction, and so, in the production of aneuploid ova and zygotes. The frequencies of the latter would be seen to increase with maternal age. Thus, the PLH provided a novel explanation for the well-known fact that most numerical chromosome anomalies in experimental animals[3-5] and in man[6,7] are maternal-age dependent. Down's syndrome with free trisomy 21 is a paragon of the situation.

Clearly, the cytological relationship between postnatal age and Xma should be mirrored by similar changes in recombination frequency of linked genes. In female mice, a few pairs of loci seemed to confirm this expectation.[8,9] However, other data showed an age effect in both sexes,[10] and ad hoc tests of the PLH showed that different loci differed in age-related changes of recombination frequencies.[11] So, there seemed to be no simple relationship with Xma frequency, and the results suggested possible age-related changes in Xma positions as well.

Among the cytological tests of the hypothesis — the validity of which in the mouse, Baker[12] has *a priori* questioned — Luthardt et al.[13] confirmed the first meiotic metaphase (MI) results on Xma and univalents. So did Polani and Jagiello[14] who, however, also studied the second meiotic metaphase (MII) and so chromosome segregation: they suggested that most of the MI univalents were not true primary asynaptic or precocious desynaptic in origin, but rather were artifactual — albeit in an age-related manner (see also Reference 1). Speed[15] studied MI in the same strains of mice used by Henderson and Edwards[1] and observed the decrease of Xma, but without a corresponding increase in the frequency of univalents. Jagiello and Fang[16] set out to test the PLH by comparing the frequencies of Xma and univalents in early (fetal day 16) vs. late (fetal day 18) diplotene oocytes, a technically demanding investigation. In their strain of Swiss mice, the day 18 cells had fewer Xma and more univalents than the early (day 16) oocytes had. By contrast, in another Swiss strain together with a further strain, Speed and Chandley[17] found that the germ cells had not reached the diplotene stage on day 16, and in the day 18 and 19 diplonemas were unable unambiguously to identify Xma. They also made a search for possible anomalies of synapsis of the axes of the synaptonemal complex, but found no evidence of pairing errors that might account for the failure to form Xma between some homologs in a gestation-age related manner. It is to be stressed that synapsis between homologs is a necessary condition but not a sufficient one for crossing over recombination, and so, for Xma formation. Finally, Tease and Fisher[18] also sought to test the PLH by looking for possible fetal age-related changes in meiotic pairing, using two different inversion heterozygous mouse strains. While progressive inversion loop disappearance, through synaptic adjustment, was correlated with pachytene age, both this, and particularly, other pairing disturbances, were consistent with a production line of oogenesis.

It seems that this work on PLH has overlooked the fact that the hypothesis consists of three distinct arms: (1) sequential prenatal production and *pari passu*, sequential postpubertal release of oocytes; (2) sequential decrease of Xma of oocytes with fetal age; and (3) consequently prenatal determination of univalents on which postnatal maternal-age related nondisjunction would depend. Proposition (1) is the key element, while propositions (2) and (3) are subordinate: they are not obligate and could be expected to be affected by environmental forces external to the fetus and especially by genetic variation. It is with the first arm of PLH, a production line of oogenesis in a narrower sense, that the present study is concerned.

II. MATERIALS AND METHODS

For the experiments, we used the fetal ovaries of syngeneic mice of the C3H/He strain. The technique employed was an in vitro/in vivo method developed for the study of meiosis in the female mouse.[19] In essence, and for all the experiments, the ovaries were explanted on day 14 of embryonic development (day of plug = day 1) and were then maintained in an organ culture system[20] in microwells for 8 days, during which time they matured at the same rate as in vivo, 1 day in culture corresponding very closely to a day of gestation. At the end of the in vitro phase, a proportion of the ovaries matured in culture was used to study the germ cells that had progressed to pachynema/diplonema. For this, single ovaries, in general one per fetus, were homogenized mechanically to make a cell suspension. To follow the subsequent maturation of the oocytes equivalent to their natural development from birth to puberty and then to ovulation, the in vitro matured ovaries were transplanted under the kidney capsule of previously spayed young females and allowed to develop there for a minimum of 19 days prior to harvesting the mature oocytes from which chromosome preparations of first and second meiotic metaphases (MI/MII) were made. The culture medium used was Ham's F10 (Gibco Biocult), supplemented with L-glutamine, 20% donor calf serum (Flow Labs, McLean, Va.) and antibiotics. In order to follow the sequential committment of oogonia to the meiotic prophase and their post-transplantation release as mature oocytes at MI/MII, the germ cells were marked at suitable times (see below) by exposing the ovaries in vitro to ^{3}HTdR (0.1 μCi/mℓ: 0.5 Ci/mmol), so that the radioactive precursor could be incorporated into their DNA during its premeiotic synthesis. The radioactivity of the labeled cells, both prophases and metaphases, was assessed using standard autoradiographic stripping film methods.[20]

In addition to data on the proportions of labeled cells at prophase and metaphase, it was necessary to have information on labeled oocytes *in situ* in the post-transplantation ovary. This was done by scoring the proportion of radiographically marked intrafollicular germinal vesicles using autoradiography and a reconstruction procedure on Bouin-fixed, paraffin wax embedded, serially sectioned ovaries. For autoradiography, we used Kodak AR10 stripping film developed for 5 weeks at 4°C, following which the sections were stained with Meyer's haemulum prior to mounting.

III. DESIGN OF EXPERIMENTS AND COMMENTS

We had previously established that exposing fetal day 14 ovaries in vitro to ^{3}HTdR yielded oocytes nearly 100% of which were labeled, whereas on days 16/17 only some 20 to 30% of MI/MII were marked (see Reference 20, especially Table 2 and Figure 4). With this information, we conducted three sets of experiments to test the PLH.

The first and most direct test was to expose the explanted fetal ovaries in vitro to ^{3}HTdR on day 14, or alternatively, but still using ovaries explanted on day 14, to the radioactive precursor on day 16/17; and then, having obtained the required baseline of labeled prophases, the object was to attempt to harvest the oocytes at increasing intervals of time after trans-

Table 1
RESULTS OF GERM CELL LABELING

In vitro ^{3}HTdR to explanted ovaries on	Oocytes at MI/MII		
	Label	No label	Total
d14			
Obs.	414	15	429
Exp.	417	12	—
d15			
Obs.	73	107	180
Exp.			
PLH valid	?	?	—
PLH not valid	149	31	—
d16/17			
Obs.	4	249	253
Exp.			
PLH valid	±0	±253	—
PLH not valid	98	155	—

Note: Germ cells were labeled within the explanted fetal ovaries with ^{3}HTdR on day 14, 15, or 16/17. The expected values are derived from the proportion of labeled prophase cells. In the case of day 16/17, two expectations can be given depending on whether or not a production line system of oogenesis (PLH) exists (see text).

plantation, namely at 19 days or at 5, 7, 9, 11, 15, and 20 weeks. We expected that nearly all harvested oocytes from ovaries exposed to ^{3}HTdR in vitro on day 14 of gestation would be labeled. By contrast, exposure of ovaries in vitro on day 16/17 would differentiate between the validity or otherwise of the hypothesis. If valid, then harvesting earlier after transplantation should yield practically only unlabeled oocytes, whereas an increasing proportion of those harvested at later times should display the label. If instead similar proportions of labeled oocytes were obtained at each harvest time, the hypothesis would be rejected, and the proportion of oocytes that were expected to be labeled by the radioactive DNA precursor supplied to the fetal ovaries in vitro, was indicated by the proportion of prophases found to be labeled.

Analysis of these first experiments showed that the method that we employed and the radiosensitivity of the oocytes would not permit the use of this direct aging approach. We found that, taking two measures of technical success, graft viability and MI/MII yield, in vitro labeled ovaries were detrimentally affected compared with unlabeled controls. However, from the test and the control ovaries the two parameters remained steady at the harvest times of 19 days and 5 and 7 weeks, while only beyond this post-transplantation period graft viability and oocyte yield deteriorated. Therefore, the results from these early harvests (204 MI/MII) were included in the final analysis (Table 1).

On the basis of these first findings, two further sets of experiments were done in which we harvested only at one time, i.e., at 19 days after heterotopic transplantation (indirect approach). The fetal ovaries were treated in vitro as in the first set of tests and the prophase cells were studied. The interpretation of the results in terms of support for or rejection of the PLH was as already outlined: if valid, we expected to find practically only unlabeled MI/MII from ovaries treated on day 16/17; by contrast, if the hypothesis were not valid, oocytes would be labeled and their proportion would be expected to correspond to the proportion of prophase cells found to be labeled at the end of the in vitro phase.

The first of these two sets of indirect experiments gave results identical to those from the early harvests in the direct experiments, which they thus confirmed, supporting strongly the

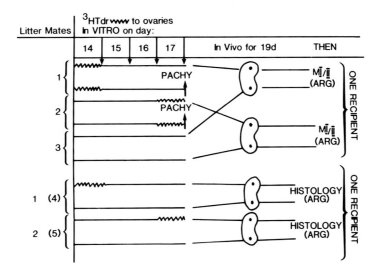

FIGURE 1. Represents the plan of the third experiment (see text) using ovaries from litter mates and a single recipient for each experiment. The position of the in vitro matured ovaries when transplanted under the kidney capsule is indicated; for each fetus one ovary was used for prophase analysis at the end of the in vitro phase and the other for MI/MII at the end of the in vivo phase. Whenever possible, litter mates were used also for the histological study.

existence of a production line system of oogenesis. Also, using this same experimental protocol, we exposed ovaries in culture to the radioactive precursor on day 15 of gestation (ovaries explanted on day 14), expecting to obtain, if the PLH was valid, results intermediate between those from day 14 and those from day 16/17, and significantly different from those dictated by the corresponding prophase labeling data.

As a final check, and in order to minimize variation, we carried out the third set of experiments, also indirect, and according to the same in vitro and in vivo protocol as the second set of tests. However, we used litter mates to supply test and control ovaries at the two different labeling times for prophase and MI/MII, and only one recipient for each group of litter mate ovaries, and we prepared ovaries for the collateral confirmatory histology study, again using the litter mates whenever possible, and again using only one recipient per set of ovaries, both test and control (Figure 1).

IV. RESULTS

Labeling information was obtained from 862 MI/MII oocytes: 204 from the first, 555 from the second, and 103 from the third set of experiments. Because the results from the three sets were concordant, they can be pooled and presented by the day of in vitro exposure to ^{3}HTdR of the explanted ovaries (Table 1). Thus, 429 MI/MII were derived from day 14, 180 from day 15, and 253 from days 16/17 exposures.

From each experiment we also obtained data on labeled prophases which were used to construct the expected values of labeling of the MI/MII on the alternative assumptions of the validity, or otherwise, of the PLH (Table 1). In all, we scored 6087 prophase cells. Pooling the results as above, there were 1963 cells to use as labeling standards for the ovaries exposed on day 14 (97.1% were labeled), 1420 as standards for day 15 (83.0% labeled), and 2704 for day 16/17 (38.7% labeled).

As can be seen, the results from day 16/17 labeling experiments, the critical late-labeling time, fit well with an expectation based on the premise that the PLH is valid, and not with

Table 2
HISTOLOGY RESULTS OF
LABELED FETAL OVARIES

In vitro ^{3}HTdR to explanted ovaries on	Number of ovaries	Germinal vesicles	
		Labeled	Not labeled
d14	5	168	34
		83%	17%
d16/17	3	64	120
		35%	65%
Controls	3	—	55

Note: The germ cells were labeled with ^{3}HTdR at two gestation ages. The germinal vesicles were scored for autoradiographic grains in serial histological sections with reconstruction in order to assess individual vesicles occupying more than one section (see text).

the alternative. The results from day 15 are intermediate, but particularly, they are significantly different from an expectation based on the view that the PLH may not be valid. The results from day 14 in vitro labeling can really be considered labeling controls. The histology results are set out in Table 2 and show that the proportions of labeled oocytes within the ovaries exposed to the radioactive precursor on day 14 or day 16/17 of gestational age are as expected from the corresponding labeling proportions of pachynemas/diplonemas. In other words, there is no deficiency of labeled germinal vesicles (see Section V).

V. DISCUSSION

It behoves us to consider first how appropriate to the testing of the key element of the PLH were the techniques we employed. Orthotopic transplantation experiments of genetically marked fetal ovaries to generate genetically marked young.[20] convinced us that the in vitro/ in vivo technique is a good mimic of natural conditions. In addition, there is no evidence from labeling experiments[20] that the germ cells in the in vitro cultured ovaries behave unusually; their commitment, entry, and transit through the meiotic prophase seem quite normal. These previous experiments, and the present results from labeling ovaries in vitro on day 14 of gestation, clearly show the absence of detrimental effects on oocyte survival. However, the first set of experiments indicates that this conclusion is only valid, given the radiation dose used, for experiments terminated not later than 7 weeks after transplantation. Also, the day 14 labeling experiments tell that, provided harvesting is done early, there is no evidence of interference with oocyte maturation through the follicular stages right up to Graafian follicle formation and ovulation (see below).

We stress again that the results are derived from three independent sets of experiments that are congruous with each other, and that their individual outcomes fit the predictions that can be made given that the PLH is valid. Taken together with the above comments on the appropriateness of the techniques used, we can confidently accept that there exists a production line system of murine oogenesis, in the narrower sense. This rests on the critical results from labeling in vitro ovaries on day 16/17 of gestation. Clearly, the reason why from these ovaries we practically obtain only unlabeled oocytes is not because of elimination or failed maturation of the labeled ones as a result of radiation damage, a possibility that had to be considered; that this is not the case is in general already clear from the day 14 labeling tests, but more specifically from the fact that at histology, with serial sections reconstruction, we find labeled mature and maturing oocytes *in situ* in precisely the proportions expected from the pachynema/diplonema labeling data.

We had set out to test the key aspect of the PLH, namely the first-in first-out, last-in last-out view of it, and, as we said, we find support for the existence of such a system. However, caution demands that the hypothesis be considered valid strictly within the limits of species, indeed strain, and experimental approach. The latter seems appropriate, and it would appear strange if such a basic phenomenon as a production line system operated within one taxon only and not in other mammals.

The two subordinate propositions of the PLH, namely the decrease in Xma frequency and increase in univalents with their nondisjunctional consequences, were not tested by these experiments. As we indicated in Section I, these features of the PLH are not straightforward or uncontroversial. However, these two other parts of the hypothesis may also prove to be correct, at least in some genetic backgrounds (strains). If so, then some maternal-age dependent numerical chromosome anomalies might arise as a byproduct of the production line system of oogenesis, as suggested by Henderson and Edwards,[1] and as a result of the influence of "gradients of development" on recombination via chiasma formation.

The question would then arise as to what proportion of trisomics, and their complementary monosomics, of maternal origin may be attributable to this prenatal mechanism, and what proportion, conversely, would be due to postnatal oocyte aging, as it seems *a priori* highly unlikely that all maternal-age dependent anomalies have the same origin. Such a problem would be especially relevant to man — assuming that a production line system operated in human oogenesis — in view of the high frequency of maternal-age dependence at conception of numerical chromosome anomalies, their proportionally frequent maternal origin at MI, the many years during which a woman may bear children in the face, for example, of decreasing oocyte availability, and a progressively diminishing efficiency of sex-hormonal homeostasis, or of other triggers of nondisjunction, both genetic and environmental.

VI. SUMMARY

By explanting and maintaining in organ culture fetal mouse ovaries, followed by transplantation into spayed females for further oocyte maturation to meiotic metaphase, we tested the PLH of Henderson and Edwards,[1] according to which germ cells, known to become committed to meiosis in fetal life, are released after puberty in the same sequence as that of meiotic entry *in fetu* ("first-in first-out; last-in last-out"). There are also subsidiary stipulations of the hypothesis related both to chiasma frequency, negatively correlated with fetal age, and to nondisjunction postnatally. We traced the sequential entry into meiosis of germ cells by radioactively labeling their DNA synthesized while in culture. At the end of this in vitro phase, prophase cells were sampled and examined, and the proportion of cells labeled noted. After the subsequent in vivo, post-transplantation phase, the proportion of labeled oocytes at MI/MII was likewise determined. By labeling, while in the cultured fetal ovaries, the germ cells entering meiosis early (day 14 of gestation) and conversely, those entering late (day 16/17), and by comparing the proportions of labeled cells at metaphase of meiosis with those at its prophase, the conclusion was reached that a production line system of oogenesis exists in the mouse.

ACKNOWLEDGMENTS

We are very grateful to Professor Martin Bobrow for his continued interest and support. P.E.P. warmly thanks Action Research for the Crippled Child, and The Mr. and Mrs. Archie

Sherman Foundation for generous financial support over the years; and The Spastics Society for laboratory and backup facilities. As always, we are grateful to Mrs. Barbara Merchant for her invaluable help with the literature and to Miss Adrianne Knight for typing the manuscript.

REFERENCES

1. **Henderson, S. A. and Edwards, R. G.,** Chiasma frequency and maternal age in mammals, *Nature (London),* 218, 22, 1968.
2. **Mather, K.,** Crossing-over, *Biol. Rev.,* 13, 252, 1938.
3. **Gosden, R. G.,** Chromosomal anomalies of preimplantation mouse embryos in relation to maternal age, *J. Reprod. Fertil.,* 35, 351, 1973.
4. **Yamamoto, M., Endo, A., and Watanabe, G.,** Maternal age dependence of chromosome anomalies, *Nature (London) New Biol.,* 241, 141, 1973.
5. **Martin, R. H., Dill, F. J., and Miller, J. R.,** Nondisjunction in aging female mice, *Cytogenet. Cell Genet.,* 17, 150, 1976.
6. **Polani, P. E.,** A review of the cause of chromosome anomalies, in *The Future of Prenatal Diagnosis,* Galjaard, H., Ed., Churchill Livingstone, Edinburgh, 1982, chap. 2.
7. **Hassold, T. J. and Jacobs, P. A.,** Trisomy in man, *Annu. Rev. Genet.,* 18, 69, 1984.
8. **Bodmer, W. F.,** Effects of maternal age on the incidence of congenital abnormalities in mouse and man, *Nature (London),* 190, 1134, 1961.
9. **Reid, D. H. and Parsons, P. A.,** Sex of parent and variation of recombination with age in the mouse, *Heredity,* 18, 107, 1963.
10. **Fisher, R. A.,** A preliminary linkage test with *Agouti* and *undulated* mice, *Heredity,* 3, 229, 1949.
11. **Wallace, M. E., MacSwiney, F. J., and Edwards, R. G.,** Parental age and recombination frequency in the house mouse, *Genet. Res.,* 28, 241, 1976.
12. **Baker, T. G.,** Oogenesis and ovulation, in *Reproduction In Mammals,* 2nd ed., Vol. 1, Austin, C. R. and Short, R. V., Eds., Cambridge University Press, New York, 1982, chap. 2.
13. **Luthardt, F. W., Palmer, C. G., and Yu, P.-L.,** Chiasma and univalent frequencies in aging female mice, *Cytogenet. Cell Genet.,* 12, 68, 1973.
14. **Polani, P. E. and Jagiello, G. M.,** Chiasmata, meiotic univalents and age in relation to aneuploid imbalance in mice, *Cytogenet. Cell Genet.,* 16, 505, 1976.
15. **Speed, R. M.,** The effects of ageing on the meiotic chromosomes of male and female mice, *Chromosoma,* 64, 241, 1977.
16. **Jagiello, G. and Fang, J. S.,** Analyses of diplotene chiasma frequencies in mouse oocytes and spermatocytes in relation to ageing and sexual dimorphism, *Cytogenet. Cell Genet.,* 23, 53, 1979.
17. **Speed, R. M. and Chandley, A. C.,** Meiosis in the foetal mouse ovary. II. Oocyte development and age-related aneuploidy. Does a production line exist?, *Chromosoma,* 88, 184, 1983.
18. **Tease, C. and Fisher, G.,** Further examination of the production-line hypothesis in mouse foetal oocytes. I. Inversion heterozygotes, *Chromosoma,* 93, 447, 1986.
19. **Polani, P. E., Crolla, J. A., Seller, M. J., and Moir, F.,** Meiotic crossing-over exchange in the female mouse visualised by BUdR substitution, *Nature (London),* 278, 348, 1979.
20. **Polani, P. E., Crolla, J. A., and Seller, M. J.,** An experimental approach to female mammalian meiosis: differential chromosome labeling and an analysis of chiasmata in the female mouse, in *Bioregulators of Reproduction,* P & S Biomedical Sciences Symposia Series, Jagiello, G. and Vogel, H. J., Eds., Academic Press, New York, 1981, 59.

Chapter 20

MOLECULAR BIOLOGY OF ANTI-MÜLLERIAN HORMONE

Jean-Yves Picard, Daniel Guerrier, Axel Kahn, and Nathalie Josso

TABLE OF CONTENTS

I. THE AMH MOLECULE

Anti-Müllerian hormone (AMH) is a disulfide-bonded glycoprotein dimer responsible for regression of the Müllerian ducts in the male fetus.[1] It is produced essentially by immature Sertoli cells[2-4] but also by granulosa cells of growing and antral follicles from birth onwards.[5-7]

Although AMH has been purified from follicular fluid,[5] immature testicular tissue, by far the most potent producer, is usually chosen as starting material. Fetal bovine testes, obtained from commercial slaughterhouses, are incubated either 4 hr[8] or 45 min,[9] and the hormone is isolated from the conditioned medium after ammonium sulfate precipitation.

A. Purified AMH

Purification of bovine AMH to homogeneity has been achieved by Picard and Josso[8] by use of a simple protocol involving ion-exchange chromatography and affinity-chromatography with a monoclonal antibody. Purified AMH is characterized by its reduction-sensitive behavior (Figure 1). Exposure to 2-mercaptoethanol reduces AMH to a subunit 72,000 in molecular weight according to Picard and Josso.[8] According to Cate et al.,[10] the dimer dissociates into a doublet, comprising subunits of 70,000 and 74,000 mol wt, respectively, the smaller species being a proteolytic product. In the absence of reducing agents, the subunits form dimers and tetramers. The microheterogeneity disclosed by polyacrylamide gel electrophoresis is not due to inadequate purification since all the protein bands, with the exception of a single minor one, are recognized by Western blotting with a monoclonal antibody. Also, the amino acid composition of the purified product[11] is similar to the amino acid composition specified by the sequence of the cloned gene.[10] Purified AMH elicits total regression of fetal rat Müllerian ducts in organ culture at a concentration of 1.5 µg/mℓ (10 nM). A polyclonal antibody developed against purified AMH was used as a probe to screen an expression library obtained in phage lambda gt-11.[12]

B. Fraction Green-3

Other investigators have submitted ammonium sulfate precipitates of bovine testicular incubation medium to sequential anion and cation exchange chromatography, followed by affinity chromatography successively on wheat germ agglutinin and Matrex-Green.[13,14] The resulting product, named Green-3, has moderate anti-Müllerian activity at a concentration of 6 µg/mℓ[15] and is endowed with activities against cancer of the female reproductive tract[16] and against meiosis of female germ cells,[15] not shared by purified AMH.[17,17a]

Electroelution of the AMH subunits contained in fraction Green-3 has allowed sequence determination, construction of a nucleotide probe, and cloning of the gene for bovine and human AMH.[10]

II. CLONING AMH cDNA AND GENE

Because AMH represents an extremely low (0.01%) proportion of the proteins synthesized by the gonads, powerful techniques, adapted to the detection of rare messenger RNAs (mRNAs), were required for successful cloning of its complementary DNA (cDNA).

A. The Immunological Approach

A polyclonal antibody to purified bovine AMH was used to screen a lambda gt-11 expression library.[12] To obtain the cDNA clone library, double-stranded DNA complementary to total poly(A) + RNA extracted from the testis of a 2-week-old calf was inserted into lambda gt-11 DNA (Protoclone kit, Promega Biotec, Madison, Wis.), and packaged using Gigapack (Vector Cloning Systems, San Diego, Calif.). A library of 1.5 million primary clones with 94% recombinants was obtained. Plated clones were screened according to Young and Davis[18]

PAGE OF PURE BOVINE AMH

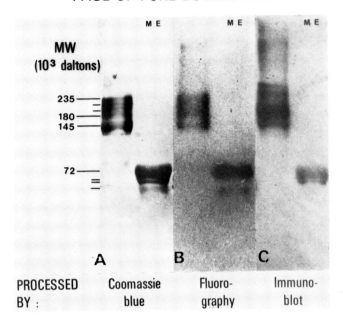

FIGURE 1.   Electrophoresis of purified AMH on polyacrylamide gradient gels in the presence of sodium dodecyl sulfate, with or without reduction of samples by 5% 2-mercaptoethanol (ME). (A) Coomassie blue stain. AMH (1.5 μg) in the unreduced state is formed by a 145,000 mol wt dimer and a series or larger polymers. After reduction by mercaptoethanol of disulfide bonds, these bands disappear, replaced by a major 72,000 mol wt band. (B) AMH, synthesized in the presence of tritiated fucose and revealed by fluorography, shows a similar pattern. (C) Western blot of purified AMH (1.5 μg) revealed by immunoblotting with a monoclonal antibody. Same pattern as A except that the minor 56,000 mol wt component of the subunit is not stained.

using anti-bovine AMH antiserum diluted 1/250 and radioiodinated protein A. Those phage plaques producing a signal on duplicate nitrocellulose filters were replated at lower density up to homogeneity. Three clones: 4, 5, and 8, positive after three rounds of replating, were identified after screening of 500,000 clones.

The specificity of the isolated inserts was tested by anti-epitope technology,[19,20] i.e., by confirming that the antibodies eluted from the fusion proteins synthesized by the clones were capable of recognizing AMH on Western blots and of precipitating radioiodinated bovine AMH in a double antibody precipitation test.[21] In both tests, only anti-epitope antibodies binding to inserts 4 and 5 gave a clearly positive signal. The anti-epitope antibody eluted from clone 8 did not precipitate iodinated AMH and required 19 days of exposure to become visible on Western blots, compared to 4 and 24 hr for antibodies eluted from the two other clones.

Two inserts, 4 and 5, were found to cross-hybridize with each other: 5, the smaller of the two, was therefore not studied further. Insert 4 was found to bind a 2.1-kb mRNA species, present only in immature testicular tissue and in ovarian follicles. The size and tissue-specific expression of this RNA are in agreement with those expected for an AMH-specific mRNA. In contrast, insert 8 hybridizes to a 1-kb ubiquitous species of mRNA, thus confirming that the insert codes for a fragment of a protein only faintly related to AMH.

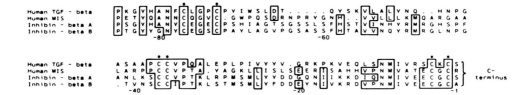

FIGURE 2.   Comparison of amino acid composition of the C-terminal domain of human AMH, human TGF-β, and the β subunits of porcine inhibin A and B. All the cysteine residues are conserved. (From Cate, R. L., et al., *Cell*, 45, 685, 1986. With permission.)

B. The Oligonucleotide Approach

The above results can be considered only preliminary, compared to the progress achieved in Boston by Biogene Corporation in collaboration with Dr. Donahoe's group. Cate and his colleagues[10] electroeluted the AMH subunit from polyacrylamide gels loaded with fraction Green-3,[13,14] sequenced it, and constructed oligonucleotide probes which they tested by hybridization to Northern blots of bovine testicular mRNA. The resultant reduction of probe degeneracy greatly facilitated the isolation of a cDNA clone contained in a lambda gt-10 library and the subsequent isolation of a genomic clone from a bovine cosmid library. The bovine cDNA clone was used as a probe to isolate a human genomic clone from a human cosmid library. The specificity of the human genomic clone was proven by the insertion of a viral promoter-driven fragment of the human genomic clone into COS cells. Conditioned medium from transfected COS cells produced regression of fetal rat Müllerian ducts in organ culture.

C. Structure of the *AMH* Gene: the Disulfide-Linked Protein Superfamily

As demonstrated by Cate et al.,[10] the human AMH gene is only 2.7 kb long and contains five exons. The promoter of the bovine AMH gene contains the sequence CCGCCC 60 nucleotides upstream of the transcription initiation site, as do the various promoters encoding particular housekeeping genes. The N-terminal region is the one showing the least homology between human and bovine AMH (62% over the first 110 amino acids). In contrast, the C-terminus shows the strongest homology with 108 of the last 112 amino acids conserved. The C-terminal region of AMH also shows some degree of homology with the C-terminal region of the beta subunit of porcine inhibin and the transforming growth factor TGF-β (Figure 2). Both these proteins happen to be disulfide-linked dimers. Inhibin has the same cellular origin as AMH, but not the same ontogeny, since it is secreted in adulthood by Sertoli cells and granulosa cells. Inhibin, formed by the association of alpha and beta subunits, decreases the secretion of follicle-stimulating hormone by the pituitary. In contrast, the association of two beta subunits yields FRP or activin, a protein which activates follicle-stimulating hormone (FSH) release.[22] TGF-β is a multifunctional peptide-controlling proliferation and differentiation of many cell types.[23] A new homodimer form of TGF-β has recently been isolated.[24]

In spite of their structural relationship, and in spite of the capacity of TGF-β to inhibit epithelial growth and stimulate endothelial proliferation[23] — precisely the effect of AMH upon the Müllerian duct — AMH and TGF-β do not share physiological effects. TGF-β has no anti-Müllerian activity detectable by bioassay (Figure 3), nor does AMH compete with TGF-β in a radioreceptor assay or have an effect on angiotensin-stimulated steroidogenesis of adrenal cells.[25]

III. LOCALIZATION OF THE HUMAN *AMH* GENE

We have recently mapped the gene coding for AMH to the extremity of the short arm of

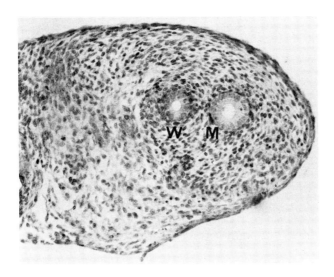

FIGURE 3.   Aspect of rat fetal Müllerian duct, 14.5 days postcoitum, exposed to 10 ng/mℓ of TGF-β, purified from human platelets (gift of C. Cochet). Note inhibition of epithelial growth. Hematoxylin-eosin × 150.

chromosome 19, sub-band p13.3, using somatic cell genetics and *in situ* hybridization.[26] The gene responsible for directing the synthesis of human TGF-β is located on the long arm of the same chromosome, sub-band q13.1.[27]

For somatic cell hybridization, genomic DNA was extracted from cell cultures of 9 man × mouse and 26 man × hamster hybrids, and from their parent cell lines. The chromosome constitution of all hybrids was ascertained by karyotype and enzyme analysis at the time of DNA extraction. DNA was digested by restriction endonuclease Taq 1, electrophoresed on a 1% agarose gel, blotted and hybridized with AMH probes under stringent conditions as described.[26] Three probes were used for this purpose: the double-stranded 1.2-kb cDNA insert isolated by Picard et al.,[12] which cross-hybridizes with human and rat AMH, and two single-stranded probes prepared by primer extension from either the 3′ or the 5′ extremity of this insert. The single hybridizing human restriction fragment has 1.65 kb and can easily be distinguished from the 1.05-kb mouse restriction fragment, and from the three 0.75 to 1.35 hamster restriction fragments. A first experiment, performed with the 3′ probe on a standard panel of previously characterized hybrids, allowed the exclusion of all chromosomes except 10, 14, and 19 (Table 1). To reach a definitive conclusion, a complementary panel was analyzed using all three probes (Figure 4). The restriction fragment hybridizing to the AMH probes consistently segregated with chromosome 19, and was discordant with chromosome 10 and chromosome 14.

The assignment of *AMH* to chromosome 19 was confirmed and refined by *in situ* hybridization. This was performed on metaphase chromosome spreads prepared from lymphocytes of a normal human male[28] using the full insert contained in plasmid pEMBL9, tritium-labelled by nick translation. Analysis of 157 metaphase spreads showed that 7.2% of the silver grains were located on chromosome 19. No other hybridization peaks were found in the genome. The distribution of grains on chromosome 19 is shown in Figure 5. Approximately 82% of the grains were clustered on bands 19p133 and 19p132, with a maximum on the former, which appears to be the most probable location of *AMH*.

IV. THE PERSISTENT MÜLLERIAN DUCT SYNDROME

Persistence of Müllerian primordia in genetic males is seen in two circumstances. Usually,

Table 1
SEGREGATION OF AMH AND CHROMOSOMES OF
25 HUMAN-RODENT HYBRIDS[a]

Chromosome	Segregation[b]				Number of discordant clones/ total number of informative hybrids	Percent discordant
	+/+	−/−	+/−	−/+		
1	5	4	12	1	13/22	59
2	4	4	13	1	14/22	64
3	10	2	8	3	11/23	48
4	9	5	4	1	5/19	26
5	10	4	8	2	10/24	42
6	11	1	6	4	10/21	48
7	6	4	11	2	13/23	56
8	8	3	4	3	7/18	39
9	5	5	11	1	12/22	54
10	5	4	10	0	10/19	53
11	10	1	7	5	12/23	52
12	12	3	6	3	9/24	37
13	8	3	8	3	11/22	50
14	8	5	7	0	7/20	35
15	5	4	11	2	13/22	59
16	8	5	6	1	7/20	35
17	5	3	13	2	15/23	65
18	8	4	7	1	8/20	40
19	10	5	3	0	3/18	17
20	13	3	4	1	5/21	24
21	11	1	5	3	8/20	40
22	9	4	9	1	10/23	43
X	6	0	2	3	5/11	45
Y	1	3	4	0	4/8	50

[a]   The presence of a human DNA restriction fragment hybridizing to the AMH probe was taken as proof of the presence of *AMH* in a given hybrid. Karyotypes were established on 12 to 20 metaphase cells.

[b]   +/+ = Both *AMH* and the chromosome are present; −/− = neither *AMH* nor the chromosome are present. +/− = *AMH* is present, but not the chromosome. −/+ = the chromosome is present, but not *AMH*. The values in these columns indicate the number of hybrids in each situation for the chromosome listed on the left. Only those hybrids where a given chromosome is present in at least 30% of the cells examined were taken into account; therefore, the total number of hybrids evaluated for each chromosome is always less than 25.

the presence of uterus and tubes is associated with defective virilization of the external genitalia and urogenital sinus. As discussed by Josso et al.,[29] this combined deficiency of testosterone and AMH-dependent steps of sex differentiation points to gonadal dysgenesis as the most probable cause of the syndrome. In such patients, namely those affected by mixed gonadal dysgenesis and dysgenetic male pseudohermaphroditism, a defect of the *AMH* locus and/or a receptor defect are probably not involved.

More rarely, persistence of Müllerian primordia occurs in otherwise normal males, and the condition is sometimes genetically transmitted in a manner suggesting either a dominant sex-limited or a recessive X-linked disorder.[30] Since the *AMH* gene is not on the X chromosome, as shown by others[10] and ourselves,[26] three possibilities are open to discussion. First, the persistent Müllerian duct syndrome is due to resistance of the end organ to the action of AMH, and AMH should be demonstrated by immunocytochemistry in testicular biopsies performed before age 5.[31] Second, the condition is due to lack of production of AMH itself and is expressed in heterozygous males (dominant sex-limited transmission) who have inherited an abnormal chromosome 19 probably from their mother. According to this

215

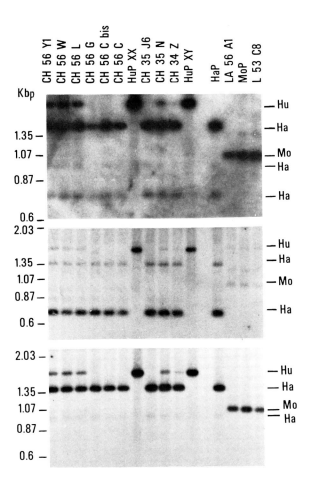

FIGURE 4. Autoradiogram of hybridization of ^{32}P-AMH probes to DNA of ten man-rodent hybrid cell-cell lines exhibiting independent segregation of chromosomes 10, 14, and 19, and to their parent cell lines (HaP = hamster parent; MoP = mouse parent; HuP = human parent). CH = man-hamster hybrid; L = man-mouse hybrid. Genomic DNA was digested with Taq 1 (Appligène, Strasbourg, France) for 6 hr at 65°C, separated on a 1% agarose gel, and transferred 6 hr to Hybond membranes (Amersham) according to the manufacturer's instructions. (A) Full insert probe, specific activity 7 × 10^7 cpm/μg, 1.5 × 10^6 cpm/mℓ; (B) 5′ probe, specific activity 2 × 10^9 cpm/μg, 2 × 10^6 cpm/μg; (C) 3′ probe, specific activity 1.6 × 10^9 cpm/μg, 1.3 × 10^6 cpm/mℓ. Size markers are lambda DNA Hind III/φ-174 RF DNA Hae III, supplied by Pharmacia. Hybridizing DNA fragments from different species can be easily distinguished. Hu = human restriction fragment, 1.65 kb; Ha = hamster restriction fragments, 1.35, 1.00, and 0.75 kb; and Mo = mouse restriction fragment, 1.05 kb. The relative intensity of the rodent signal varies with the different probes used.

hypothesis, AMH should not be detectable in testicular tissue by immunocytochemistry — unless a biologically inactive, but immunologically reactive AMH is being produced! (Study of DNA of family members might be rewarding in this case.) The third possibility is that the persistent Müllerian duct syndrome is heterogeneous, due sometimes to lack of production of normal AMH and sometimes to a receptor defect.

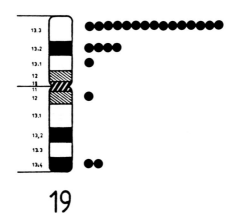

FIGURE 5.   G-band diagram of chromosome 19
(ISCN, 1981) showing distribution of silver grains
on 23 labeled chromosomes with most of the grains
localized to sub-band p133 and somewhat weaker
hybridization to sub-band p132.

REFERENCES

1. **Jost, A.,** Problems of fetal endocrinology: the gonadal and hypophyseal hormones, *Rec. Prog. Horm. Res.,* 8, 379, 1953.
2. **Blanchard, M. G. and Josso, N.,** Source of the anti-Müllerian hormone synthesized by the fetal testis: Müllerian-inhibiting activity of fetal bovine Sertoli cells in tissue culture, *Pediatr. Res.,* 8, 968, 1974.
3. **Tran, D. and Josso, N.,** Localization of anti-Müllerian hormone in the rough endoplasmic reticulum of the developing bovine Sertoli cell using immunocytochemistry with a monoclonal antibody, *Endocrinology,* 111, 1562, 1982.
4. **Hayashi, H., Shima, H., Hayashi, K., Trelstad, R. L., and Donahoe, P. K.,** Immunocytochemical localization of Müllerian inhibiting substance in the rough endoplasmic reticulum and Golgi apparatus in Sertoli cells of the neonatal calf testis using a monoclonal antibody, *J. Histochem. Cytochem.,* 32, 649, 1984.
5. **Vigier, B., Picard, J. Y., Tran, D., Legeai, L., and Josso, N.,** Production of anti-Müllerian hormone: another homology between Sertoli and granulosa cells, *Endocrinology,* 114, 1315, 1984.
6. **Takahashi, M., Hayashi, M., Manganaro, T. F., and Donahoe, P. K.,** The ontogeny of Müllerian inhibiting substance in granulosa cells of the bovine ovarian follicle, *Biol. Reprod.,* 35, 447, 1986.
7. **Bézard, J., Vigier, B., Tran, D., Mauléon, P., and Josso, N.,** Immunocytochemical study of anti-Müllerian hormone in sheep ovarian follicles during fetal and postnatal development, *J. Reprod. Fertil.,* 80, 509, 1987.
8. **Picard, J. Y. and Josso, N.,** Purification of testicular anti-Müllerian hormone allowing direct visualization of the pure glycoprotein and determination of yield and purification factor, *Mol. Cell. Endocrinol.,* 34, 23, 1984.
9. **Budzik, G. P., Swann, D. A., Hayashi, A., and Donahoe, P. K.,** Enhanced purification of Müllerian inhibiting substance by lectin affinity chromatography, *Cell,* 21, 909, 1980.
10. **Cate, R. L., Mattaliano, R. J., Hession, C., Tizard, R., Farber, N. M., Cheung, A., Ninfa, E. G., Frey, A. Z., Gash, D. J., Chow, E. P., Fisher, R. A., Bertonis, J. M., Torres, G., Wallner, B. P., Ramachandran, K. L., Ragin, R. C., Manganaro, T. F., MacLaughlin, D. T., and Donahoe, P. K.,** Isolation of the bovine and human genes for Müllerian inhibiting substance and expression of the human gene in animal cells, *Cell,* 45, 685, 1986.
11. **Picard, J. Y., Goulut, C., Bourrillon, R., and Josso, N.,** Biochemical analysis of bovine testicular anti-Müllerian hormone, *FEBS Lett.,* 195, 73, 1986.
12. **Picard, J. Y., Benarous, R., Guerrier, D., Josso, N., and Kahn, A.,** Cloning and expression of cDNA for anti-Müllerian hormone, *Proc. Natl. Acad. Sci. U.S.A.,* 83, 5464, 1986.

13. **Budzik, G. P., Powell, S. M., Kamagata, S., and Donahoe, P. K.,** Müllerian inhibiting substance fractionation by dye affinity chromatography, *Cell,* 34, 307, 1983.

14. **Budzik, G. P., Donahoe, P. K., and Hutson, J. M.,** A possible purification of Müllerian inhibiting substance and a model for its mechanism of action, *Prog. Clin. Biol. Res.,* 171, 207, 1985.

15. **Takahashi, M., Koide, S. S., and Donahoe, P. K.,** Müllerian inhibiting substance as oocyte meiosis inhibitor, *Mol. Cell Endocrinol.,* 47, 225, 1986.

16. **Fuller, A. F., Krane, I. M., Budzik, G. P., and Donahoe, P. K.,** Müllerian inhibiting substance reduction of colony growth of human gynecologic cancers in a stem cell assay, *Gynecol. Oncol.,* 22, 135, 1985.

17. **Rosenwaks, Z., Liu, H. C., Picard, J. Y., and Josso, N.,** Anti-Müllerian hormone is not cytotoxic to human endometrial cancer in tissue culture, *J. Clin. Endocrinol. Metab.,* 59, 166, 1984.

17a. **Tsafriri, A., Picard, J. Y., and Josso, N.,** Immunopurified anti-Müllerian hormone does not inhibit spontaneous resumption of meiosis in vitro of rat oocytes, *Biol. Reprod.,* 38, 481, 1988.

18. **Young, R. A. and Davis, R. W.,** Efficient isolation of genes by using antibody probes, *Proc. Natl. Acad. Sci. U.S.A.,* 80, 1194, 1983.

19. **Weinberger, C., Hollenberg, S., Ong, E. S., Harmon, J. M., Brower, S. T., Cidlowski, J., Thompson, E. B., Rosenfeld, M. G., and Evans, R. M.,** Identification of human glucocorticoid receptor complementary DNA clones by epitope selection, *Science,* 228, 740, 1985.

20. **Govindan, M. V., Devic, M., Green, S., Gronemeyer, H., and Chambon, P.,** Cloning of the human glucocorticoid receptor cDNA, *Nucleic Acids Res.,* 13, 8293, 1985.

21. **Vigier, B., Picard, J. Y., Campargue, J., Forest, M. G., Heyman, Y., and Josso, N.,** Secretion of anti-Müllerian hormone by immature bovine Sertoli cells in primary culture, studied by a competition type radio-immunoassay: lack of modulation by either FSH or testosterone, *Mol. Cell Endocrinol.,* 43, 141, 1985.

22. **Tsonis, C. G. and Sharpe, R. M.,** Dual gonadal control of follicle-stimulating hormone, *Nature (London),* 321, 724, 1986.

23. **Sporn, M. B., Roberts, A. B., Wakefield, L. M., and Assoian, R. K.,** Transforming growth factor-beta: biological function and chemical structure, *Science,* 233, 532, 1986.

24. **Cheifetz, S., Weatherbee, J. A., Tsang, M. L. S., Anderson, J. K., Mole, J. E., Lucas, R., and Massagué, J.,** The transforming growth factor-beta system, a complex pattern of cross-reactive ligands and receptors, *Cell,* 48, 409, 1987.

25. **Cochet, C.,** personal communication, 1987.

26. **Cohen-Haguenauer, O., Picard, J. Y., Mattei, M. G., Serero, S., Nguyen, V. C., de Tand, M. F., Guerrier, D., Hors-Cayla, M. C., Josso, N., and Frézal, J.,** Mapping of the gene for anti-Müllerian hormone to the short arm of human chromosome 19, *Cytogenet. Cell Genet.,* 44, 2, 1987.

27. **Fujii, D. M., Brissenden, J. E., Derynck, R., and Francke, U.,** Transforming growth factor beta gene maps to human chromosome 19 long arm and to mouse chromosome 7, *Somatic Cell. Mol. Genet.,* 12, 281, 1986.

28. **Mattei, M. G., Philip, N., Passage, E., Moisan, J. P., Mandel, J. L., and Mattei, J. F.,** DNA probe localization at 18pp113 band by in situ hybridization and identification of a small supernumerary chromosome, *Hum. Genet.,* 69, 268, 1985.

29. **Josso, N., Fékété, C., Cachin, O., Nézelof, C., and Rappaport, R.,** Persistence of Müllerian ducts in male pseudohermaphroditism, and its relationship to cryptorchidism, *Clin. Endocrinol. (Oxford),* 19, 247, 1983.

30. **Sloan, W. R. and Walsh, P. C.,** Familial persistent Müllerian duct syndrome, *J. Urol.,* 115, 459, 1976.

31. **Tran, D., Picard, J. Y., Campargue, J., and Josso, N.,** Immunocytochemical detection of anti-Müllerian hormone in Sertoli cells of various mammalian species, including man, *J. Histochem. Cytochem.,* 35, 733, 1987.

Chapter 21

NON-HORMONE-MEDIATED SEX CHROMOSOMAL EFFECTS IN DEVELOPMENT: ANOTHER LOOK AT THE Y CHROMOSOME-TESTICULAR HORMONE PARADIGM

Stan R. Blecher and Laura J. Wilkinson

TABLE OF CONTENTS

I. INTRODUCTION

Mammals have evolved an ingenious mechanism of sexual development that combines sex chromosomal control of sex determination (the commitment of the indifferent gonad to testicular or ovarian development) with the use of hormones for secondary sex differentiation (the development of internal organs of reproduction and external genitalia). In a system in which hormone-sensitive stages of development occur when individuals of both sexes are in a maternal environment, total dependency on hormones for primary determination of sex would clearly be untenable.[1]

The relationship of genetic and chromosomal factors to hormonal factors in sexual development has been well worked out during the last 50 years, thanks largely to the classical work of Greene and co-workers,[2] Jost,[3] and Josso.[4] The major features of the paradigm developed on the basis of these studies remain supportable. However, new data on some aspects of sexual development require some modification of the general theory. Several of the exceptions, including some studied in our laboratory, will be mentioned below.

II. THE Y CHROMOSOME-TESTICULAR HORMONE PARADIGM

In mammals, a genetic factor or factors on the Y chromosome determines maleness, since XO mice[5] and humans[6] (individuals with Turner Syndrome) have predominantly female phenotypes, whereas male patients with the Klinefelter Syndrome are XXY.[7] Evidently independent of the number of X chromosomes present, the presence of one Y chromosome leads to testis formation. Having served this purpose, Y-chromosomal sex-determining genes are believed to have no further role to play.[1] Subsequent development, called sex differentiation, is caused by hormones produced by the testes. As is the case in sex determination, the male state is the induced and the female the noninduced. Early work of Greene et al.[2] and Jost[3] showed that in the presence of testes, internal and external genitalia are masculinized, and in the absence of testes, they are feminized. Müllerian-inhibiting hormone[4] (MIH), produced in the testis, prevents the further development of the paramesonephric (Müllerian) duct in the male; in the absence of this hormone, the duct serves as precursor for the development of the uterine tubes, the uterus, and the upper portion of the vagina. Testosterone and its metabolite dihydrotestosterone promote the development of the mesonephric (Wolffian) duct system (MDS) and the urogenital sinus into male internal organs, and masculinize the external genitalia. Like the testes, the urogenital sinus and the external genitalia go through an "indifferent" phase; that is, the same precursor can be induced to become male or female, in contrast to the situation pertaining to the Wolffian and Müllerian duct systems.

According to the hypothesis of Wachtel et al.,[8] the gonadal inducer may be the male-specific H-Y antigen originally discovered in transplantation experiments.[9] When studied by serological techniques, male-specific antigen is called, by some workers, serologically detectable male (SDM) antigen.[10] The existing paradigm, then, is that gonadal sex is determined by sex-determining genes of the Y chromosome, and that all subsequent development of sexually dimorphic organs is hormonal. Any Y-chromosomal effect other than that mentioned above, and any effect of the X chromosome, on sexual dimorphism, would constitute exceptions to this rule. X-chromosomal effects could be due to noninactivated chromosomes, segments, or genes. X and Y effects could also be due to inactivated or active regions functioning by mechanisms other than gene transcription, e.g., as binding sites for regulatory molecules.[11] Normal action of X-linked genes in the differential segment, undergoing dosage compensation by the inactivation process of the Lyon hypothesis, are excluded from this discussion. As will be seen in the following, there appear to be several exceptions to the paradigm.

III. NON-SEX-DETERMINING Y-CHROMOSOMAL GENES

Aside from the role of sex determination, the existence of genes on the Y chromosome related, in various ways, to other aspects of testicular structure and function has been suggested several times.

Sex reversed (Sxr),[12] originally believed to be an autosomal dominant, sex-limited mimic gene of sex determination in the mouse, is now thought[13] to be a duplication of the Y-chromosomal testis-determining gene *(Tdy),*[14] which in the mouse appears to be situated on the long arm of the Y, close to the centromere. A transpositioning of *Sxr,* probably through an ancestral paracentric inversion, evidently resulted in *Sxr* being situated at the distal tip of the Y chromosome. In every meiosis in the testis of such *Sxr* carrier males, *Sxr* transfers, through crossover, to an X chromatid. Fertilization by such an *Sxr* chromatid of a normal, X-bearing ovum results in an XX*Sxr* "sex-reversed" mouse in which testes develop. Evidently normal masculinization occurs through androgen production. The testes, however, possess no germ cells in adults, evidently because of the inability of XX germ cells to proceed through gametogenesis in a testicular environment.[15]

In contrast to XO humans, XO mice are fertile. X*Sxr*O animals do undergo spermatogenesis, but the sperm are immotile.[12] It has accordingly been postulated that although *Sxr* represents a duplication of the testis-determining factor of the Y chromosome, it lacks a determinant for sperm motility.

McLaren and co-workers[16] presented evidence that the testis-determining and H-Y transplantation antigen-coding properties of *Sxr* are now separable; an H-Y negative variant has been called *Sxr'.* The component that confers H-Y positivity also acts as a "spermatogenesis gene", since spermatogenesis is blocked in its absence, as seen in X*Sxr'*O animals. Evidence for a Y-chromosomal spermatogenesis gene also comes from studies on XO/XY/XYY mosaic mice.[17] Finally, Hayward[18] described a Y-chromosomal gene that affects testis size in the mouse. In summary, it is possible that the mouse Y chromosome contains, aside from the *Tdy* locus, separate genetic loci for testis size, spermatogenesis, sperm motility, H-Y transplantation antigen, and SDM antigen.

Elizabeth Simpson and co-workers[19] recently presented evidence from studies on Y chromosome-deletion XY female and XX male individuals that the gene for H-Y transplantation antigen in the human is on the long arm of the Y chromosome. This may correspond to the site at which a postulated spermatogenesis-related Y-chromosomal gene in the human is situated.[20] Evidence has been presented[21] that testis size in humans may also be Y determined. The same may apply to other primates.[22]

In humans and mice, the male-determining locus of the Y chromosome is close to the centromere, on the short and long arms, respectively. Pairing and crossing-over between X and Y chromosomes occur in both species, near the tips of the X and Y short arms (humans) and long arms (mice). It is evident that genes situated distal to the site of crossover, on both X and Y chromosomes, will not be permanently linked to the sex-determining loci or centromeres of their respective chromosomes, and accordingly, will be genetically indistinguishable from autosomal loci. Accordingly, such loci have been named *pseudoautosomal* by Burgoyne.[23] *Sxr* is the prototypical pseudoautosomal gene; examples of pseudoautosomal genes in the human are also known.[24,25]

The gene locus for the enzyme steroid sulfatase *(STS)* in the mouse presented a paradox for some time, since some authors found evidence that it is X linked,[26] and others found evidence that it is autosomal.[27] The paradox has now been resolved; *STS* in the mouse is evidently pseudoautosomal, and both X and Y loci exist.[28] It had been known, prior to this discovery, that in the human, *STS* is in the noninactivated segment of the X.[29] Thus, this locus is just within the pairing segment in the human X-Y pair, and just distal to it in the mouse. Furthermore, *STS* is an example of a sex chromosomal gene that in the human produces a sexually dimorphic effect that is neither concerned with testis determination nor is androgen dependent.

The first X-linked gene locus shown to escape inactivation in either X chromosome, in XX individuals, was *Xg*.[30] Goodfellow and Tippett[31] studied the 12E7 antigen, coded for at the *MIC2* locus, and showed that this locus was either identical to *Xg* (suggesting the existence of *Yg* alleles) or extremely closely linked to *Xg*. Subsequently, *MIC2* was shown to be pseudoautosomal.[32] The role of this sexually dimorphic, non-sex-determining androgen independent, sex chromosomal gene is unknown.

A sexual dimorphism of tooth size (male teeth larger than female teeth) exists in humans,[33] other primates,[34] and rats.[35] Alvesalo and co-workers[36,37] have shown that in the human, a Y-linked gene situated on the long arm at Yq11 determines male sexual tooth size predominance, in both deciduous and permanent teeth. In contrast, Heller and Blecher[38] showed that male mice at puberty (35 ± 1 days of age) have very much smaller molar teeth than females have. This reversed sexual dimorphism was shown to be hormonal in origin rather than sex chromosomal, since in XO females and XY animals hemizygous for *Tfm*, (testicular feminization syndrome, see below), in both of which mutants hormonal status is female, large molars were observed. XX*Sxr* animals, which are chromosomally female but hormonally masculinized, had small molars.[38]

The sex chromosomes carry genes for body height. Patients showing features of the Turner Syndrome, with chromosome constitution 46,XXq⁻ (deletion of the long arm of an X chromosome), are of normal height, whereas patients with 46,Xi(Xq) (isochromosome of a long arm) show the short stature typical of Turner Syndrome; from this it has been deduced that the short stature is the result of hemizygosity for a gene in the noninactivated segment of Xp. A patient with Klinefelter Syndrome and 47,Xi(Xq)Y karyotype showed the sexual features but not the tall stature characteristic of this condition.[39] From this and other data, it is concluded that the tall stature in Klinefelter Syndrome is due to an additive effect of genes on Xp and Yp.

Rosenfeld et al.[40] described a phenotypically female infant with symptoms of the Turner Syndrome and with a karyotype of 46,XYp⁻. The deleted short arm of the Y chromosome accounted for the absence of the testis-determining factor of that chromosome, and possibly, the phenotypic features suggestive of Turner Syndrome. Most XO human fetuses are inviable. It is enigmatic that a segment of the X chromosome when in hemizygous state appears to be usually lethal or, when not lethal, dysmorphogenic, but when homozygous, appears to be viable and normally morphogenic, However, paradoxically, early stillborn XO fetuses have less of the Turner stigmata than late or surviving fetuses.[41,42] Possibly, the critical segment of the X is essential in homozygous state for normal development, is lethal in hemizygous state, and becomes viable (and dysmorphogenic) in a variant or mutated state. This suggests that living Turner Syndrome patients are carriers of a "survival mutation" on their monosomic X chromosome.

IV. WHOLE SEX CHROMOSOMES AND ANDROGEN SENSITIVITY

In a challenging paper on sex determination and sex differentiation in the human, Opitz[43] pointed out that Ohno,[44] in his development of the data derived by Jost[3] and others, draws the conclusion that extragonadal sexual dimorphism in mammals is due to prenatal or postnatal androgen effect. However, Opitz argued that a primary genetic effect of the Y chromosome in some nongenital, somatic sex differences has not been excluded. Opitz proposed that a test of the hypothesis that the Y chromosome affects somatic traits of sexual dimorphism would be an examination of dermatoglyphic similarity in male and female sibs of XY patients with Testicular Feminization Syndrome (TFS). In this condition, although the testes appear to be normally induced, target organs and tissues normally sensitive to androgens fail to respond to these hormones because of absence or abnormality of the hormone receptor.[44] In conformity with the paradigm described above, external genitalia remain in the uninduced feminine state. Androgen production proceeds vigorously because of loss of feedback control. Excess testosterone gets converted by aromatization to estrogen with resultant feminization: external genitalia mature as in a normal female (except that the vagina ends as a blind pouch where the Müllerian duct component would normally contribute to this organ), and feminine breast development occurs. The gene for TFS is X linked;[45] in conformity with Ohno's principle of conservation of the mammalian X chromosome,[46] homologous X-linked TFS exists in the mouse[47] and other mammals.

Penrose[48] showed that sex chromosomes influence dermatoglyphic total ridge count (TRC) in a manner suggestive of a whole chromosome effect, rather than one that could be ascribed to sex-linked genes. Sexual dimorphism in TRC is well-known; in Caucasian populations, for example, males have higher TRC (approximately $\bar{x} \sim 147$, cf. females $\bar{x} \sim 127$). In patients with additional X chromosomes, such as XXX and XXY, TRC is lowered, and in patients with XO, TRC is elevated to above the level of normal males.[48]

The X and Y chromosomes each appear to influence the ridge count in the same direction, but X chromosomes appear to have a greater effect. Penrose[48] suggests that the effect is dependent not on sex-linked genes, but on some as yet undetermined effect of whole sex chromosomes or segments.

Polani and Polani[49] studied TRC and palm-print patterns in families of patients with TFS. Their results indicated that palm-print patterns are influenced by sex chromosomal constitution. TRC in TFS propositae was the same as in female controls, suggesting hormonal influence. However, in secondary cases identified through TFS families, TRC was lower than in control females, and in both fathers and mothers of propositae, it showed a tendency to be lower than controls. These latter data might be compatible with an X-chromosomal effect (see below).

Blecher[50] studied TRC and palm patterns in a population of hybrid San (Bushman) peoples from the Pandamatenga area of Botswana, and found no significant differences between male and female TRC. However, TRC was lower in males and females than in Caucasoid peoples. Curiously, the San peoples display several traits of interest in this context. They have short physical stature, and the men are relatively ahirsute and have infantile genitalia, with the penis in a semierect position. These traits as well as the quite pronounced verticality of the forehead and "bossing" of the parietal prominences seen in these people have been interpreted as features of pedomorphism or neoteny (retention into adult life of infantile features),[51] a process believed to have played a role in human evolution.[51,52] The female of the San demonstrate extreme deposition of adipose tissue in the buttocks, producing the characteristic trait of steatopygia. Also characteristic of the San female is the globose areolar of the breast and the tablier,[53] a pronounced enlargement of the labia minora described by the early explorers of Africa as the "Hottentot apron", and for many years considered to be the result of ritual manipulation.

Steatopygia, globose areolar, and tablier of the female are all extremes of feminine, estrogen-dependent traits. Ahirsutism, infantile genitalia, and short stature of males, may also represent feminization; feminization may be a component of pedomorphism. The San allele at the X-linked locus for the androgen receptor may code for a receptor which is less sensitive to circulating androgens than is the receptor of other races. Development of external genitalia, hirsutism, body stature, and fat distribution are all androgen dependent. Androgens and estrogens appear to act in opposition in some situations, possibly including development of the breast.[54] In TFS, breast development may be even more pronounced than it is in normal females, presumably because of the unopposed action of estrogens. The globose areolar and the tablier of the San female may be similarly explained by estrogen activity that is less opposed by androgens than it is in females of other races. The low TRC in males of the Pandamatenga peoples, predominantly of San origin, may relate in part to hormonal and in part to X-chromosomal status. It appears possible that all the dermatoglyphic data cited above[48-50] could be accounted for if the segment of the X chromosome responsible for the effect proposed by Penrose,[48] and the TFS locus, (Tfm) were within the same region. (A partially homologous Y segment would also be postulated.) We envisage a variant within this segment in the San X chromosome causing a partial form of androgen insensitivity, and an extreme form of the TRC-lowering property.

Mittwoch[55] has developed the hypothesis that sex determination is dependent upon a "whole-chromosome" effect of the sex chromosomes, specifically an effect of the Y chromosome in speeding up cell division. Gonadal volumes, protein, and DNA content of male rat embryos are greater than those of their female litter mates; a similar situation pertains in humans and other mammals. In human true hermaphrodites, ovaries are situated more commonly on the left and testes more commonly on the right. In normal human fetuses, right gonads exceed left gonads in weight, protein and DNA content, and in numbers of cells. Thus, ovaries are smaller and grow slower than testes, and left gonads are smaller and grow slower than right gonads. Mittwoch[55] concludes that the observed lateral asymmetry of gonadal growth in human fetuses provides an explanation of the asymmetrical distribution of ovaries and testes in human hermaphrodites.

Sex determination and differentiation in birds is characterized by bilateral asymmetry. In the females of most species of birds, only the left gonad becomes a functional ovary, while the right gonad remains rudimentary. In comparing human sex determination with that of birds, Mittwoch concludes[55] that there is a tendency in both for left gonads to become ovaries and for right gonads to be testes; in humans, this tendency manifests itself in hermaphroditism, whereas in birds, the left gonadal tendency to be an ovary is seen in the normal female. In the human, the asymmetry is overridden by the phylogenetically more recent mechanism of Y-chromosomal male determination.

If whole-chromosome effects of X and Y chromosomes do play a role in sex determination through mechanisms relating to rate of cell division as suggested by Mittwoch,[55] or through any other process, it would be necessary to postulate means by which these general effects of the sex chromosomes become translated into organ-specific expression. Testis-specific proteins[56] and potential coding sequences[57] are known. Tissue-specific gene expression may involve local modification of existing regulatory proteins, rather than synthesis of new, specific proteins.[48] No mechanism has yet been proposed by which organ-specific modification can be applied to the Mittwoch theory of sex determination.

A generalized whole-body effect of sex chromosome differences must be presumed to account for the greater body weight of male than female 12-day rat embryos,[59] in which gonadal development is rudimentary. Sexual dimorphism of body size in human fetuses, recorded by ultrasonic scanning, has also been regarded as being nonhormonal,[60] though on a less well-documented basis.

Schmahl and co-workers[61] reported experiments in which male and female mice were exposed to X-irradiation at 11 to 13 days of postcoital age. Brains were fixed, sectioned, and stained for histological study at 18 days post conception and differences between male and female responses were reported. Schmahl and co-workers concluded[61] that, since the radiation took place prior to hormone production by the testis, the sexually dimorphic response must be due to chromosomal rather than hormonal effects. This result is interesting, but fails to take into account the possibility that substances released by the radiation damage may be initially dormant, and subsequently, only affect the brain after androgens are being produced in the male. Duplication of Schmahl's experiments using appropriate mutants[38] (see above) might resolve this question.

TFS, discussed above, may be partial or complete. In the murine *Tfm*/Y hemizygote, no trace of MDS remnants were described in the original report.[47] In a strain carrying *Tfm*, some unusual features of *Tfm* and the closely linked X chromosome coat-color marker gene Blotchy (*Blo*) were observed in Ohno's laboratory. In heterozygotes, the effect of *Blo* was very pronounced, producing an almost homogeneously light color instead of the expected blotches of light and agouti pigmentation. In addition, sexual development was affected, as evidenced in animals in which *Sxr* had been introduced. It was suggested that a hypothetical gene close to *Tfm* and *Blo* was affecting both these genes in some type of *cis*-acting position effect, possibly through disturbance of normal X-chromosomal inactivation. It was proposed that this putative controlling gene, named O^{hv}, might be an allele at the Xce locus.[62]

XY animals hemizygous for O^{hv}, Blo, and Tfm were found to possess microscopical epididymides.[62] Since Tfm/Y hemizygotes were expected to demonstrate no remnants of the MDS, the possibility existed that O^{hv} was acting directly on the MDS. However, the original study on Tfm in the mouse had not reported microscopical examination for epididymal remnants. Blecher[63] performed this study on Tfm/Y animals not carrying O^{hv} and showed that microscopical epididymides are a trait of Tfm and occur even in the absence of O^{hv}. Thus, these microorgans either are not dependent for their survival on androgens, or are not entirely insensitive to androgens. Subsequently, Scott and Blecher[64] examined the structures for the existence of histologically demonstrable β glucuronidase, which this laboratory had previously shown to be androgen dependent in the epididymis.[65] The enzyme is present at low levels in the microscopical epididymides of the Tfm/Y mouse, induced by endogenous androgens, but exogenous androgens fail to induce further activity.[64]

V. THE XXSxr GENOTYPE AND ANDROGEN SENSITIVITY

Because Tfm/Y hemizygote mice were, contrary to expectation, found to possess microscopical epididymides, we tested another prediction of the "Y chromosome-androgen paradigm". We examined whether, in XX animals masculinized by high levels of androgens, the male derivatives of the MDS so produced are entirely normal. Epididymides of XXSxr mice were studied histologically. We call these animals pseudomales because they are neither true males nor truly "sex reversed". They were found to lack the major epithelial cell type characteristic of the Initial Segment (IS) of the epididymis, called cell type EH 9 according to the scheme of Blecher and Kirkeby.[66] To determine whether the IS itself was absent, or whether the epithelium was transformed into one of a different histological appearance, serial sections through entire epididymal heads were examined and the continuity of the epididymal duct studied, and it was shown that the IS is indeed absent in XXSxr pseudomales.[67] In addition, vascularization of the epididymis is abnormal.[68] Androgen levels in epididymis, testis, and serum of these animals are normal, and exogenous testosterone failed to induce normal development of the IS.[69]

We report here preliminary electron microscopical findings on epididymides of 47-day-old XXSxr pseudomales and their normal, wild-type XY litter mate controls. Figures 1 and 2 depict normal and mutant development, respectively. Mutants showed major disruption of epithelial-mesenchymal development. Epithelial cells contained vesicles, often in mitochondria, such that normal mitochondria were rarely, if ever, seen. Basal lamina separating epithelial cells from underlying mesenchyme was thickened and abnormally infolded. Mesenchymal connective tissue was quite abnormal; collagen fibers were virtually absent, and replaced by thin, thread-like structures which might represent immature collagen molecules. Connective tissue cells were disorganized with respect to their position and form, contained vesicles, and were abnormally large. Similarly, smooth muscle cells were abnormal in shape, size, and position.

Since there is no evidence to associate the abnormalities observed with hormonal effects, the disruption of differentiation of this MDS derivative might be due to abnormal developmental signals resulting from the XXSxr genotype. We conclude that loci on the sex chromosomes may directly affect epithelial-mesenchymal interactions without mediation of androgens. The relationship of this putative function with sex determination of the gonad remains to be elucidated.

VI. SUMMARY

Mammalian sexual development is believed to take place through a cascade of events. Sex chromosomes, through genic and/or whole chromosome effects, determine the fate of

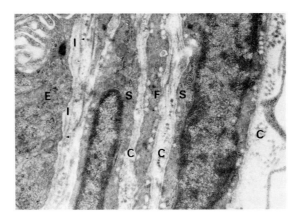

FIGURE 1. Electron micrograph of epididymis in 47-day-old wild-type male mouse. Note the epithelial cell (E) surrounded by a compact layer of smooth muscle cells (S) and fibrocytes (F). A basal lamina (arrows) with few infoldings (I) lies deep to the epithelial cell. Closely packed collagen fibrils (C) in longitudinal and cross-sectional arrangement are found between the smooth muscle cells and fibrocytes. Magnification × 15,140.

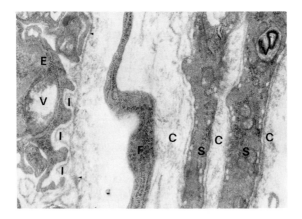

FIGURE 2. Electron micrograph of epididymis in 47-day-old XXSxr sex-reversed mouse. Note the abnormally shaped basal surface of the epithelial cell (E). Note also the thickened basal lamina (arrows) with numerous infoldings (I), and the abnormally shaped fibrocytes (F) and smooth muscle cells (S) surrounding the epithelial cell. All cell types contain abnormal vesicles (V). Collagen fibrils (C) are diffuse and irregular in shape. The entire layer of smooth muscle cells and fibrocytes is thickened due to the large extracellular spaces and abnormal shape of the cells. Magnification × 15,140.

an initially indifferent gonad to become either a testis or an ovary. It is believed that the Y chromosome causes the gonad to become a testis, and that, once determined, the testis directs the rest of sexually dimorphic development through production of hormones, and neither the Y nor X chromosome plays any further role in sexual development. MIH produced by the testis leads to degeneration of the Müllerian duct in males; absence of the testis and therefore, of MIH leads to development of the Müllerian duct into organs of the internal female genitalia. Androgens, also produced by the testis, promote the development of the

MDS and urogenital sinus into internal organs of the male genital tract, and in addition, androgens masculinize the external genitalia. Absence of androgens leads to degeneration of the MDS in females, and retention of the uninduced, feminine state of the external genitalia.

Exceptions to the above scheme exist at the levels of primary gonadal sex determination, secondary sexual differentiation, and the development of somatic, nongonadal sexually dimorphic structures. At the level of primary sex determination, XO Turner Syndrome and XXY Klinefelter Syndrome are examples of enigmas in sexual development not yet explained.

In secondary sex differentiation, the existing paradigm accounts in general terms for the observed phenomena; however, the persistence of microscopical epididymides in mice with testicular feminization, the absence of an IS of the epididymis in XX*Sxr* pseudomales, and the major disruptions observed in epithelial-mesenchymal developmental interactions in this organ are examples of developmental features not accounted for by the hormonal theory of control of secondary sexual development. Of somatic nongenital traits, dermatoglyphics, genes for stature, and, in the human, Y-linked genes for tooth size are examples of sexual dimorphism that appear to present a strong case for direct influence of sex chromosomes on normal somatic development.

NOTE ADDED IN PROOF

A recent paper by W.-S. O. et al. (*Nature (London)*, 331, 716, 1988) presents further evidence in support of the viewpoint that the "the classical view of mammalian sexual differentiation may have over-emphasized the role of testicular hormones, and overlooked earlier genetic effects". These authors report differentiation of the scrotum, mammary glands, gubernaculum, and processus vaginalis in the tammar wallaby, well before gonadal differentiation. The authors add that "there are also some examples of apparent sex differences which precede gonadal differentiation in eutherian mammals" and conclude that "the conventional view that all somatic sexual dimorphisms in mammals are a consequence of gonadal hormone secretion can no longer be supported".

REFERENCES

1. **Ohno, S., Nagai, Y., Ciccarese, S., and Iwata, H.,** Testis-organizing H-Y antigen and the primary sex-determining mechanism of mammals, *Recent Prog. Horm. Res.,* 35, 449, 1979.
2. **Greene, R. R., Burrill, M. W., and Ivy, A. C.,** Experimental intersexuality: the effect of antenatal androgens on sexual development of female rats, *Am. J. Anat.,* 65, 415, 1939.
3. **Jost, A.,** The role of fetal hormones in prenatal development, *Harvey Lecture Series 55,* Academic Press, New York, 1961.
4. **Vigier, B., Picard, J.-Y., Bezard, J., and Josso, N.,** Anti-Müllerian hormone: a local or long-distance morphogenetic factor?, *Hum. Genet.,* 58, 85, 1981.
5. **Welshons, W. J., and Russell, L. B.,** The Y-chromosome as the bearer of male determining factors in the mouse, *Proc. Natl. Acad. Sci. U.S.A.,* 45, 560, 1959.
6. **Ford, C. E., Jones, K. W., Polani, P. E., Almeida, J. C., and Briggs, J. H.,** A sex-chromosome anomaly in a case of gonadal dysgenesis (Turner's Syndrome), *Lancet,* I/7075, 711, 1959.
7. **Ford, C. E., Polani, P. E., Briggs, J. H., and Bishop, P. M. F.,** A presumptive human XXY/XX mosaic, *Nature (London),* 183, 1030, 1959.
8. **Wachtel, S. S., Ohno, S., Koo, G. C., and Boyse, E. A.,** Possible role for H-Y antigen in the primary determination of sex, *Nature (London),* 257, 235, 1975.
9. **Gasser, D. L. and Silvers, W. K.,** Genetics and immunology of sex-linked antigens, *Adv. Immunol.,* 15, 215, 1972.
10. **Silvers, W. K., Gasser, D. L., and Eicher, E. M.,** H-Y antigen, serologically detectable male antigen and sex determination, *Cell,* 28, 439, 1982.

11. **Chandra, H. S.,** A model for mammalian male determination based on a passive Y chromosome, *Mol. Gen. Genet.,* 193, 384, 1984.
12. **Cattanach, B. M., Pollard, C. E., and Hawkes, S. G.,** Sex-reversed mice: XX and XO males, *Cytogenetics,* 10, 318, 1971.
13. **Evans, E. P., Burtenshaw, M. D., Cattanach, B. M.,** Meitoic crossing-over between the X and Y chromosomes of male mice carrying the sex-reversing (*Sxr*) factor, *Nature (London),* 300, 443, 1982.
14. **Washburn, L. L. and Eicher, E. M.,** Sex reversal in XY mice caused by dominant mutation on chromosome 17, *Nature (London),* 303, 338, 1983.
15. **McLaren, A.,** Sex reversal in the mouse, *Differentiation,* 23, S93, 1983.
16. **McLaren, A., Simpson, E., Tomonari, K., Chandler, P., and Hogg, H.,** Male sexual differentiation in mice lacking H-Y antigen, *Nature (London),* 312, 552, 1984.
17. **Levy, E. R. and Burgoyne, P. S.,** The fate of XO germ cells in the testes of XO/XY and XO/XY/XYY mouse mosaics: evidence for a spermatogenesis gene on the mouse Y chromosome, *Cytogenet. Cell Genet.,* 42, 208, 1986.
18. **Hayward, P.,** Y chromosome effect on adult testis size, *Nature (London),* 250, 499, 1974.
19. **Simpson, E., Chandler, P., Goulmy, E., Disteche, C. M., Ferguson-Smith, M. A., and Page, D. C.,** Separation of the genetic loci for the H-Y antigen and for testis determination on human Y chromosome, *Nature (London),* 326, 876, 1987.
20. **Tiepolo, L. and Zuffardi, O.,** Localization of factors controlling spermatogenesis in the nonfluorescent portion of the human Y chromosome long arm, *Hum. Genet.,* 34, 119, 1976.
21. **Diamond, J. M.,** Variation in human testis size, *Nature (London),* 320, 488, 1986.
22. **Harcourt, A. H., Harvey, P. H., Larson, S. G., and Short, R. V.,** Testis weight, body weight and breeding system in primates, *Nature (London),* 293, 55, 1981.
23. **Burgoyne, P. S.,** Genetic homology and crossing over in the X and Y chromosomes of mammals, *Hum. Genet.,* 61, 85, 1982.
24. **Cooke, H. J., Brown, W. R. A., and Rappold, G. A.,** Hypervariable telomeric sequences from the human sex chromosomes are pseudoautosomal, *Nature (London),* 317, 687, 1985.
25. **Simmler, M. C., Rouyer, F., Vergnaud, G., Hystrom-Lahti, M., Ngo, K. Y., de la Chapelle, A., and Weissenbach, J.,** Pseudoautosomal DNA sequences in the pairing region of the human sex chromosomes, *Nature (London),* 317, 692, 1985.
26. **Gartler, S. M. and Riverst, M.,** Evidence for X-linkage of steroid sulfatase in the mouse: steroid sulfatase levels in oocytes of XX and XO mice, *Genetics,* 103, 137, 1983.
27. **Erickson, R. P., Harper, K., and Kramer, J. M.,** Identification of an autosomal locus affecting steroid sulfatase activity among inbred strains of mice, *Genetics,* 105, 181, 1983.
28. **Keitges, E., Rivest, M., Siniscalco, M., and Gartler, S. M.,** X-linkage of steroid sulphatase in the mouse is evidence for a functional Y-linked allele, *Nature (London),* 315, 226, 1985.
29. **Muller, C. R., Migl, B., Traupe, H., and Ropers, H. H.,** X-linked steroid sulfatase: evidence for different gene-dosage in males and females, *Hum. Genet.,* 54, 197, 1980.
30. **Race, R. R.,** Is the Xg blood group locus subject to inactivation?, *Human Genetics,* Elsevier, Amsterdam, 1972, 311.
31. **Goodfellow, P. N. and Tippett, P.,** A human quantitative polymorphism related to Xg blood groups, *Nature (London),* 289, 404, 1981.
32. **Goodfellow, P. J., Darling, S. M., Thomas, N. S., and Goodfellow, P. N.,** A pseudoautosomal gene in man, *Science,* 234, 740, 1986.
33. **Garn, S. M., Lewis, A. B., and Walenga, A. J.,** Crown-size profile pattern comparisons of 14 human populations, *Arch. Oral Biol.,* 13, 1235, 1968.
34. **Garn, S. M., Kerewsky, R. S., and Swindler, D. R.,** Canine "field" in sexual dimorphism of tooth size, *Nature (London),* 212, 1501, 1966.
35. **Riesenfeld, A.,** Relationship between facial protrusion and tooth length in four strains of rats, *Acta Anat.,* 97, 118, 1977.
36. **Alvesalo, L. and de la Chapelle, A.,** Tooth sizes in two males with deletions of the long arm of the Y-chromosome, *Ann. Hum. Genet.,* 45, 49, 1981.
37. **Alvesalo, L. and Kari, M.,** Sizes of deciduous teeth in 47,XYY males, *Am. J. Hum. Genet.,* 29, 486, 1977.
38. **Heller, N.-H. and Blecher, S. R.,** Reverse, hormone-dependent sex difference in molar tooth mass in pubertal mice, *Arch. Oral Biol.,* 27, 325, 1982.
39. **Gardiner, A., Brown, M. M., and Gray, J. E.,** Unusual chromosomal variant in Klinefelter's syndrome, *Br. Med. J.,* 4, 1123, 1978.
40. **Rosenfeld, R. G., Luzzatti, L., Hintz, R. L., Miller, O. J., Koo, G. C., and Wachtel, S. S.,** Sexual and somatic determinants of the human Y chromosome: studies in a 46,XYp⁻ phenotypic female, *Am. J. Hum, Genet.,* 31, 458, 1979.

41. **Singh, R. P. and Carr, D. H.,** The anatomy and histology of XO human embryos and fetuses, *Anat. Res.,* 155, 369, 1966.
42. **Carr, D. H.,** personal communication, 1986.
43. **Opitz, J. M.,** Comments on some genetic abnormalities of sex determination and sex differentiation in *Homo sapiens, Eur. J. Pediatr.,* 133, 77, 1980.
44. **Ohno, S.,** *Major Sex-Determining Genes,* Vol. 2, Springer-Verlag, Berlin, 1979.
45. **Meyer, W. J., III, Migeon, B. R., and Migeon, C. J.,** Locus on human X chromosome for dihydro-testosterone receptor and androgen insensitivity, *Proc. Natl. Acad. Sci. U.S.A.,* 72, 1469, 1975.
46. **Ohno, S.,** *Sex Chromosomes and Sex-Linked Genes,* Vol. 1, Springer-Verlag, Berlin, 1967.
47. **Lyon, M. F. and Hawkes, S. G.,** X-linked gene for testicular feminization in the mouse, *Nature (London),* 227, 1217, 1970.
48. **Penrose, L. S.,** Finger-print pattern and the sex chromosomes, *Lancet,* 1, 298, 1967.
49. **Polani, P. E. and Polani, N.,** Dermatologyphics in the testicular feminization syndrome, *Ann. Hum. Biol.,* 6, 417, 1979.
50. **Blecher, S. R.,** Dermatoglyphics of the Pandamatenga Bush-Bantu hybrids, *Hum. Hered.,* 22, 149, 1972.
51. **Tobias, P. V.,** Bushmen of the Kalahari, *Man (London).,* 57, 33, 1957.
52. **Bolk, L.,** Origin of racial characteristics in man, *Am. J. Phys. Anthropol.,* 13, 1, 1929.
53. **Orkin, C. J.,** The tablier in Griqua women, *Leech,* 50, 42, 1970.
54. **Griffin, J. E. and Wilson, J. D.,** The syndromes of androgen resistance, *N. Engl. J. Med.,* 302, 198, 1980.
55. **Mittwoch, U.,** Males, females and hermaphrodites, *Ann. Hum. Genet.,* 50, 103, 1986.
56. **McCarrey, J. R. and Thomas K.,** Human testis-specific PGK gene lacks introns and possesses characteristics of a processed gene, *Nature (London),* 326, 501, 1987.
57. **Bishop, C. E. and Hatat, D.,** Molecular cloning and sequence analysis of a mouse Y chromosome RNA transcript expressed in the testis, *Nucleic Acids Res.,* 15, 2959, 1987.
58. **Maniatis, T., Goodbourn, S., and Fischer, J. A.,** Regulation of inducible and tissue-specific gene expression, *Science,* 236, 1237, 1987.
59. **Scott, W. J. and Holson, J. F.,** Weight differences in rat embryos prior to sexual differentiation, *J. Embryol. Exp. Morphol.,* 40, 259, 1977.
60. **Pedersen, J. F.,** Ultrasound evidence of sexual difference in fetal size in first trimester, *Br. Med. J.,* 281, 1253, 1980.
61. **Schmahl, W., Weber, L., and Kriegel, H.,** Sexual dimorphism of mouse fetal brain lesions after X-irradiation prior to gonadal differentiation, *Experientia,* 35, 1653, 1979.
62. **Drews, U., Blecher, S. R., Owen, D. A., and Ohno, S.,** Genetically directed preferential X-activation seen in mice, *Cell,* 1, 3, 1974.
63. **Blecher, S. R.,** Microscopic epididymides in testicular feminisation, *Nature (London),* 275, 748, 1978.
64. **Scott, J. E. and Blecher, S. R.,** β-Glucuronidase activity is present in the microscopic epididymis of the *Tfm/*Y mouse, *Dev. Genet.,* 8, 11, 1987.
65. **Blecher, S. R. and Kirkeby, S.,** Histochemical studies on genetical control of hormonal enzyme inducibility in the mouse V. Histochemical evidence for androgen inducibility of β-glucuronidase in the epididymis, *Acta Histochem.,* 70, 8, 1982.
66. **Blecher, S. R. and Kirkeby, S.,** Histochemical studies on genetical control of hormonal enzyme inducibility in the mouse. I. Non-specific esterase activity and regional histology of the epididymis, *J. Anat.,* 125, 247, 1978.
67. **LeBarr, D. K. and Blecher, S. R.,** Epididymides of sex-reversed XX mice lack the initial segment, *Dev. Genet.,* 7, 109, 1986.
68. **LeBarr, D. K. and Blecher, S. R.,** Decreased arterial vasculature of the epididymal head in XX*Sxr* pseudomale ("sex-reversed") mice, *Acta Anat.,* 129, 123, 1987.
69. **LeBarr, D. K., Blecher, S. R., and Moger, W. H.,** Androgen levels and androgenization in sex-reversed (XX*Sxr* pseudomale) mouse: absence of initial segment of epididymis is independent of androgens, *Arch. Androl.,* 17, 195, 1986.

Part VI: Molecular Biology in Medicine and Agriculture

Chapter 22

REVERSE GENETICS AND THE
DUCHENNE MUSCULAR DYSTROPHY LOCUS

Anthony P. Monaco and Louis M. Kunkel

TABLE OF CONTENTS

I. INTRODUCTION

The study of inherited human diseases by "reverse genetics" has had recent advances in chronic granulomatous disease (CGD), Duchenne muscular dystrophy (DMD), and retinoblastoma.[1] As the human genome becomes increasingly accessible by molecular cloning technology,[2] many diseases caused by defects in unknown gene products will be amenable to similar reverse genetic approaches. To begin, the map position of an inherited human disorder is usually established by several methods. The two most common approaches include linkage analysis with DNA markers that detect restriction fragment-length polymophisms (RFLPs)[3] and cytological detection of structural abnormalities (deletions and translocations). These two mapping methods can narrow a disease locus to a region of roughly several million base pairs.

To understand regions of the genome that are structurally altered or mutated to give rise to disease, it is first necessary to physically isolate and map DNA sequences from within these regions, and to characterize them at a molecular level. One approach used to clone the gene responsible for X-linked CGD was subtractive hybridization of CGD-enriched cDNA with RNA from a patient bearing a large Xp21 deletion.[4] The subtracted cDNA was used as a probe to detect an exon in cloned genomic DNA around the pERT379 locus. The hybridizing pERT379 genomic fragment identified a 4.7-kb mRNA that was structurally altered or not detectable in the monocyte RNA of several CGD patients. A different approach that was successful in defining exons for the genes involved in DMD[5] and retinoblastoma[6] was to search cloned genomic DNA for nucleotide sequence conservation across mammalian species. This strategy was based on the assumption that DNA sequences encoding amino acids (exons) were conserved during evolution, while noncoding sequences (introns) diverged over time.[7]

II. DUCHENNE MUSCULAR DYSTROPHY

The *DMD* locus has the highest mutation rate among X-linked genetic disorders (1/3000);[8] about one third of cases are new or sporadic. The disease is characterized clinically by the early onset of progressive muscle weakness and degeneration of muscle fibers. Affected males are usually confined to a wheelchair for mobility by age 12 with death by the end of the second decade. Becker muscular dystrophy (BMD) is a less severe X-linked myopathy with a later age of onset and milder clinical progression of muscle weakness and degeneration. The site for DMD and BMD mutations was pinpointed to the band Xp21 of the human X chromosome short arm by genetic linkage analysis[9-12] and detection of structural alterations.[13-15]

III. ISOLATION OF DNA PROBES IN Xp21

The physical isolation of DNA sequences from within the Xp21 region was accomplished using three independent approaches. The compiled data are represented in Figure 1 which summarizes our current understanding of the genomic structure of the *DMD/BMD* locus. Recombinant genomic-phage libraries constructed from flow-sorted X chromosome DNA yielded several random X-specific probes that map close to and flank the band Xp21.[9,10,16-18] The genomic loci defined by the X-specific DNA probes were found to be closely linked to DMD and BMD mutations segregating in families, but were not close enough to directly detect structural alterations within the affected gene itself.[9-11,19,20] To obtain DNA sequences within the gene responsible for DMD and BMD, two different structural alterations of the band Xp21 were manipulated. In rare cases of females with muscular dystrophy, balanced X; autosome translocations were detected with one breakpoint

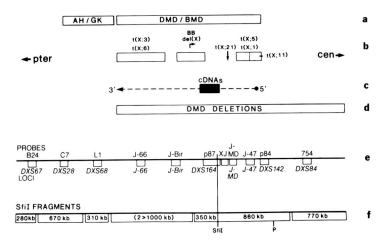

FIGURE 1. The Xp21 *DMD/BMD* locus. The different levels of the diagram illustrate features of the Xp21 region that have emerged from various types of study. (a) Approximate extent of the *DMD/BMD* locus and its position relative to the adrenal hypoplasia (*AH*) and glycerol kinase (*GK*) deficiency loci. (b) Regions identified as containing translocation breakpoints in DMD patients; the blocks indicate the region of uncertainty to which the breakpoints map as determined by hybridization experiments on translocation chromosomes segregated in mouse-human somatic cell hybrids.[32] (c) The black box shows the region containing sequences that hybridize to the cDNA clones that have so far been derived from a 16-kbp transcript found in skeletal muscle. The dotted line and arrow indicate the predicted extent of genomic DNA (approximately 2,000 kilobase pairs) corresponding to the whole of this transcript, and the predicted direction of transcription. (d) The block shows the extent of DNA that is covered by DMD deletions the endpoints of which have been mapped to date. (e) Positions of loci throughout Xp21 and designations of some of the probes that identify the loci. For example, the locus *DXS164* was originally identified by the 200-bp probe, pERT87 (p87); the locus has been extended by chromosome walking to cover 220 kbp of DNA as shown. The J-66 and J-Bir regions (containing the outlying endpoints of two long DMD deletions, the other endpoints of which are within the *DXS164* locus) are positioned on the distal side of the *DXS164* locus and the B.B. deletion breakpoint, with J-66 more distal than J-Bir. The relative order for the loci identified by probes B24, C7, and L1 (*DXS67, DXS28,* and *DXS68,* respectively) was assigned by deletion analysis of male patients with combinations of AH, GK, and DMD.:[30,35] The J-MD and J-47 regions (outlying ends of long DMD deletions beginning in the *DXS164* locus) lie on the distal side of the X;11 translocation. A DNA fragment (HIP25) generated by competitive reassociation with B.B.'s DNA also maps to sequences between these two translocation breakpoints.[37] The probes pERT84 (p84) and 754 identify loci *DXS142* and *DXS84,* which are on the centromere side of the X;21 and X;11 translocations, with the latter being the closer to the centromere. The X;5 and X;1 translocations are positioned between sequences identified by J-MD and pERT84.[32] (f) Order of Sfi I restriction fragments that have been identified by pulsed field-gel electrophoresis distributed over approximately 4500 kbp of DNA in the Xp21 region.[35-37] Additional Sfi I fragments may exist between some of the DNA clones, and this is indicated by spaces between the fragments. The position of a Sfi I site within the *DXS164* locus is shown. The letter P indicates a restriction site that is only partially cleaved, presumably because of methylation. (From Monaco, A. P. and Kunkel, L. M., *Trends Genet.,* 3, 34, 1987. With permission.)

in the band Xp21. In one case, the autosomal breakpoint was in the ribosomal gene locus on chromosome 21.[21] A ribosomal gene segment was used as a probe to isolate the breakpoint from this translocation chromosome thus yielding an Xp21-specific DNA fragment on the other side.[22] The Xp21-specific fragment (XJ) detected deletions in unrelated DMD and BMD males and was tightly linked to mutations segregating in families.

A different approach to isolate DNA fragments specifically within the Xp21 region was to use a visible interstitial deletion identified in a male patient, B. B., with multiple X-linked disorders: DMD, CGD, retinitis pigmentosa, and the McLeod red blood cell phenotype.[15] DNA isolated from a lymphoblastoid cell line established from B. B. was used in a competition, reassociation, and cloning strategy to yield nine DNA fragments absent from the Xp21 megabase deletion.[23] One of these nine DNA fragments, pERT87, was found to identify deletions in unrelated DMD males and was also tightly linked to mutations segregating in families.[24]

IV. EXPANSION OF Xp21 LOCI

A. Chromosome Walking

The small pERT87 probe was used to initiate chromosome "walking"[25] in genomic-phage libraries to isolate DNA sequences involved in DMD and BMD. 220 kb of genomic DNA were isolated (designated the *DXS164* locus)[12] containing sequences structurally altered in DMD and BMD males. In an international study of 1346 DMD and BMD males, 88 (6.5%) were found to have complete or partial deletions of the *DXS164* locus.[26] Most deletions were large (>100 kb), with 40% having one breakpoint in cloned DNA. The deletion endpoints were overlapping and nonoverlapping, with deletions extending toward both the centromere and telomere of the short arm.[26,27] Four deletions were identified with both endpoints in the *DXS164* locus, one of which was a small 6-kb deletion detected in a BMD patient.[27,28] Common sequences of the *DXS164* locus absent in DMD and BMD patients bearing deletions provided further evidence for the allelic nature of the two clinical phenotypes.

Single-copy genomic clones from the *DXS164* locus were used to search for RFLPs to follow the segregation of mutation and recombination events in DMD families. Many RFLPs were found that have been used in prenatal diagnosis and carrier detection of DMD. Linkage data from many laboratories have indicated a substantial recombination rate (4 to 6%) between *DXS164* subclones and DMD mutations segregating in families.[26,29,30] This indicated that the *DMD/BMD* gene locus was extremely large or that the genomic DNA containing it was recombining with mutations at a high rate.

B. Deletion Junction Cloning

Since the Xp21 locus involved in muscular dystrophy seemed to be larger than, or outside of, the *DXS164* locus, several deletion junction fragments were cloned to obtain DNA on the other side of the breakpoint, an unknown distance from the *DXS164* locus.[27] Deletion endpoints were mapped with single-copy clones from the *DXS164* locus and junction fragments were identified on Southern blots. Four deletions were chosen to construct libraries for isolation of breakpoints from deletions extending toward the centromere and telomere from the *DXS164* locus. (See Figure 1 for location relative to the *DXS164* locus and various translocation breakpoints.) Two junctions were isolated (J-MD and J-47) with their endpoints on the centromere side of both the *DXS164* locus and the X;21 translocation, yet on the distal side of the X;11 translocation. Two other deletion junctions were isolated (J-BIR and J-66) with their endpoints on the telomere side of the *DXS164* locus, yet present in the DNA of the large Xp21 deletion of B. B., a patient with DMD and multiphenotype disease.[15] The junction loci were each expanded by chromosome walking to about 50 kb, and were examined for RFLP detecting probes. A polymorphism detected by a probe from the J-BIR region was recombinant with a DMD mutation segregating in a family that was also recombinant for the *DXS164* locus and the more proximal *DXS84* (754) locus.[26,27] The inheritance pattern of Xp21 probes in this family placed the mutation and recombination event as distal in the *DMD* locus and confirmed the large region in which mutations were thought to give rise to DMD and BMD.

V. SEARCH FOR THE *DMD* GENE

In addition, the *DXS164* locus was shown to be centrally located to the *DMD/BMD* gene by deletion analysis and physical mapping with respect to translocation breakpoints.[26,27] To find potential exons in the 220 kb of the *DXS164* locus, single-copy genomic clones were tested by hybridization on Southern blots containing various mammalian DNA samples.[5] Significant nucleotide conservation was defined by cross-species hybridization on Southern blots at high stringency. Two DNA fragments, separated by more than 70 kb in the *DXS164* locus, were found to be conserved among all mammalian species tested. The equivalent genomic loci were isolated from the mouse and mapped to the mouse X chromosome in a somatic cell hybrid bearing the mouse X on a hamster DNA background.[5] The nucleotide sequences of the human and mouse conserved DNA fragments were determined. This indicated intron: exon borders with short, open reading frames coding for amino acids. One of the human conserved fragments (pERT87-25) identified a large 16-kb transcript in fetal skeletal muscle and was used as a probe to isolate cDNA clones constructed from polyadenylated RNA.[5] The cDNA clones (1.0 kb) were mapped to Xp21 on genomic Southern blots and hybridized to many small restriction fragments in phage clones spread across 110 kb of the *DXS164* locus. The exons were predicted to be quite small (150 base pairs) and the introns large (average 16 kb); the ratio of cDNA clone size to genomic DNA predicted a *DMD/BMD* locus of 1 to 2 million base pairs.

A large size for the *DMD/BMD* locus is consistent with many aspects of the disease, especially the high mutation frequency.[31] Most deletions identified in DMD and BMD patients were found to be large in extent (>100 kb), and their breakpoints heterogeneously located in the Xp21 region.[26,27] Translocations giving rise to the disease in females also have their breakpoints widely spaced over the Xp21 region at the cytological and molecular mapping level.[13,14,32] Most recently, the advent of pulsed-field gel electrophoresis (PFGE)[33,34] has allowed several groups to construct a megabase map of the Xp21 region.[35-37] This map also indicates that deletions and translocation breakpoints involved in DMD and BMD are spread over a 1500 to 2000-kb region (see Figure 1). The large size for the *DMD/BMD* locus explains the high recombination rate (4 to 6%) of RFLP-detecting probes within the gene with widely spaced mutations giving rise to disease. This will obviously complicate prenatal diagnosis and carrier detection and will necessitate the use of many DNA markers within and flanking the gene locus.

VI. CONCLUSION

The molecular cloning of the complete cDNA representation of the large *DMD/BMD* gene is now in progress. The nucleotide sequence and predicted amino acid sequence from mouse and human should enlighten investigators as to a possible function for such a large protein. Focused study of this large muscle protein by molecular biologists, biochemists, and physiologists may one day lead to a treatment for the X-linked muscular dystrophies. In general, approaches used to isolate and characterize DNA sequences from the Xp21 region in search for the *DMD/BMD* gene may be useful models for research attempts in other loci related to human disease.

ACKNOWLEDGMENTS

This work was supported by the Muscular Dystrophy Association of America, the NIH (RO1 NS23740 and HD18658) and PHS NRSA (2T 32 GM0775307) from the National Institute for General Medical Sciences.

REFERENCES

1. **Orkin, S. A.,** Reverse genetics and human disease, *Cell,* 47, 845, 1986.
2. **Poustka, A. and Lehrach, H.,** Jumping libraries and linking libraries: the next generation of molecular tools in mammalian genetics, *Trends Genet.,* 2, 174, 1986.
3. **Botstein, D., White, R. L., Skolnick, M., and Davis, R. W.,** Construction of a genetic linkage map in man using restriction fragment length polymorphisms, *Am. J. Hum. Genet.,* 32, 314, 1980.

4. **Royer-Pokora, B., Kunkel, L. M., Monaco, A. P., Goff, S. C., Newburger, P. E., Baehner, R. L., Cole, F. S., Curnette, J. T. and Orkin, S. H.,** Cloning the gene for an inherited human disorder (chronic granulomatous disease) on the basis of its chromosomal location, *Nature (London)*, 322, 32, 1986.

5. **Monaco, A. P., Neve, R., Colletti-Feener, C., Bertelson, C. J., Kurnit, D. M., and Kunkel, L. M.,** Isolation of candidate cDNAs for portions of the Duchenne muscular dystrophy gene, *Nature (London).*, 323, 646, 1986.

6. **Friend, S. H., Bernards, R., Rogelj, S., Weinberg, R. A., Rapaport, J. M., Albert, D. M., and Dryja, T. P.,** A human DNA segment with properties of the gene that predisposes to retinoblastoma and osteosarcoma, *Nature (London)*, 323, 643, 1986.

7. **Perler, F., Efstratiadis, A., Lomedico, P., Gilbert, W., Kolodner, R., and Dodgson, J.,** The evolution of genes: the chicken preproinsulin gene, *Cell,* 20, 555, 1980.

8. **Moser, H.,** Duchenne muscular dystrophy: pathogenetic aspects and genetic prevention, *Hum. Genet.,* 66, 17, 1984.

9. **Davies, K. E., Pearson, P. L., Harper, P. S., Murray, J. M., O'Brien, T., Sarfarazi, M., and Williamson, R.,** Linkage analysis of two cloned DNA sequences flanking the Duchenne muscular dystrophy locus on the short arm of the human X chromosome, *Nucleic Acids Res.,* 11, 2303, 1983.

10. **Aldridge, J. A. Kunkel, L., Bruns, G., Tantravahi, U., Lalande, M., Brewster, T., and Moreau, E.,** A strategy to reveal high frequency RFLPs along the human X chromosome, *Am. J. Hum. Genet.,* 36, 546, 1984.

11. **de Martinville, B., Kunkel, L. M., Bruns, G., Morle, F., Koenig, M., Mandel, J. L., Horwich, A., Latt, S. A., Gusella, J. F., Housman, D., and Francke, U.,** Localization of DNA sequences in the region Xp21 of the human X chromosome: search for molecular markers close to the Duchenne muscular dystrophy locus, *Am. J. Hum. Genet.,* 37, 235, 1985.

12. **Goodfellow, P. N., Davies, K. E., and Ropers, H. H.,** Report of the committee on the genetic constitution of the X and Y chromosomes, *Cytogenet. Cell Genet.,* 40, 296, 1985.

13. **Boyd, Y. and Buckle, V. J.,** Cytogenetic heterogeneity of translocations associated with Duchenne muscular dystrophy, *Clin. Genet.,* 29, 108, 1986.

14. **Boyd, Y., Buckle, V., Holt, S., Munro, E., Hunter, D., and Craig, I.,** Muscular dystrophy in girls with X;autosome translocations, *J. Med. Genet.,* 23, 484, 1986.

15. **Francke, U., Ochs, H. D., de Martinville, B., Giacalone, J., Lindgren, V., Disteche, C. M., Pagon, R. A., Hofker, M. H., van Ommen, G. J. B., Pearson, P. L., and Wedgwood, R. J.,** Minor Xp21 chromosome deletion in a male associated with expression of Duchenne muscular dystrophy, chronic granulomatous disease, retinitis pigmentosa, and McLeod syndrome, *Am. J. Hum. Genet.,* 37, 250, 1985.

16. **Davies, K. E., Young, B. D., Elles, R. G., Hill, M. E., and Williamson, R.,** Cloning of a representative genomic library of the human X chromosome after sorting by flow cytometry, *Nature (London)*, 293, 374, 1981.

17. **Kunkel, L. M., Tantravahi, U., Eisenhard, M., and Latt, S. A.,** Regional localization on the human X of DNA segments cloned from flow-sorted chromosomes, *Nucleic Acids Res.,* 10, 1557, 1982.

18. **Kunkel, L. M., LaLande, M., Monaco, A. P., Flint, A., Middlesworth, W., and Latt, S. A.,** Construction of a human X-chromosome-enriched phage library which facilitates analysis of specific loci, *Gene,* 33, 251, 1985.

19. **Dorkins, H., Junien, C., Mandel, J. L., Wrogemann, K., Moison, J. P., Martinez, M., Old, J. M., Bundey, S., Schwartz, M., Carpenter, N., Hill, D., Lindlof, M., de la Chapelle, A., Pearson, P. L., and Davies, K. E.,** Segregation analysis of a marker localised Xp21.2-Xp21.3 in Duchenne and Becker muscular dystrophy families, *Hum. Genet.,* 71, 103, 1985.

20. **Hofker, M. H., Wapenaar, M. C., Coor, N., Bakker, E., van Ommen, G. J. B., and Pearson, P. L.,** Isolation of probes detecting restriction fragment length polymorphisms from X-chromosome specific libraries: potential use for diagnosis of Duchenne muscular dystrophy, *Hum. Genet.,* 70, 148, 1985.

21. **Verellen-Dumoulin, Ch., Freund, M., DeMeyer, R., Laterre, Ch., Frederic, J., Thompson, M. W., Markovic, V. D., and Worton, R. G.,** Expression of an X-linked muscular dystrophy in a female due to translocation involving Xp21 and non-random inactivation of the normal X chromosome, *Hum. Genet.,* 67, 115, 1984.

22. **Ray, P. N., Belfall, B., Duff, C., Logan, C., Kean, V., Thompson, M. W., Sylvester, J. E., Gorski, J. L., Schmickel, R. D., and Worton, R. G.,** Cloning of the breakpoint of an X;21 translocation associated with Duchenne muscular dystrophy, *Nature (London)*, 318, 672, 1985.

23. **Kunkel, L. M., Monaco, A. P., Middlesworth, W., Ochs, H. D., and Latt, S. A.,** Specific cloning of DNA fragments absent from the DNA of a male patient with an X chromosome deletion, *Proc. Natl. Acad. Sci. U.S.A.,* 82, 4778, 1985.

24. **Monaco, A. P., Bertelson, C. J., Middlesworth, W., Colletti, C.-A., Aldridge, J., Fischbeck, K. H., Bartlett, R., Pericak-Vance, M. A., Roses, A. D., and Kunkel, L. M.,** Detection of deletions spanning the Duchenne muscular dystrophy locus using a tightly linked DNA segment, *Nature (London)*, 842, 1985.

25. **Bender, W., Arkam, M., Karch, F., Beachy, P. A., Peifer, M., Spierer, P., Lewis, E. B., and Hogness, D. S.,** Molecular genetics of the bithorax complex in Drosophila melanogaster, *Science,* 221, 23, 1983.
26. **Kunkel, L. M., et al.,** Analysis of deletions in the DNA of patients with Becker and Duchenne muscular dystrophy, *Nature (London),* 322, 73, 1986.
27. **Monaco, A. P., Bertelson, C. J., Colletti-Feener, C., and Kunkel, L. M.,** Localization and cloning of Xp21 deletion breakpoints involved in muscular dystrophy, *Hum. Genet.,* 75, 221, 1987.
28. **Hart, K. A., Hodgson, S., Walker, A., Cole, C. G., Johnson, L., Bubowitz, V., and Bobrow, M.,** DNA deletions in mild and severe Becker muscular dystrophy, *Hum. Genet.,* 75, 281, 1987.
29. **Fischbeck, K. H., Ritter, A. W., Tirschwell, D. L., Kunkel, L. M. Bertelson, C. J., Monaco, A. P., Hejtmancik, J. F., Boehm, C., Ionasescu, V., Ionasescu, R., Pericak-Vance, M., Kandt, R., and Roses, A. D.,** Recombination with PERT87 (DXS164) in families with X-linked muscular dystrophy, *Lancet,* 2, 104, 1986.
30. **Bertelson, C. J., Bartley, J. A., Monaco, A. P., Colletti-Feener, C., Fischbeck, K., and Kunkel, L. M.,** Localization of Xp21 meiotic exchange points in DMD families, *J. Med. Genet.,* 23, 531, 1986.
31. **Monaco, A. P. and Kunkel, L. M.,** A giant locus for the Duchenne and Becker muscular dystrophy gene, *Trends Genet.,* 3, 33, 1987.
32. **Boyd, Y., Munro, E., Ray, P., Worton, R., Monaco, T., Kunkel, L., and Craig, I.,** Molecular heterogeneity of translocations associated with muscular dystrophy, *Clin. Genet.,* 31, 265, 1987.
33. **Schwartz, D. C. and Cantor, C. R.,** Separation of yeast chromosome-sized DNAs by pulsed field gel electrophoresis, *Cell,* 37, 67, 1984.
34. **Carle, G. F., Frank, M., and Olson, M. V.,** Electrophoretic separations of large DNA molecules by periodic inversion of the electric field, *Science,* 4, 65, 1986.
35. **van Ommen, G. J. B., Verkerk, J. M. H., Hofker, M. H., Monaco, A. P., Kunkel, L. M., Ray, P., Worton, R., Wieringa, B., Bakker, B., and Pearson, P. L.,** A physical map of 4 million bp around the Duchenne muscular dystrophy gene on the human X-chromosome, *Cell,* 47, 499, 1986.
36. **Burmeister, M. and Lehrach, H.,** Long-range restriction map around the Duchenne muscular dystrophy gene, *Nature (London),* 324, 582, 1986.
37. **Kenwrick, S., Patterson, M., Speer, A., Fischbeck, K., and Davies, K.,** Molecular analysis of the Duchenne muscular dystrophy region using pulsed field gel electrophoresis, *Cell,* 48, 351, 1987.

Chapter 23

THE MOLECULAR BIOLOGY OF VISION

Larry A. Donoso

TABLE OF CONTENTS

I. INTRODUCTION

Blindness is feared by more people than any other physical affliction except cancer. The biochemical mechanisms by which light energy is converted into an electrical impulse and subsequently processed by the brain are highly complex and have been investigated for over a century. Consequently, in this article, it is possible only to highlight a few aspects that relate to the molecular biology of vision, and only rhodopsin and S-antigen will be discussed in any detail. The reader is referred to other reviews for additional information.[1,2]

II. ANATOMY OF THE EYE

The eye (Figure 1) is composed of three tissue layers: the outer fibrous tunic, the middle uveal tract, and the inner retina. The fibrous tunic, which maintains the shape of the eye, is composed of the white opaque sclera and the clear cornea. The cornea is a curved surface forming a convex lens of approximately 43 diopters of refractive power, thus, constituting the main refractive element of the eye. Light refracted by the cornea passes through the anterior chamber formed by the iris posteriorly and the cornea anteriorly. Light rays are further refracted by the lens, a biconvex structure, which on occasion may become cloudy forming a cataract.

Refracted light finally passes through the vitreous, the posterior chamber of the eye, to impinge on the retina. The second tissue layer is composed of the uveal tract consisting of the iris, choroid, and ciliary body. This layer of the eye is highly vascularized and serves a nutritive function. The innermost layer of the eye is the retina. The retina represents an extension of the brain with neurons eventually terminating in the occipital lobe. There are at least ten well-defined layers of the retina including neural, glial, and vascular tissue elements (Figure 2).

III. THE PHOTOTRANSDUCTION OF VISION

The photoreceptor cell layer of the retina contains some 130 million specialized rods and cones which are the principal cells that interact with light and initiate the visual process. These cells are composed of a synaptic terminal, an inner segment, and an outer segment containing one of four visual pigments (Figure 3). The rod photoreceptor cell, which absorbs light maximally at 495 nm, is responsible for monochromatic vision in dim light and contains the visual pigment rhodospin.[3] In contrast, the cone cells contain one of three visual pigments and are responsible for color vision (trichromatic vision) and function in normal light intensity. Each visual pigment absorbs a particular range of wavelengths corresponding to the electromagnetic spectrum between 400 and 700 nm. For example, the red-sensitive pigment absorbs maximally at 560 nm, whereas the green and blue pigments absorb maximally at 530 and 420 nm, respectively.[4]

The complex biochemical mechanisms of the visual process are mediated in part by a series of well-studied photoreceptor cell proteins including rhodopsin, guanine nucleotide-binding protein (G-protein, transducin), guanosine $3'5'$ monophosphate phosphodiesterase (GMP), rhodopsin kinase, and S-antigen (48 K protein).[5] The process by which a single photon of light is converted into an electrical impulse (Figure 4) is highly efficient; each molecule of light-activated rhodopsin initiates a cascade of biochemical reactions involving transducin and GMP and results in an overall amplification of 10^5.

IV. VISUAL PIGMENT OF ROD PHOTORECEPTOR CELLS

The first important clues concerning the role of rhodopsin in vision were recorded by the German biologist Franz Boll, who observed differences in color changes in the frog eye

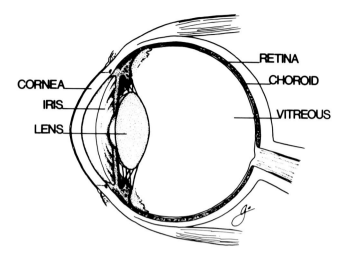

FIGURE 1. Diagram of the eye showing the cornea, iris, lens, retina, sclera, choroid, and vitreous.

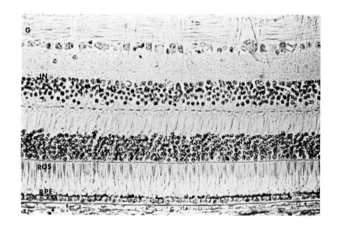

FIGURE 2. Anatomy of the retina showing: G, ganglion cell layer; IN, inner nuclear layer; ROS, rod outer segments of photoreceptor cell layer; and RPE, single cell layer of the retinal pigment epithelium.

under dark and light conditions.[6] Shortly thereafter, in 1877, Kuhne extracted rhodopsin.[6] In more recent years, rhodopsin from bovine and human retinas has been extensively characterized. The primary structure of rhodopsin has been determined by a variety of conventional biochemical techniques.[7] The bovine protein contains 348 amino acids and has a molecular weight of 41,399, including two oligosaccharides. Rhodopsin also contains the chromophore 11-*cis*-retinaldehyde which is linked to the protein through a Schiff base. The action of light in initiating the visual process results in the isomerization of 11-*cis*-retinaldehyde to the all-*trans*-form. The genes encoding bovine and human rhodopsin also have been identified, isolated, and sequenced. The human rhodopsin gene, located on chromosome 3, contains four introns and the deduced amino acid sequence is more than 90% homologous with bovine rhodopsin.[8]

A knowledge of the primary amino acid sequence of rhodopsin has led to predictions concerning the structural organization of the protein. According to the available biophysical and chemical data, rhodopsin is a transmembrane protein containing as many as seven helical

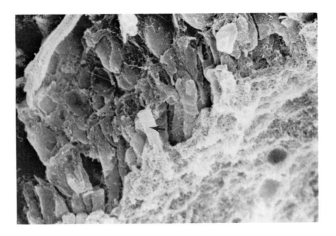

FIGURE 3.   Scanning electron micrograph of the retina showing rod outer segments (arrow) of the photoreceptor cells.

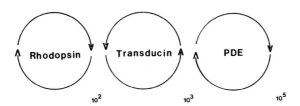

FIGURE 4.   Phototransduction of vision illustrating amplifying cascade whereby one photon of light interacts with rhodopsin which eventually results in the activation of 10^5 molecules of phosphodiesterase (PDE).[5]

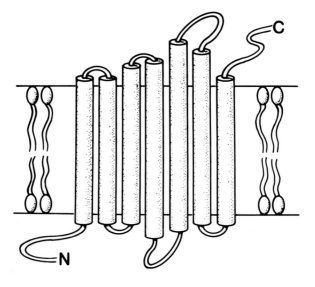

FIGURE 5.   Diagramatic representation of rhodopsin molecule in rod outer segment membrane showing seven transmembrane domains.[3]

columns which span the rod outer segment membrane (Figure 5). The amino-terminal portion of the molecule is oriented toward the intradisc surface, and the carboxy-terminal portion is oriented toward the cytoplasmic surface and contains several potential sites subject to light-dependent phosphorylation. The chromophore 11-*cis*-retinaldehyde is situated in a hydrophobic cleft of the molecule and is aligned parallel to the plane of the lipid bilayer.

Of recent interest is the finding that rhodopsin shares sequence homologies with the β-adrenergic receptor (BAR).[9] The BAR is part of the adenylate cyclase system which provides the effector mechanism for the interaction of many hormones and drugs. The BAR consists of several subunits including a catalytic moiety and a regulatory G-protein. This sequence homology parallels similarities in function since rhodopsin and BAR both are involved in signal transductions that involve interactions with other guanine nucleotide regulatory proteins.

V. VISUAL PIGMENT OF CONE PHOTORECEPTOR CELLS

The perception of color is a cortical response to specific stimuli received by the cone cells of the retina. In an elegant study, Nathans and associates,[3,4] isolated and sequenced genomic and cDNA clones corresponding to each of the color visual pigments. The deduced amino acid sequences showed three unique proteins with sequence homology with rhodopsin, including transmembrane binding domains. Of additional interest was the assignment of the green and red pigment to genes on the X chromosome, consistent with the observation that most inherited red/green color deficiencies are X linked.

VI. TRANSDUCIN

Transducin, which interacts with photoexcited rhodopsin, is a member of a family of guanine nucleotide-binding proteins (GTP) collectively referred to as G-proteins.[10] All GTP-binding proteins are composed of three subunits α, β, and γ. Although the β-subunits of the G-proteins appear similar or identical, the α-subunits are unique. In this regard, Lochrie and associates[10] isolated and sequenced cDNA clones encoding the α-subunit of transducin from bovine retina. Comparison of the deduced amino acid sequence of the retinal protein revealed sequence homologies to several proteins including yeast elongation factors. Another striking feature is the similarity between α-transducin and the *ras* gene products, G-proteins that may be important in regulating cell growth and oncogenesis.

VIII. S-ANTIGEN

Another protein intimately involved in the visual process is S-antigen. S-antigen is a 50,000-mol-wt protein that has been extensively purified and characterized from both bovine[11] and human retina.[12] Although the exact function of S-antigen is unknown, it binds to photo-excited rhodopsin and is involved in the quenching of light-induced cGMP, suggesting an important role in the phototransduction of vision.[13] In addition, when S-antigen is injected into susceptible experimental animals, it causes an experimental autoimmune uveitis (EAU).[11] EAU is characterized, in part, as a T cell-mediated disorder. T cells play a significant role in the pathogenesis of this disease since EAU develops following the adoptive transfer of lymph node cells or spleen cells from animals previously immunized with S-antigen.[14] Furthermore, athymic nude rats, deficient in T cells, fail to develop EAU following the immunization with S-antigen. EAU can be induced, however, in athymic nude rats following the adoptive transfer of lymph node cells or spleen cells from heterozygous nude rats previously immunized with S-antigen. In humans, T cells have been implicated in some forms of uveitis, such as sympathetic ophthalmia,[15] and in some patients with uveitis, T cell-mediated responses to S-antigen can be elicited.[16]

Table 1
AMINO ACID SEQUENCE OF S-ANTIGEN

```
M K A N K P A P N H V I F K K I S R D K S V T I Y L
G K R D Y I D H V E R V E P V D G V V L V D P E L V
K G K R V Y V S L T C A F R Y G Q E D I D V M G L S
F R R D L Y F S Q V Q V F P P V G A S G A T T R L Q
E S L I K K L G A N T Y P F L L T F P D Y L P C S V
M L Q P A P Q D V G K S C G V D F E I K A F A T H S
T D V E E D K I P K K S S V R L L I R K V Q H A P R
D M G P Q P R A E A S W Q F F M S D K P L R L A V S
L S K E I Y Y H G E P I P V T V A V T N S T E K T V
K K I K V L V E Q V T N V V L Y S S D Y Y I K T V A
A E E A Q E K V P P N S S L T K T L T L V P L L A N
N R E R R G I A L D G K I K H E D T N L A S S T I I
K E G I D K T V M G I L V S Y Q I K V K L T V S G L
L G E L T S S E V A T E V P F R L M H P Q P E D P D
T A K E S F Q D E N F V F E E F A R Q N L K D A G E
Y K E E K T D Q E A A M D E
```

Note: The single-letter abbreviation code for amino acids is used as follows: A, alanine; C, cysteine; D, aspartic acid; E, glutamic acid; F, phenylalanine; G, glycine; H, histidine; I, isoleucine; K, lysine; L, leucine; M, methionine; N, asparagine; P, proline; Q, glutamine; R, arginine; S, serine; T, threonine; V, valine; W, tryptophan; and Y, tyrosine.

Using traditional biochemical and cDNA cloning and sequencing techniques, the amino acid sequence of S-antigen has been determined (Table 1).[17] Analysis of the amino acid sequence data indicates that S-antigen contains 404 amino acids and codes for a protein of approximately 45,000 mol wt. Based on the determined sequence, S-antigen contains two potential carbohydrate attachment sites, potential cholera and pertussis toxin-binding sites, and a potential phosphorylation site, all features consistent with the notion of a protein involved in the visual pathway.

In order to identify pathogenic sites in the molecule, peptides have been synthesized which correspond to the entire length of S-antigen. One peptide, peptide M, 18 amino acids in length, has been found to be highly pathogenic for the induction of EAU in Lewis rats.[18] Clinically, the disease that develops is characterized by hyperemia of the iris and cornel limbus followed by inflammatory exudates within the anterior chamber of the eye. Histopathologically, a severe inflammatory response is observed that results in the complete destruction of the photoreceptor cell layer of the retina (Figure 6). Furthermore, the EAU induced by the peptide is identical to the disease caused by the immunization with native S-antigen.

VIII. RETINOBLASTOMA

It is also possible that tumors can arise from the photoreceptor cell layer of the retina. The most common intraocular tumor of childhood is retinoblastoma, occurring with a frequency of approximately 1 in 20,000 live births.[19] Clinically, patients with small tumors may present with visual difficulty or strabismus ("crossed-eyes"). In more advanced cases, the retinal tumor may be so large that it can occupy the entire globe (Figure 7). Often, in these cases, the pupil appears white, "leukokoria". In the past, the histogenesis of these tumors was controversial, but a significant body of evidence indicates that they arise from primitive retinoblasts. Popott and Ellsworth[20] noted that retinoblastomas resemble human fetal retina. In this regard, both S-antigen and rhodopsin have been identified in human fetal retina.[21,22] Although the origin of retinoblastomas has been disputed, it is generally accepted that these tumors are of neuronal origin. Well-differentiated tumors may contain rossettes

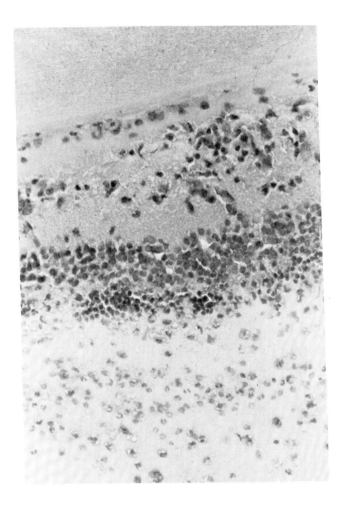

FIGURE 6. Complete destruction of the photoreceptor cell layer of the retina resulting from the immunization of Lewis rats with S-antigen.

or fluerettes, features attributed to represent an "attempt" by the tumor to form photoreceptor elements. Further support for the neuronal origin of these tumors has been the demonstration of neuronal-associated antigens, such as neuron-specific enolase, rhodopsin, and S-antigen in these tumors.[23-25] Using monoclonal antibody, MAbA9-C6, Donoso and associates demonstrated S-antigen immunoreactivity in 19 cases of retinoblastoma.[26] In these cases, S-antigen immunoreactivity was most pronounced in those areas of the tumor containing well-differentiated rosettes or fleurettes (Figure 8). In five of these cases, an intense well-circumscribed "halo" staining pattern was observed which corresponded to rhodopsin immunoreactivity.

The identification and characterization of the retinoblastoma gene has been a major goal of molecular biologists concerned with ocular oncology. Recent studies have shown that recessive oncogenes that initiate certain childhood tumors, such as retinoblastoma, may normally regulate the expression of other genes during development. Failure of the regulating gene could thus allow for the development of cancer. In that regard, retinoblastoma appears as a prototypic model for the study of such recessive oncogenes.

Previous studies have shown that the genetic locus determining the susceptibility to retinoblastoma resides in the q14 band of chromosome 13. Other studies have shown that the gene for the enzyme esterase D is in close proximity to the locus for the retinoblastoma

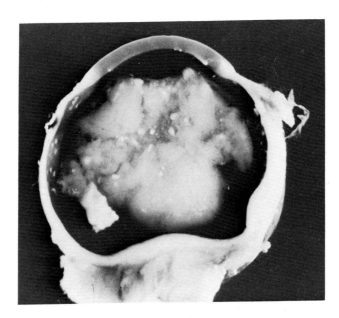

FIGURE 7. Low-power photomicrograph of globe containing a retinoblastoma. Entire globe is filled with tumor.

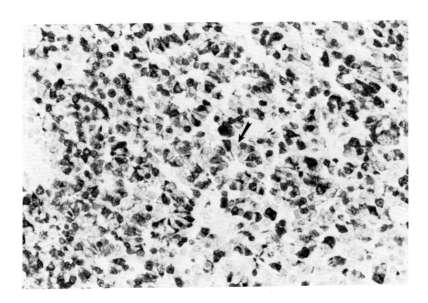

FIGURE 8. S-antigen immunoreactivity in retinoblastoma. Tumor is composed of well-differentiated containing "rosettes" (arrow). S-antigen immunoreactivity is indicated by dark-brown immunoprecipitate.

gene.[27] Friend and associates[28] recently isolated a cDNA segment that detected a chromosome segment having the properties of the retinoblastoma gene, spanning at least 70 kb in human chromosome band 13q14. This initial finding was refined and expanded in an elegant study by Lee and associates[29] who identified a gene encoding a mRNA of 4.6 kb located in the proximity of the *esterase D* locus and identified it as the retinoblastoma susceptibility gene on the basis of its chromosomal location, homozygous deletion, and the finding of tumor-specific alteration in expression (Figure 9). Transcription of this gene was abnormal in all

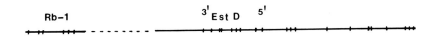

FIGURE 9. Organization of the retinoblastoma gene in chromosome.[29]

six retinoblastomas examined, but was unaltered and present in fetal human retina and placenta. Their studies indicate that the retinoblastoma gene contains at least 12 exons and is distributed over 100 kb.

Furthermore, sequence analysis of cDNA clones prediced a hypothetical protein of 816 amino acids with features found in nucleic acid-binding proteins. These studies provide the basis for the continued study of recessive genetic mechanisms in human cancers.

IX. PINEAL GLAND

It is of additional interest to note that the eye is not the only organ of the body that contains photoreceptor cells. In lower vertebrates the pineal gland is structurally and functionally a photoreceptive organ.[30] The immunologic association of the pineal and retina is further supported by the finding of S-antigen in both organs. It is perhaps not unexpected to find S-antigen immunoreactivity in tumors arising from both tissues.[31-33] On rare occasions retinoblastoma may occur in association with pineal tumors. Most of these cases have been patients with bilateral tumors of the eye which invariably represent the genetic form of the disease. These cases are now considered to be another expression of the retinoblastoma gene.

The molecular biology of vision is complex. Despite this, great progress has been made in recent years concerning the phototransducion of vision and the study of tumors arising from the retina. Future studies will almost certainly identify additional photoreceptor cell proteins which participate in the visual process. Similar studies concerning the retinoblastoma gene should allow for the identification of patients at risk genetically for the development of this malignant ocular cancer.

ACKNOWLEDGMENTS

Supported in part by NIH grant EY5095, the Pennsylvania Lions Sight Conservation and Eye Research Foundation, and the Crippled Childrens Vitreo-Retinal Research Foundation of Memphis (David Meyer, M.D., Director). Larry A. Donoso, M.D., Ph.D. is the Thomas D. Duane Research Professor of Ophthalmology, Jefferson Medical College, Thomas Jefferson University, Philadelphia, Pa.

REFERENCES

1. **Pober, J. S. and Bitensky, M. W.,** Light-regulated enzymes of vertebrate retinal rods, *Advances in Cyclic Nucleotide Research,* Raven Press, New York, 1979, 265.
2. **Venkateswara, R. P., Helson, L., Ellsworth, R. M., Reid, T., and Gilbert, F.,** Chromosomal abnormalities in human retinoblastoma, *Cancer,* 58, 663, 1986.
3. **Nathans, J., Thomas, D., and Hogness, D. S.,** Molecular genetics of human color vision: the genes encoding blue, green and red pigments, *Science,* 232, 193, 1986.
4. **Nathans, J., Piantanida, T. P., Eddy, R. L., Shows, T. B. and Hogness, D. S.,** *Science,* 232, 203, 1986.
5. **Lewin, R.,** Unexpect progress in photoreception, *Science,* 227, 500, 1985.
6. **Duke-Elder, S. S.,** *The Physiology of the Eye and of Vision in System of Ophthalmology,* Duke-Elder, S. S., Ed., C. V. Mosby, St. Louis, Mo., 1976, 476.

7. **Hargrave, P. A., McDowell, J. H., Curtis, D. R., Wang, J. K., Juszcak, E., Fong, S. L., Rao, J. K. M., and Argos, P.,** The structure of bovine rhodopsin, *Niophys. Struct. Mech.,* 9, 235, 1983.

8. **Nathans, J. and Hogness, D. S.,** Isolation and nucleotide sequence of the gene encoding human rhodopsin, *Proc. Natl. Acad. Sci. U.S.A.,* 81, 4851, 1984.

9. **Dixon, R. A. F., Kobilka, B. K., Strader, D. J., Benovic, J. L., Dohlman, H. G., Frielle, T., Baoanowski, M. A., et al.,** Cloning of the gene and cDNA for mammalian B-adrenergic receptor and homology with rhodopsin, *Nature (London),* 321, 75, 1986.

10. **Lochrie, M. A., Hurley, J. B. and Simon, M. I.,** Sequence of the alpha subunit of photoreceptor G protein: homologues between transducin, ras, and elongation factors, *Science,* 228, 96, 1985.

11. **Wacker, W. B., Donoso, L. A., Kalsow, C. M., Yankeelow, J. A., Jr., and Organisciak, D. T.,** Experimental allergic uveitis, isolation, characterization, and localization of a soluble uveitopathogenic antigen from bovine retina, *J. Immunol.,* 119, 1949, 1977.

12. **Beneski, D. A., Donoson, L. A., Edleberg, K. E., Magargal, L. E., Folberg, R., and Merryman, C.,** Human retinal S-antigen, *Invest. Ophthalmol. Vis. Sci.,* 6, 686, 1984.

13. **Pfister, C., Chabre, M., Plouet, J., Tuyen, V. V., De Kozak, Y., Faure, J. P., and Kuhn, H.,** Retinal S-antigen identified as the 48K protein regulating light-dependent phosphodiesterase in rods, *Science,* 228, 891, 1985.

14. **Merryman, C. F., Donoso, L. A., Sery, T. W., Bauer, A., Shinohara, T., and Sciutto, E.,** S-antigen: adoptive transfer of experimental autoimmune uveitis following immunization with a small synthetic peptide, *Arch. Ophthalmol.,* 105, 841, 1987.

15. **Jakobiec, F. A., Marboe, C. C., Knowles, D. M., Iwamoto, T., Harrison, W., Chang, S., and Coleman, D. J.,** An analysis of the inflammatory infiltrate by hybridoma monoclonal antibodies, immunochemistry and correlative electron microscopy, *Ophthalmology,* 90, 76, 1983.

16. **Nussenblatt, R. B., Gery, I., Ballintine, E. J., and Wacker, W. B.,** Cellular immune responsiveness of uveitis patients to retinal S-antigen, *Am. J. Ophthalmol.,* 89, 173, 1980.

17. **Yamaki, K., Takahashi, Y., Sakuragi, S., and Matsubara, K.,** Molecular cloning of the S-antigen cDNA from bovine retina, *Biochem., Biophys. Res. Commun.,* 142, 904, 1987.

18. **Donoso, L. A., Merryman, C. F., Shinohara, T., Dietzschold, B., Wistow, G., Craft, C., Morley, W., and Henry, R. T.,** S-antigen: identification of the MAbA9-C6 monoclonal antibody binding site and the uveitopathogenic sites, *Curr. Eye Res.,* 5, 995, 1986.

19. **Shields, J. A.,** *Diagnosis and Management of Intraocular Tumors,* C. V. Mosby, St. Louis, Mo., 1983.

20. **Popott, N. and Ellsworth, R. M.,** The fine structure of nuclear alterations in retinoblastoma and developing human retina: in vivo and in vitro observations, *J. Ultrastruct. Res.,* 29, 535, 1969.

21. **Donoso, L. A., Merryman, C. F., Edelberg, K. E., Naids, R., and Kalsow, C.,** S-antigen in the developing retina and pineal gland: a monoclonal antibody study, *Invest. Ophthal. Vis. Sci.,* 4, 561, 1985.

22. **Donoso, L. A., Hamm, H., Dietzschold, B., Augusburger, J. J., Shields, J. A., and Arbizo, V.,** Rhodopsin and retinoblastoma. A monoclonal antibody histopathologic study, *Arch. Ophthalmol.,* 104, 111, 1986.

23. **Donoso, L. A., Folberg, R., and Arbizo, V.,** Retinal S-antigen and retinoblastoma, *Arch. Ophthalmol.,* 103, 855, 1986.

24. **Perentes, E., Herbert, C. P., Rubinstein, L. J., Herman, M. M., Uffer, S., Donoso, L. A., and Collins, V. P.,** Immunohistochemical characterization of human retinoblastomas in situ: an analysis with multiple markers, *Am. J. Ophthalmol.,* 103, 647, 1987.

25. **Rodrigues, M. M., Wiggert, B., Shields, J. A., Donoso, L. A., Bardenstein, D., Katz, N., Friendly, D., Chader, G.,** Retinoblastoma: immunohistochemistry and cell differentiation, *Ophthalmology,* 94, 378, 1987.

26. **Donoso, L. A., Folberg, R., and Arbizo, V.,** Retinal S-antigen and retinoblastoma, *Arch. Ophthalmol.,* 103, 855, 1985.

27. **Squire, J., Dryja, T. P., Dunn, J., Goddard, A., Hofmann, T., Musarella, M., Huntington, F. W., Becker, A. J., Gallie, B. L., and Phillips, R. A.,** Cloning of the esterase D gene: a polymorphic gene probe closely linked to the retinoblastoma locus on chromosome 13, *Proc. Natl. Acad. Sci. U.S.A.,* 83, 6573, 1986.

28. **Friend, S. H., Bernards, R., Rogelj, S., Weinberg, R. A., Rapaport, J. M., Albert, D. M., and Dryja, T. P.,** A human DNA segment with properties of the gene that predisposes to retinoblastoma and osteosarcoma, *Nature (London),* 323, 643, 1986.

29. **Lee, W. H., Bookstein, R., Hong, F., Young, L. J., Shew, J. Y., Eva, Y. H., and Lee, P.,** Human retinoblastoma susceptibility gene: cloning, identification, and sequence, *Science,* 235, 1394, 1987.

30. **Wurtman, R. J., Axelrod, J., and Kelly, D. E.,** *The Pineal,* Academic Press, New York, 1960.

31. **Donoso, L. A., Rorke, L. B., Shields, J. A., Augusburger, J. J., Brownstein, S., and Lahoud, S.,** *Am. J. Ophthalmol.,* 103, 57, 1987.

32. **Donoso, L. A., Shields, J. A., Felberg, N. T., Martyn, L. J., Truex, K. C., and D'Cruz, C. A.,** Intracranial malignancy in patients with bilateral retinoblastoma, *Retina,* 1, 67, 1981.

33. **Perentes, E., Rubinstein, L. J., Herman, M. M., and Donoso, L. A.,** S-antigen immunoreactivity in human pineal glands and pineal parenchymal tumors: a monoclonal antibody study, *Acta Neuropathol.,* 71, 224, 1986.

Chapter 24

NONCLASSICAL STEROID 21-HYDROXYLASE DEFICIENCY AS A CAUSE OF REPRODUCTIVE DYSFUNCTION

Maria I. New, Susan Drucker, and Phyllis W. Speiser

TABLE OF CONTENTS

I. INTRODUCTION

21-Hydroxylase deficiency (21-OHD) is an autosomal recessive disorder of adrenal steroidogenesis that is resolvable by clinical and hormonal criteria into classical and nonclassical forms. Classical 21-OHD is the most frequently occurring adrenal enzymatic defect causing congenital adrenal hyperplasia. Affected females present at birth with ambiguous genitalia resulting from high adrenal androgen levels during fetal development, and affected neonates of either genetic sex may often present with salt wasting resulting from deficient adrenal synthesis of aldosterone. Nonclassical 21-OHD produces virilization at any developmental stage postnatally; excess adrenal androgen levels may cause no clinically apparent effect in some cases, whereas in others, symptoms may appear with rapid onset in late childhood, peripubertally, or in adulthood.

In this chapter we will focus on the clinical and hormonal manifestations of the nonclassical disorder, and outline current methods of diagnosis and management. Specific effects on gonadal function will then be discussed, and recent advances in classical and molecular genetics will be reviewed in brief.

II. CLINICAL PROFILE OF NONCLASSICAL 21-OHD

A. Clinical Findings

Nonclassical 21-OHD (NC 21-OHD) may be manifest in children as virilization, premature pubarche, and advanced bone age with resultant short stature due to early epiphyseal fusion. In female adolescents and adults, the clinical picture may resemble polycystic ovarian disease; diminished fertility may occur in either sex.[1,2] Female infants with NC 21-OHD, unlike those with classical 21-OHD, show no evidence of virilization *in utero*, and are born with normal external genitalia. In certain individuals, the disorder is without clinical stigmata; these cases are, however, characterized by the presence of hormonal abnormalities identical to those that occur in patients with overt signs of androgen excess.[3] Table 1 provides the terms used to distinguish the various forms of classical and NC 21-OHD. The classical and nonclassical disorders are human major-histocompatibility-complex-linked allelic variants of a monogenic disease.[4-8]

The first NC 21-OHD patients were described in 1957.[9] Since a variety of symptoms at different stages of development may be attributable to an androgen excess syndrome, and since individual cases may vary greatly in the presentation and degree of such symptoms (Figure 1), the pediatrician must be aware of this phenotypic variability. In one patient, the symptoms of NC21-OHD were present at 6 months of age with pubic hair and accelerated growth. These symptoms gradually disappeared, but the biochemical defect persisted. The young female child may present with any of these signs: precocious adrenarche without thelarche, mild clitoromegaly and labial fusion, acne, advanced bone age, or early growth spurt. The male child may present similarly with evidence of adrenarche without gonadarche (i.e., sexual hair and phallic enlargement with small testes).

In adolescent girls, hirsutism, severe acne, menstrual abnormalities, male-pattern balding, or short stature warrants evaluation to rule out an adrenal steroidogenic defect. In boys, signs of adrenal androgen excess may include acne, advanced stature and bone age, or adrenarche (i.e., sexual hair growth) without gonadarche.

The symptomatology of adult women can be similar to that of girls in adolescence. Difficulty conceiving or carrying a pregnancy to term also may be observed. Hyperandrogenism from a deficiency of adrenal 21-hydroxylase in males is a potential causal factor in dysregulation of the hypothalamic-pituitary-gonadal axis by adrenal androgens, with resultant oligospermia and subfertility.[2]

Table 1
GENOTYPIC CHARACTERIZATION OF THE FORMS OF 21-HYDROXYLASE DEFICIENCY

Form of 21-OHase deficiency	Clinical phenotype	Hormonal phenotype (in response to ACTH)	Genotype
Classical	Prenatal virilization; fully symptomatic	Marked elevation of precursors (serum 17-OHP and Δ^4-A)	$\dfrac{\text{21-OH def}^{\text{Severe}}}{\text{21-OH def}^{\text{Severe}}}$
Nonclassical	Symptomatic: later development of virilization; milder symptoms Asymptomatic: no virilization of other symptoms	Moderate elevation of precursors	$\dfrac{\text{21-OH def}^{\text{Severe}}}{\text{21-OH def}^{\text{Mild}}}$ or $\dfrac{\text{21-OH def}^{\text{Mild}}}{\text{21-OH def}^{\text{Mild}}}$
Carrier	Asymptomatic	Precursor level greater than normal	$\dfrac{\text{21-OH def}^{\text{Severe}}}{\text{21-OHase}^{\text{Normal}}}$ or $\dfrac{\text{21-OH def}^{\text{Mild}}}{\text{21-OHase}^{\text{Normal}}}$
Normal	(Asymptomatic)	Lowest levels — some overlap seen with carriers	$\dfrac{\text{21-OHase}^{\text{Normal}}}{\text{21-OHase}^{\text{Normal}}}$

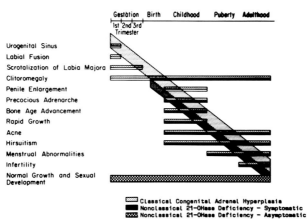

FIGURE 1. Clinical spectrum of HLA-linked steroid 21-hydroxylase deficiency. (From New, M. I. et al., *The Metabolic Basis of Inherited Disease*, 5th ed., Stanbury, J. B., et al., Eds., McGraw-Hill, New York, 1983, 973. With permission.)

B. Hormonal Characterization and Diagnosis

Biochemically, reduced 21-hydroxylase activity is most evident as accumulation of the cortisol precursor 17α-hydroxyprogesterone (17-OHP), the principal substrate for the steroid 21-hydroxylase enzyme. The accepted standard diagnostic test is the 60-min ACTH (adrenocorticotropic hormone) stimulation test. Serum 17-OHP radioimmunoassay[10] is performed before and after the intravenous (i.v.) administration of 0.25 mg Cortrosyn (synthetic ACTH_{1-24}, Organon, West Orange, N.J.). The basal and stimulated 17-OHP levels are plotted on a reference nomogram (Figure 2) in order to assign the subject a 21-hydroxylase deficiency genotype. The nomogram shows a distribution of hormonal responses along a regression line reflecting a spectrum of enzymatic deficiency.[11] The hormonal values aggregate along

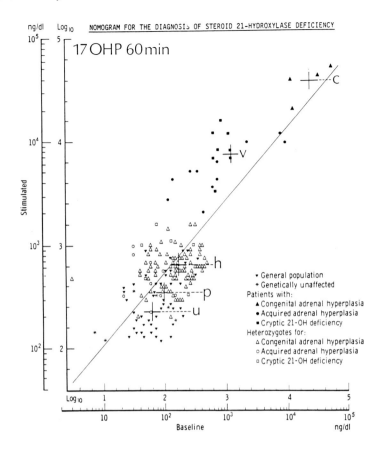

FIGURE 2.   Nomogram relating baseline to ACTH-stimulated serum concentrations of 17-OHP. The scales are logarithmic. A regression line for all data points is shown. The mean for each group is indicated by a large cross and adjacent letter: c denotes classical 21-OHD; v, variant or NC 21-OHD (combined mean of values in patients with asymptomatic and symptomatic disease); h, heterozygotes for all forms of 21-hydroxylase deficiency; p, general population; and u, persons known to be unaffected (e.g., siblings of 21-OHD patients, who carry neither affected parental haplotype as determined by HLA typing).

the regression line into the following easily distinguishable groups: (1) patients with classical 21-OHD adrenal hyperplasia, who show the highest baseline values and the most marked elevations of ACTH-stimulated 17-OHP; (2) patients affected with NC 21-OHD, exhibiting less elevated 17-OHP in both states; (3) heterozygote carriers of a 21-OHD, who have a mild enzyme deficiency unmasked only upon ACTH stimulation; and (4) the general population. Hormonal values in family members predicted by HLA genotyping to be unaffected for 21-OHD fall at the lowest point of the regression line and serve as the best control population for normal 21-hydroxylase activity. Those members of the general population whose responses are in the heterozygote range may actually be carriers of a gene for 21-OHD. This can be corroborated by examining HLA linkage markers (see below).

Heterozygotes for classical and NC 21-OHD demonstrate indistinguishable 17-OHP responses to ACTH stimulation. In contrast with the clear 17-OHP elevations uniformly seen in classical 21-OHD, random base line 17-OHP values are not always diagnostic in NC 21-OHD affected subjects, and abnormally high base line 17-OHP levels are most consistently seen in early morning (0800 hr).[12]

Attempts to detect heterozygosity for the *21-hydroxylase* gene hormonally prior to the discovery of HLA linkage were only partially successful due to incomplete separation of

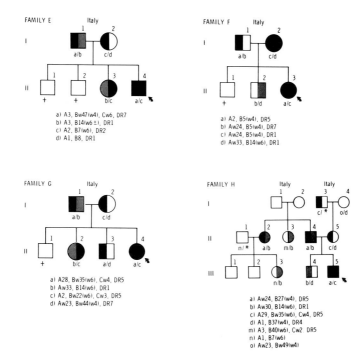

FIGURE 3.   Family pedigrees (E-H) with nonclassical symptomatic adrenal hyperplasia. ○/□, unaffected individual; ●/■, infant with classical 21-OHD; ◐/◪, heterozygous carriers of the severe deficiency gene for CAH; and ◐, patient with nonclassical adrenal hyperplasia having both a severe 21-OHD (classical) gene and a mild deficiency (nonclassical) gene. (From New, M. I., et al., *The Metabolic Basis of Inherited Disease*, 5th ed., Stanbury, J. B. et al., Eds., McGraw-Hill, New York, 1983, 973. With permission.)

21-OHD carriers and unaffected subjects of ACTH testing.[13] The method for utilizing HLA genotyping to determine heterozygosity is shown in Figure 3. The HLA prediction may be confirmed by hormonal testing. Improved distinction between heterozygotes and unaffected subjects in adult test populations has been reported following overnight adrenal suppression with single-dose dexamethasone administration in the late evening of the day preceding the ACTH stimulation test.[14,15]

C. Medical Therapy

Low-dose glucocorticoid replacement therapy is employed to treat all classical and symptomatic NC 21-OHD patients. By providing cortisol (or the equivalent in glucocorticoid activity) in physiological range, this treatment suppresses endogenous ACTH release and adrenocortical activity, thereby lessening the accumulation of 17-OHP as available substrate for the overproduction of androgens. Hydrocortisone is administered to children in the range 10 to 20 mg/m²/day divided equally in two daily doses. Older patients, who are not at risk for growth suppression, may be treated with a longer-acting synthetic steroid such as dexamethasone or prednisone in the appropriate dose. Patients are monitored by measuring serum and urinary 21-hydroxylase precursor levels, growth velocity, and bone age over time to ensure good androgen suppression without signs of hypercortisolism.

Serum RIAs (radioimmunoassays) are used to assess the adequacy of hormonal replacement regimens. Normative RIA standards for adrenal steroids in normal children and patients with congenital adrenal hyperplasia (CAH) have been established.[16-20] 17-OHP and its direct product in the androgen pathway, Δ^4-androstenedione (Δ^4-A), provide the most sensitive index of control in serum; testosterone (T) is useful only in females and prepubertal males.

Salivary levels correlate well with serum free steroid hormone levels. The effectiveness of early morning salivary assay for 17-OHP in hormonal evaluation for NC 21-OHD has been proven,[21] and a very recent report relates salivary T to the bioavailable fraction of this hormone in plasma.[22] The high correlation with serum value, and ease of collection and noninvasive nature makes salivary assays optimal for clinical evaluations for steroid function. In addition, use of this type of test is economically feasible in large-scale population screenings in determining the true prevalence of 21-OHD and other late-onset adrenal steroidogenic disorders.

III. NONCLASSICAL 21-OHD AND GONADAL FUNCTION

Gynecologists in clinical practice have long recognized the efficacy of cortisone treatment in selected oligomenorrheic and infertile women.[23-25] Recent retrospective studies have reported widely divergent estimates of the frequency of NC 21-OHD in girls and women evaluated for hirsutism and oligomenorrhea ranging from 1.2 to 30%.[6,26-33] The incidence of NC 21-OHD in our population of hirsute, oligomenorrheic women is 14%.[31] The incidence of impaired fertility in this group of patients is unknown. Published results still do not permit analysis of fertility rates of female patients with proven NC 21-OHD. Given the high frequency of NC 21-OHD and its variability by ethnic group, it is possible that this particular disorder of adrenal steroidogenesis may correlate with fertility problems in females in ethnic populations at high risk. By comparison, in classical 21-OHD congenital adrenal hyperplasia, the fertility rate was estimated at 36% in those women who began treatment between 6 and 20 years of age, while no woman beginning treatment after the age of 20 years conceived.[34] A recent survey of case histories at the same institution reported divergent values of 60% and 0 to 2.5% for the fertility rates of patients with the simple virilizing and salt-wasting forms, respectively.[35] Vaginal adequacy for coitus was indicated as the primary determinant, although sexual preference must also be considered important.[36] Clearly, such factors make any specific assessment of gonadal function difficult in the classical disorder.

The incidence of NC 21-OHD in infertile men has not been established, but is estimated to be low based on one relatively small test population of men with idiopathic oligospermia.[37] Conversely, among men confirmed to be affected with NC 21-OHD, there is currently no estimate of the incidence of infertility. There have been case reports of normal fertility with the classical disorder[38,39] as well as reports of gonadal dysfunction and infertility.[39] In spite of a recent claim that low-dose glucocorticoid treatment is not beneficial,[40] at least one case of successful reversal of infertility from oligospermia with glucocorticoids has been reported,[11] indicating that this treatment (with careful monitoring of dosage) may be effective. It is therefore plausible that a subgroup of individuals with asymptomatic NC 21-OHD are men in whom subfertility has gone undetected and who are capable of normal fertility pending reversal of the hormonal imbalance induced by excessive adrenal sex steroids.

IV. GENETICS OF 21-OHD

A discussion of the genetics of NC 21-OHD would be incomplete without an account of the more general investigations on 21-OHD which led us to determine that NC 21-OHD was one of the most frequent autosomal recessive disorders in man.[41]

A. Epidemiology of Classical 21-OHD

Several surveys have established that the 21-OHD is transmitted as an autosomal recessive trait.[42-45] Estimates of the prevalence of salt wasting in the classical disorder have increased,[46,47] perhaps related to the diminishing incidence of missed diagnosis (although fatalities still occur), and salt wasting is now known to occur in more than 66% of classical cases; preliminary analysis of worldwide screening data indicates even higher rates of occurrence.[70]

In Europe and the U.S., recent estimates of the incidence of congenital adrenal hyperplasia have been between 1:5000 and 1:15,000 giving a gene frequency of approximately 1:100 in the general Caucasian population.[48,49]

A reliable and valid test for classical 21-OHD in newborns by measuring for 17-OHP in a heel-prick capillary blood specimen impregnated on filter paper was developed in 1977 in our laboratory.[10] Screening programs using this technique, conducted in 1982 and 1985, established values of 1:280 and 1:684 for the unusually high frequency of the salt-wasting form among the Yup'ik Eskimos of southwestern Alaska.[50,51]

B. Population Genetics of NC 21-OHD

Screening of infants or children for serum 17-OHP elevations with tests such as the microfilter paper technique is not informative in NC 21-OHD. Frequency studies have therefore relied on typing for HLA antigens as genetic markers in conjunction with ACTH stimulation hormonal testing in obligate heterozygote parents. The reliability of this approach was confirmed through analysis of pedigree data using the affected sib pair method of Thomson and Bodmer.[52] (The premise of this method is the observation that as a disease gene becomes increasingly rare, the proportion of affected sibs sharing linkage markers on both haplotypes increases.) It was found that NC 21-OHD is far more common than the classical disorder, and also that its frequency of occurrence is ethnic group-specific.[41] The greatest prevalence was seen in Ashkenazic Jews (19.1%); high gene frequencies were also seen among Hispanics (13.6%), Yugoslavs (12.5%), and Italians (5.8%). In other Caucasians, 41% of whom had Anglo-Saxon ancestry, the gene frequency was 3.2%. The corresponding heterozygote frequencies were 1:3 for Ashkenazic Jews, 1:4 for Hispanics, 1:5 for Yugoslavs, 1:9 for Italians, and 1:14 for other Caucasians. Disease frequencies were 1:27 for Ashkenazi Jews, 1:53 for Hispanics, 1:63 for Yugoslavs, 1:333 for Italians, and 1:1000 for other Caucasians (Figure 4).[41]

Among the patient population referred to this clinic for premature adrenarche, we found a 32% incidence of NC 21-OHD.[32] Of the women referred to us for hirsutism, the incidence of NC 21-OHD was 14%.[31] In other series, the incidence of NC 21-OHD among women with hirsutism and oligomenorrhea ranged from 1.2 to 30%.[6,26-30,33] None of these studies specified the ethnic group composition of the sample.

C. HLA Linkage

The HLA system is located on the short arm of chromosome 6. The genes for these cell-surface antigens (termed A, B, C, and DR) serve as useful markers for closely linked disease genes. The linkage between HLA-B and congenital adrenal hyperplasia due to 21-OHD was first reported in 1977[53] and subsequently confirmed in a larger series of families.[54] In 1980, linkage was also established between NC 21-OHD and the *HLA-B* locus.[55]

Linkage disequilibrium refers to a preferential gametic association or nonassociation between specific allelic pairs at two linked loci.[56] HLA-Bw47; DR7 was the first haplotype segment recognized to be in positive linkage disequilibrium with classical 21-OHD.[57-59] The antigen B60(40) occurs in association with both salt wasting and simple virilizing classical 21-OHD, but the increased relative risk is greater with the salt-wasting form.[60] Simple virilizing 21-OHD is associated with HLA-B51(5) in some ethnic populations.

NC 21-OHD was first associated with HLA-B14 in 1980 in Ashkenazi Jews.[55,60,61] The study of the high frequency of the nonclassical disorder revealed linkage disequilibrium between HLA-B14 and NC 21-OHD in Hispanics and Italians, as well as in Ashkenazi Jews, and a 32-fold increased relative risk for this antigen in all ethnic groups surveyed.[41] Different HLA markers are associated with 21-OHD in Yugoslavs, suggesting a separate mutation in this population.[62]

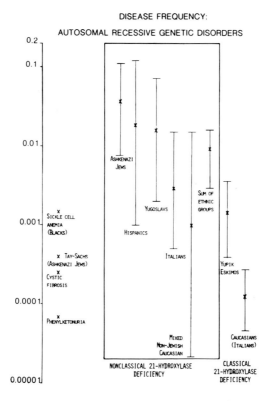

FIGURE 4.   The disease frequencies of NC 21OHD and classical 21-OHD relative to other common autosomal recessive disorders. Bars represent 5 to 95% confidence levels. (From Speiser, P.W. Dupont, B., Rubinstein, P., Piazza, A., Kastelan, A., and New, M. I., *Am. J. Hum. Genet.*, 37, 650, 1985. With permission.)

The antigen HLA-B35 occurs more frequently in association with classical 21-OHD patients, but is negatively associated with nonclassical disease, while the nonclassical-associated B14 shows a decreased frequency of occurrence in the classical deficiency.[41] Notably, HLA-B8, more frequent in many autoimmune disorders, is decreased in both the classical and nonclassical forms of 21-OHD.[40,59]

D. Molecular Genetics

It is known from earlier biochemical studies that a microsomal cytochrome, P-450 is the enzyme active for steroid 21-hydroxylation (P450c21). Molecular genetic studies have demonstrated that there are two structural genes encoding P450c21, each located immediately adjacent to one of the two *C4* genes encoding the fourth component of serum complement.[63,64] Deletion of the 21-OHase *B* gene (two gene names: *OH21B*;*CYP21B*) and neighboring *C4B* gene is found in association with the haplotype HLA-A3; Bw47; DR7 in 21-OHD patients with the salt-wasting phenotype. In contrast, 21-OHase A (*OH21A*, *CYP21A*) and *C4A* are deleted in the haplotype HLA-A1; B8; DR3, and individuals homozygous for this deletion are capable of normal rates of cortisol biosynthesis. On this basis, it was proposed that the *B* gene (*CYP21B*) is active and that the *A* gene (*CYP21A*) is inactive.[63] DNA sequence analysis has confirmed that the *CYP21B* gene encodes the active enzyme, and that the *CYP21A* gene is a pseudogene.[65,66]

Study of individuals with the HLA-B14:DR1 haplotype associated with nonclassical disorder has identified a third expressed C4 protein by gel electrophoresis. Molecular genetic analysis has demonstrated that this genotype indeed carries a third (duplicated *C4B*) gene[67]

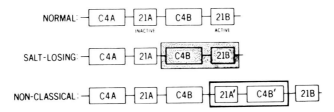

FIGURE 5. Schematic of haplochromosomal arrangements of the genes encoding the fourth component of serum complement (two isozymes, C4A and C4B) and steroid 21-hydroxylase (pseudogene *OH21A* and active gene *OH21B*). All genes in this segment are transcribed in the same direction (left to right, as shown here); orientation of the segment (between HLA-B and HLA-DR) on the chromosome has not been confirmed. Top: normal gene arrangement. Middle: pattern occurring on the haplotype HLA-Bw47; DR7 with a null allele at the *C4B* locus and absent 3.7 kb Taq 1 band. This normally present restriction fragment is missing from other genotypes causing the severe, salt-wasting form of the disease; thus the extent of what is here diagrammed as a single deletion probably varies (in some cases the *C4B* gene is completely unaffected). Bottom: in cases of NC 21-OHD associated with the haplotype HLA-B14; DR1, the *C4B* gene and *OH21A* (or *OH21A*-like) gene appear to be duplicated as suggested by the presence of increased intensity of a 3.2 Taq 1 band cut from the normally occurring *OH21A* gene. (From Werkmeister, J. W., New, M. I., Dupont, B., and White, P. C., *Am. J. Hum. Genet.*, 39, 461, 1986. With permission.)

and an extra *CYP21A* or *CYP21A*-like gene, based on the size of restriction enzyme fragments[68,69] (Figure 5). It is not known whether the duplicated pseudogene contributes to the development of the 21-OHD phenotype or simply is a marker in association with a separate B gene mutation.

Approximately one fourth of the 21-OHD alleles examined from patients not carrying the HLA-Bw47 haplotype are also deletions of the active B gene, and the majority of these alleles also have a *C4B* deletion. To date, all patients with homozygous deletions of the *CYP21B* gene have salt-wasting classical 21-OHD.

The remaining three fourths of classical disease alleles do not have deletions, but presumably represent point mutations, which may not yield informative RFLPs (restriction fragment length polymorphisms) with a full-length cDNA probe on Southern blot hybridization.

REFERENCES

1. **Kohn, B., Levine, L. S., Pollack, M. S., Pang, S., Lorenzen, F., Levy, D., Lerner, A., Rondanini, G. F., Dupont, B., and New, M. I.,** Late-onset steroid 21-hydroxylase deficiency: a variant of classical congenital adrenal hyperplasia, *J. Clin. Endocrinol. Metab.*, 55, 817, 1982.
2. **Wischusen, J., Baker, H. W. G. and Hudson, B.,** Reversible male infertility due to congenital adrenal hyperplasia, *Clin. Endocrinol.*, 14, 571, 1981.
3. **Levine, L. S., Dupont, B., Lorenzen, F., Pang, S., Pollack, M., Oberfield, S., Kohn, B., Lerner, A., Cacciari, E., Mantero, F., Cassio, A., Scaroni, C., Chiumello, G., Rondanini, G. F., Gargantini, L., Giovannelli, G., Virdis, R., Bartolotta, E., Migliori, C., Pintor, C., Tato, L., Barboni, F., and New, M. I.,** Cryptic 21-hydroxylase deficiency in families of patients with classical congenital adrenal hyperplasia, *J. Clin. Endocrinol. Metab.*, 51, 1316, 1980.
4. **Rosenwaks, Z., Lee, P. A., Jones, G. S., Migeon, C. J., and Wentz, A. C.,** An attenuated form of congenital virilizing adrenal hyperplasia, *J. Clin. Endocrinol. Metab.*, 49, 335, 1979.
5. **Lobo, R. A. and Goebelsmann, U.,** Adult manifestation of congenital adrenal hyperplasia due to incomplete 21-hydroxylase deficiency mimicking polycystic ovarian disease, *Am. J. Obstet. Gynecol.*, 138, 720, 1980.

6. **Migeon, C. J., Rosenwaks, Z., Lee, P. A., Urban, M. D., and Bias, W. B.,** The attenuated form of congenital adrenal hyperplasia as an allelic form of 21-hydroxylase deficiency, *J. Clin. Endocrinol. Metab.,* 51, 647, 1980.

7. **Pollack M. S., Levine, L. S., O'Neill, G. J., Pang, S., Lorenzen, F., Kohn, B., Rondanini, F. G., Chiumello, G., New M. I., and Dupont, B.,** HLA linkage and B14, DR1, BfS haplotype association with the genes for late onset and cryptic 21-hydroxylase deficiency, *Am. J. Hum. Genet.,* 33, 540, 1981.

8. **New, M. I., Dupont, B., Pang, S., Pollack, M. S., and Levine, L. S.,** An update of congenital adrenal hyperplasia, *Recent Prog. Horm. Res.,* 37, 105, 1981.

9. **Decourt, J., Jayle, M. F., and Baulieu, E.,** Virilisme cliniquement tardif avec excretion de pregnanetriol et insuffisance de la production du cortisol., *Ann. Endocrinol.,* 18, 416, 1957.

10. **Pang, S., Hotchkiss, J., Drash, A. L., Levine, L. S., and New, M. I.,** Microfilter paper method for 17α-progesterone radioimmunoassay: its application for rapid screening for congenital adrenal hyperplasia, *J. Clin. Endocrinol. Metab.,* 45, 1003, 1977.

11. **New, M. I., Lorenzen, F., Lerner, A. J., Kohn, B., Oberfield, S. E., Pollack, M. S., Dupont, B., Stoner, E., Levy, D. J., Pang, S., and Levine, L. S.,** Genotyping steroid 21-hydroxylase deficiency: hormonal reference data, *J. Clin. Endocrinol. Metab.,* 57, 320, 1983.

12. **Kuttenn, F.,** Late-onset adrenal hyperplasia (letter), *N. Engl. J. Med.,* 314, 451, 1986.

13. **Gutai, J. P., Kowarski, A. A., and Migeon, C. J.,** The detection of the heterozygous carrier for congenital virilizing adrenal hyperplasia, *J. Pediatr.,* 90, 924, 1977.

14. **Grosse-Wilde, H., Weil, J., Albert, E., Scholz, S., Bidlingmaier, F., Sippel, W. G., and Knorr, D.,** Genetic linkage studies between congenital adrenal hyperplasia and the HLA blood group system, *Immunogenetics,* 8, 41, 1979.

15. **Lejeune-Lenain, C., Cantraine, F., Dufrasnes, M., Prevot, F., Wolter, R., and Franckson, J. R. M.,** An improved method for the detection of heterozygosity of congenital virilizing adrenal hyperplasia, *Clin. Endocrinol.,* 12, 525, 1980.

16. **Solomon, I. L., and Schoen, E. J.,** Blood testosterone values in patients with congenital virilizing adrenal hyperplasia, *J. Clin. Endocrinol. Metab.,* 45, 355, 1975.

17. **Hughes, I. A. and Winter, J. S. D.,** The relationship between serum concentrations of 17-hydroxyprogesterone and other serum and urinary steroids in patients with congenital adrenal hyperplasia, *J. Pediatr.,* 88, 766, 1976.

18. **Pang, S., Levine, L. S., Chow, D. M., Faiman, C., and New, M. I.,** Serum androgen concentrations in neonates and young infants with congenital adrenal hyperplasia due to 21-hydroxylase deficiency, *Clin. Endocrinol.,* 11, 575, 1979.

19. **Korth-Schutz, S., Virdis, R., Saenger, P., Chow, D. M., Levine, L. S., and New, M. I.,** Serum androgens as a continuing index of adequacy of treatment of congenital adrenal hyperplasia, *J. Clin. Endocrinol. Metab.,* 46, 452, 1978.

20. **Winter, J. S. D.,** Current approaches to the treatment of congenital adrenal hyperplasia, *J. Pediatr.,* 97, 81, 1980.

21. **Zerah, M., Pang, S., and New, M. I.,** Morning salivary 17-hydroxyprogesterone is a useful screening test for nonclassical 21-hydroxylase deficiency, *J. Clin. Endocrinol. Metab.,* 65, 227, 1987.

22. **Ruutiainen, K., Sannikka, E., Santti, R., Erkkola, R., and Adlercreutz, H.,** Salivary testosterone in hirsutism: correlations with serum testosterone and the degree of hair growth, *J. Clin. Endocrinol. Metab.,* 64, 1015, 1987.

23. **Jones, H. W. and Jones, G. E. S.,** The gynecological aspects of adrenal hyperplasia and allied disorders, *Am. J. Obstet. Gynecol.,* 68, 1330, 1954.

24. **Greenblatt, R. B., Barfield, W. E., and Lampros, C. P.,** Cortisone in the treatment of infertility, *Fertil. Steril.,* 7, 203, 1956.

25. **Birnbaum, M. D. and Rose, L. I.,** The partial adrenocortical hydroxylase deficiency syndrome in infertile women, *Fertil. Steril.,* 32, 536, 1979.

26. **Child, D. R., Bu'lock, D. E., Hillier, V. F. and Anderson, D. C.,** Adrenal steroidogenesis in heterozygotes for 21-hydroxylase deficiency, *Clin. Endocrinol.,* 12, 595, 1980.

27. **Gibson, M., Lackritz, R., Schiff, I., and Tulchinsky, D.,** Abnormal adrenal responses to adrenocorticotropic hormone in hyperandrogenic women, *Fertil. Steril.,* 33, 43, 1980.

28. **Chetkowski, R. J., DeFazio, J., Shamonki, I., Judd, H. L., and Chang, R. J.,** The incidence of late-onset congenital adrenal hyperplasia due to 21-hydroxylase deficiency among hirsute women, *J. Clin. Endocrinol. Metab.,* 58, 595, 1984.

29. **Chrousos, G. P., Loriaux, D. L., Mann, D. L., and Cutler, G. B., Jr.,** Late-onset 21-hydroxylase deficiency mimicking idiopathic hirsutism or polycystic ovarian disease: an allelic variant of congenital virilizing adrenal hyperplasia with a milder enzymatic defect, *Ann. Intern. Med.,* 96, 143, 1982.

30. **Kuttenn, F., Couillin, P., Girard, F., Billaud, L., Vincens, M., Boucekkine, C., Thalabard, J.-C., Maudelonde, T., Spritzer, P., Mowszowicz, I., Boue, A., and Mauvais-Jarvis, P.,** Late-onset adrenal hyperplasia in hirsutism, *N. Engl. J. Med.,* 313, 224, 1985.

31. **Pang, S., Lerner, A. J., Stoner, E., Levine, L. S., Oberfield, S. E., Engel, I., and New, M. I.,** Late-onset adrenal steroid 3βHSD deficiency. I. A cause of hirsutism in pubertal and postpubertal women, *J. Clin. Endocrinol. Metab.,* 60, 428, 1985.

32. **Temeck, J. W., Pang, S., Nelson, C., and New, M. I.,** Genetic defects of steroidgenesis in premature pubarche, *J. Clin. Endocrinol. Metab.,* 64, 609, 1987.

33. **Baskin, H. J.,** Screening for late-onset congenital adrenal hyperplasia in hirsutism or amenorrhea, *Arch. Intern. Med.,* 147(5), 847, 1987.

34. **Klingensmith, G. J., Garcia, S. C., Jones, H. W., Jr., Migeon, C. J., and Blizzard, R. M.,** Glucocorticoid treatment of girls with congenital adrenal hyperplasia: effects on height, sexual maturation, and fertility, *J. Pediatr.,* 90, 996, 1977.

35. **Mulaikal, R. M., Migeon, C. J., and Rock, J. A.,** Fertility rates in female patients with congenital adrenal hyperplasia due to 21-hydroxylase deficiency, *N. Engl. J. Med.,* 316, 178, 1987.

36. **Federman, D. D.,** Psychosexual adjustment in congenital adrenal hyperplasia (editorial), *N. Engl. J. Med.,* 316, 209, 1987.

37. **Ojeifo, J. O., Winters, S. J., and Troen, P.,** Basal and ACTH-stimulated serum 17α-hydroxyprogesterone in men with idiopathic infertility, *Fertil. Steril.,* 42, 97, 1984.

38. **Prader, A., Zachmann, M., and Illig, R.,** Fertility in adult males with congenital adrenal hyperplasia due to 21-hydroxylase deficiency, *Acta Endocrinol. (Copenhagen),* Suppl. 177, 57, 1977.

39. **Urban, M. D., Lee, P. A., and Migeon, C. J.,** Adult height and fertility in men with congenital virilizing adrenal hyperplasia, *N. Engl. J. Med.,* 299, 1392, 1978.

40. **Burger, H. G., and Baker, H. W. G.,** The treatment of infertility, *Annu. Rev. Med.,* 38, 29, 1987.

41. **Speiser, P. W., Dupont, B., Rubinstein, P., Piazza, A., Kastelan, A., and New, M. I.,** High frequency of nonclassical steroid 21-hydroxylase deficiency, *Am. J. Hum. Genet.,* 37, 650, 1985.

42. **Wilkins, L.,** Adrenal disorders. II. Congenital virilizing adrenal hyperplasia, *Arch. Dis. Child.,* 37, 231, 1962.

43. **Prader, A.,** Die Häufigkeit des Kongenitalen Adrenogenitalen Syndroms, *Helv. Paediatr. Acta,* 13, 426, 1958.

44. **Childs, B., Grumbach, M. M., and van Wyk, J. J.,** Virilizing adrenal hyperplasia: a genetic and hormonal study, *J. Clin. Invest.,* 35, 213, 1956.

45. **Baulieu, E. E., Peillon, F., and Migeon, C. J.,** Adrenogenital syndrome, in *The Adrenal Cortex,* Eisenstein, A. B., Ed., Little, Brown, Boston, 1967, chap. 15.

46. **Fife, D. and Rappaport, E. B.,** Prevalence of salt-losing among congenital adrenal hyperplasia patients, *Clin. Endocrinol.,* 19, 259, 1983.

47. **New, M. I. and Levine, L. S.,** Recent advance in 21-hydroxylase deficiency, *Annu. Rev. Med.,* 35, 649, 1984.

48. **Rimoin, D. L. and Schimke, R. N.,** *Genetic Disorders of the Endocrine Glands,* C. V. Mosby, St. Louis, Mo., 1971, 230.

49. **Müller, W., Prader, M., Kofler, J., Glatzl, J., and Geir, W.,** Zur Häufigkeit des Kongenitalen Adrenogenitalen Syndroms, *Paediatr. Paedol.,* 14, 151, 1979.

50. **Pang, S., Murphey, W., Levine, L. S., Spence, D. A., Leon, A., LaFranchi, S., Surve, A. S., and New, M. I.,** A pilot newborn screening for congenital adrenal hyperplasia in Alaska, *J. Clin. Endocrinol. Metab.,* 55, 413, 1982.

51. **Pang, S., Spence, D. A., and New, M. I.,** Newborn screening for congenital adrenal hyperplasia with special reference to screening in Alaska, *Ann. N.Y. Acad. Sci.,* 458, 90, 1985.

52. **Thomson, G. and Bodmer, W.,** The genetic analysis of HLA and disease associations, in *HLA and Disease,* Dausset, J. and Svejgaard, A., Eds., Williams & Wilkins, Baltimore, 1977, 84.

53. **Dupont, B., Oberfield, S. E., Smithwick, E. M., Lee, T. D., and Levine, L. S.,** Close genetic linkage between HLA and congenital adrenal hyperplasia (21-hydroxylase deficiency), *Lancet,* 2, 1309, 1977.

54. **Levine, L. S., Zachmann, M., New, M. I., Prader, A., Pollack, M. S., O'Neill, G. J., Yang, S. Y., Oberfield, S. E., and Dupont, B.,** Genetic mapping of the 21-hydroxylase deficiency gene within the HLA linkage group, *N. Engl. J. Med.,* 299, 911, 1978.

55. **Laron, Z., Pollack, M. S., Zamir, R., Roitman, A., Dickerman, Z., Levine, L. S., Lorenzen, F., O'Neill, G. J., Pang, S., New, M. I., and Dupont, B.,** Late onset 21-hydroxylase deficiency and HLA in the Ashkenazi population: a new allele at the 21-hydroxylase locus, *Hum. Immunol.,* 1, 55, 1980.

56. **Dausset, J.,** The major histocompatibility complex in man, *Science,* 223, 1469, 1981.

57. **Klouda, P. T., Harris, R., and Price, D. A.,** HLA and congenital adrenal hyperplasia, *Lancet,* 2, 1046, 1978.

58. **Pucholt, B., Fitzsimmons, J. S., Gelsthorpe, K., Pratt, R. F., and Doughty, R. W.,** HLA and congenital adrenal hyperplasia, *Lancet,* 2, 1046, 1978.

59. **Pollack, M. S., Levine, L. S., Zachmann, M., Prader, A., New, M. I., Oberfield, S. E., and Dupont, B.,** Possible genetic linkage disequilibrium between HLA and 21-hydroxylase deficiency gene, *Transplant. Proc.,* 11, 1315, 1979.

60. **Dupont, B., Virdis, R., Lerner, A. J., Nelson, C., Pollack, M. S., and New, M. I.,** Distinct HLA-B antigen associations for the salt-wasting and simple virilizing forms of congenital adrenal hyperplasia due to 21-hydroxylase deficiency, in *Histocompatibility Testing 1984,* Albert, E. D. et al., Eds., Springer-Verlag, Berlin, 1984, 660.

61. **Blankstein, J., Faiman, C., Reyes, F. I., Schroeder, M. L., and Winter, J. S. D.,** Adult-onset familial adrenal 21-hydroxylase deficiency, *Am. J. Med.,* 68, 441, 1980.

62. **Dumić, M., Brkljačić, L. J., Mardešić, D., Plavšić, V., Lukenda, M., and Kaštelan, A.,** Cryptic form of congenital adrenal hyperplasia due to 21-hydroxylase deficiency in the Yugoslav population, *Acta Endocrinol. (Copenhagen),* 109, 386, 1985.

63. **White, P. C., Grossberger, D., Onufer, B. J., New, M. I., Dupont, B., and Strominger, J. L.,** Two genes encoding steroid 21-hydroxylase are located near the genes encoding the fourth component of complement in man, *Proc. Natl. Acad. Sci., U.S.A.,* 82, 1089, 1985.

64. **Carroll, M. C., Campbell, R. D., and Porter, P. R.,** Mapping of steroid 21-hydroxylase genes adjacent to complement component C4 genes in HLA, the major histocompatibility complex in man, *Proc. Natl. Acad. Sci. U.S.A.,* 82, 521, 1985.

65. **White, P. C., New M. I., and Dupont, B.,** Structure of the human steroid 21-hydroxylase genes, *Proc. Natl. Acad. Sci. U.S.A.,* 83, 5111, 1986.

66. **Higashi, Y., Yoshioka, H., Yamane, M., Gotoh, O., and Fujii-Kuriyama, Y.,** Complete nucleotide sequence of two steroid 21-hydroxylase genes tandemly arranged in human chromosome: a pseudogene and a genuine gene, *Proc. Natl. Acad. Sci. U.S.A.,* 83, 2841, 1986.

67. **Carroll, M. C., Campbell, R. D., Bentley, D. R., and Porter, R. P.,** A molecular map of the major histocompatibility complex class III region linking the complement genes C4, C2, and factor B, *Nature (London),* 307, 237, 1984.

68. **Garlepp, M. J., Wilton, A. N., Dawkins, R. L., and White, P. C.,** Rearrangement of 21-hydroxylase genes in disease-associated MHC supratypes, *Immunogenetics,* 23, 100, 1986.

69. **Werkmeister, J. W., New M. I., Dupont, B., and White, P. C.,** Frequent deletion and duplication of the steroid 21-hydroxylase genes, *Am. J. Hum. Genet.,* 39, 461, 1986.

70. **Pang, S., Wallace, M. A., Hofman, L., Thuline, H. C., Dorche, C., Lyon, I. C. T., Dobbins, R. H., Kling, S., Fujieda, K., and Suwa, S.,** Worldwide experience in newborn screening for classical congenital adrenal hyperplasia due to 21-hydroxylase deficiency, *Pediatrics,* 81, 866, 1988.

71. **New, M. I., Dupont, B., Grumbach, K., and Levine, L. S.,** *The Metabolic Basis of Inherited Disease,* Stanbury, J. B., Wyngaarden, J. B., Fredrickson, D. S., Goldstein, J., and Brown, M. S., Eds., 1983.

Chapter 25

GENETIC HETEROGENEITY IN XY SEX REVERSAL: POTENTIAL PITFALLS IN ISOLATING THE TESTIS-DETERMINING FACTOR (TDF)

Joe Leigh Simpson

TABLE OF CONTENTS

I. INTRODUCTION

A current goal in the field of human sex differentiation is isolation of the gene(s) responsible for testicular differentiation (testis-determining factor, *TDF*).[1] This approach has become possible through the availability of cloned DNA sequences (probes). Presence or absence of cloned sequences can be correlated with presence or absence of certain clinical features.[2,3] Especially attractive in exploiting this approach are sex-reversal disorders such as XY gonadal dysgenesis. Failure of a given Y-DNA sequence to hybridize with DNA of individuals having XY gonadal dysgenesis would be taken to indicate that the hybridizing sequence is identical to the locus integral for *TDF* or at least nearby. Even if the latter were true, one could still "walk the chromosome" to isolate the locus in question.

Despite the attractiveness of this approach, pitfalls remain. The major one is genetic and even etiologic heterogeneity. Suppose we are studying DNA from a group of individuals believed to have the same disorder. Suppose further, however, that the individuals studied actually do not have the same disorder but rather several different disorders that are merely clinically similar. Experimental data derived from analysis of a heterogeneous sample could be confusing, potentially causing an investigator to abandon a fundamentally solid lead because of ostensibly contradictory results. Therefore, clinical, as well as laboratory, investigators must be cognizant of genetic (causal) heterogeneity. To emphasize this concept, we shall review XY gonadal dysgenesis and other disorders of sex reversal that may yield confusion.

II. XY GONADAL DYSGENESIS

A. Phenotype

XY gonadal dysgenesis is a rare sex-reversal condition first described in 1956 by Swyer,[4] a British gynecologist. Affected individuals are 46,XY, yet show female external genitalia, a well-differentiated but infantile vagina, and Müllerian derivatives (uterus, fallopian tubes, cervix).[5,6] In lieu of gonads, fibrous streaks are located along the posterior aspects of the broad ligament. Although testicular remnants occasionally are identified, the histologic appearance of the streaks is more often indistinguishable from that of 45,X individuals. Most individuals with XY gonadal dysgenesis show no somatic anomalies. They are of normal stature, and lack the features of the Turner stigmata. At puberty, affected individuals display secondary sexual development — breast development, pubic hair, or axillary hair. However, affected individuals respond to exogenous hormones. Although obviously infantile, the uterus is potentially functional. Embryo transfer or in vitro fertilization technology with donor oocytes should produce pregnancies, as already shown in individuals who lack ovaries but possess a uterus.[7,8]

The XY gonadal dysgenesis phenotype is accepted as the human equivalent of embryonic gonadal castration in XY rabbits and other mammals. Failure of testicular differentiation during early embryogenesis has long been known to lead to the female phenotype. Although the status of the early fetal gonads in XY gonadal dysgenesis is not certain, embryonic status would appear to be ovarian. Cussen and MacMahon[9] reported an affected female who showed oocytes at birth; however, no oocytes were present at 5 years of age. Other evidence confirms that oocytes can be XY in mice.[10] Thus, one can logically conclude that ovarian differentiation is constitutive, requiring merely a single X chromosome and occurring despite presence of a Y chromosome.

B. Molecular Studies

That XY gonadal dysgenesis could potentially be useful in deletion mapping is obvious. One might logically postulate that a DNA sequence coding for *TDF* could be deleted. The experimental approach would involve searching for a Y-DNA sequence that is absent in XY gonadal dysgenesis, yet present on Yp of normal males. Such a sequence could be a candidate for *TDF*, or at least closely linked to the *TDF*.

This approach would, then, be the converse of that successfully employed by Vergnaud et al.[2] in XX males. Vergnaud et al.[2] studied 19 XX males, utilizing various DNA sequences known to be present yet not necessarily restricted to the Y short arm. In six XX males, none of these sequences was present; thus, no sequence corresponded to *TDF*. In a single XX male, sequence DXYS5 was present. In six additional XX males, both DXYS5 and a second group of sequences (including DXYS7) were present. In five other XX males, not only were DXYS5 and DXYS7 present but, so, too, was a third group of sequences signified by DYS8. The DYS8 group was never present without the DXYS7 group, nor the DXYS7 group without the DXYS5. Thus, nested analysis could be performed. The DNA sequence closest to *TDF* must be DXYS5; other sequences signified intervals progressively distal from *TDF*.

If the interval containing DXYS5 indeed contains *TDF*, one might predict that the interval containing DXYS5 would be absent in individuals with XY gonadal dysgenesis. Indeed. Disteche et al.[11] studied two individuals known to have small yet cytologically detectable deletions of the Y chromosome. DXYS5 was absent. Additional cases were later studied by Page.[12] Thus, complementary data revealed presence of sequences in XX males that are absent in XY females. Given that the *TDF* must lie close to DXYS5, accepted strategies for walking-the-chromosome are being employed.

C. X-Linked Recessive Inheritance

The experimental approach described above is based upon the premise that XY gonadal dysgenesis is caused by deletion of a submicroscopic portion of the Y chromosome. That is, the etiology is of cytogenetic origin, presumably sporadic in occurrence. To the contrary, it is clear that XY gonadal dysgenesis can on occasion be caused by a mutant X-linked recessive gene.

The first family showing X-linked inheritance was the well-cited family of Sternberg et al.[13] Other families were later reported by Espiner et al.,[14] by Simpson et al.,[15] by German et al.,[16] and by Mann et al.[17] Individuals having the X-linked recessive form of XY gonadal dysgenesis are indistinguishable from individuals with XY gonadal dysgenesis who have no other affected family members. Because the Y chromosome is completely normal in X-linked recessive kindreds, it follows that performing DNA studies on individuals having XY gonadal dysgenesis due to an X-linked mutation could be misleading. One would expect no abnormalities if DNA hybridization involving Y-specific clones were tested.

Existence of an X-linked recessive form of XY gonadal dysgenesis is relevant not only to deletion mapping, but also to H-Y antigen. Overall, approximately 75% of individuals with XY gonadal dysgenesis show H-Y antigen.[18] The pathogenesis of H-Y positive cases is assumed to involve receptor defects, whereas in H-Y negative cases the locus coding for H-Y antigen is assumed to be defective or absent. Most, if not all, of the 25 to 30% individuals with XY gonadal dysgenesis who develop germ cell neoplasia[19,20] (gonadoblastomas or dysgerminomas) show H-Y antigen, although the manner by which H-Y antigen enhances the development of neoplasia remains uncertain.

That H-Y positive individuals originally had testicular tissue, an earlier possibility raised by this author,[21] now seems less likely because of the findings of Cussen and MacMahon.[9] Although H-Y antigen studies have not been performed in all families showing X-linked recessive inheritance, the five affected individuals in the family of Mann et al.[17] were characterized by both H-Y antigen positivity and neoplasia. Neoplasia also occurred in several other X-linked recessive kindreds.[14,16] Thus, the X-linked recessive form of XY gonadal dysgenesis is characterized by H-Y antigen positivity and by predisposition toward neoplasia (Table 1). Other potential forms of XY gonadal dysgenesis, to be discussed below, may or

Table 1
FORMS OF XY GONADAL DYSGENESIS

	H-Y Antigen	Gonadal neoplasia	Inheritance
X linked	+	+	XLR
Autosomal (unproved)	− (?)	− (?)	AR (?)
Campomelic dysplasia	−	−	AR
Brosnan	Unknown	Unknown	?AR ?XLR

Note: +, Present; −, absent; (?), postulated; XLR, X-linked recessive; and AR, autosomal recessive.

may not show the same characteristics. In particular, there are indications that neoplasia is less frequent in 45,X/46,XY individuals who lack fluorescent Yq.[22]

D. Autosomal Inheritance

A second form of XY gonadal dysgenesis was postulated in 1981 by Simpson et al.[21] to be due to an autosomal recessive mutation. This arose from observations that XY gonadal dysgenesis had occurred in individuals whose parents were consanguineous.[23] In a kindred reported by Allard et al.[24] the disorder appeared to be transmitted through males, a mode of inheritance obviously incompatible with X-linked recessive inheritance. Sex-limited autosomal dominant inheritance is one possible explanation. Another is autosomal recessive inheritance, based upon the assumption that spouses of several family members were heterozygous for the same mutant allele. Such a phenomenon is not necessarily improbable if a kindred is inbred. In fact, the family of Allard et al.[24] was French Canadian (Quebec).

Based upon these leads, our group performed segregation analysis to test the hypothesis of autosomal recessive inheritance.[21] Our strategy was first to exclude families in which the disorder was either (1) known to be segregating as an X-linked recessive trait, or (2) potentially caused by other factors, as evidenced by unusual coexisting somatic features. All remaining individuals were subjected to segregation analysis. If a sex-limited autosomal recessive form of XY gonadal dysgenesis exists, 25% of males should be affected. If the disorder were due exclusively to an X-linked recessive gene, 50% of males should be affected.

Our segregation analysis was compatible with a 25% recurrence rate, favoring the hypothesis of autosomal recessive inheritance.[21] It would be tempting also to postulate that individuals with autosomal recessive XY gonadal dysgenesis were characterized not only by H-Y antigen negativity but also by lack of neoplasia (Table 1). However, the latter would be inconsistent with observations of Lusuka et al.[22] who correlated presence of fluorescent Yq with presence of gonadal neoplasia. Further testing awaits linkage studies or analysis of kindreds born to consanguineous parents. (One could not, or course, merely study families with multiple affected sibs because such an occurrence would be compatible with either X-linked recessive or autosomal recessive inheritance.)

Existence of an autosomal recessive form of XY gonadal dysgenesis would be compatible with an autosomal location for a *TDF*. That autosomal loci are integral for testicular differentiation implies that the Y may merely exert a regulatory role in testes differentiation. Such ideas are consistent with studies of Eicher et al.[25] who identified an autosomal locus integral for murine testicular determination. In their studies, C57BL/6 females were mated with *Mus poschiavinus* males. Females of the hybrid were then backcrossed to another *M. poschiavinus* male. When now placed on a C57BL/6 background, the *M. poschiavinus* Y sometimes proved incapable of directing testicular differentiation.[25] Hybrid XY mice were true hermaphrodites or sex-reversed females. The key autosomal locus lies on chromosomes no. 17,[26] which contains H-2, and thus is, presumably homologous to human chromosome no. 6.

E. Conclusion

Existence of genetic heterogeneity can be taken as evidence that the TDF sequence will not hybridize to DNA of all XY females. Indeed, failure of hybridization should be expected as more cases are analyzed. Occurrence of such a phenomenon need not detract from the usefulness of searching for sequences that fail to hybridize.

III. XY GONADAL DYSGENESIS AND CAMPOMELIC DYSPLASIA

Campomelic dysplasia is characterized by bowing of the long bones, cleft palate, micrognathia, low-set ears, brachydactyly, clinodactyly, bell-shaped chest, and small scapuli. Affected infants show postnatal growth retardation.[27] They often expire within several months of birth because of respiratory difficulties (apnea).[27] There are probably several different disorders, but at least one is inherited in autosomal recessive fashion.[27]

In addition to the features described above, it is well accepted that at least one form of this condition is associated with XY sex reversal.[28,29] Some affected 46,XY individuals also show normal *female* external genitalia, *Müllerian* derivatives (uterus, cervix, tubes), and *ovaries*. Sometimes ovarian follicles are evident, whereas on other occasions the gonads consist of ovarian-like stroma with occasional oocytes. H-Y antigen is not present.[28,29] Given the presumptive autosomal recessive etiology for campomelic dysplasia, analysis of the DNA of these individuals might prove helpful if one wishes to search for an autosomal location for *H-Y* genes or testicular determination.

IV. GENITO-PALATO-CARDIAC (GARDNER-SILENGO-WACHTEL) SYNDROME

Familiar to students of sex differentiation is the report of Bernstein et al.[30] The phenotypic female proband showed enlargement of the cranial ventricles, was mentally retarded, and manifested myoclonic jerks. Her height, weight, and head circumference were far below the third percentile, and she displayed many dysmorphic features: "transparent" skin, facial and skull asymmetry, low-set ears, prominent frontal bossing, hypertelorism, downward slanting palpebral fissures, depressed nasal bridge, prognathism, narrow lips, carp-shaped mouth, cleft palate, malaligned teeth, clinodactyly, and prominent hyperextended heels. Internal anomalies included ventricular (cardiac) septal defect, fatty metamorphosis of the liver, and fibrotic gall bladder. Serologic studies revealed hypercholesterolemia and hyperbilirubinemia. Of particular interest in the current context are genital abnormalities: female external genitalia with a normal vagina in an XY individual; Müllerian derivatives (cervix, uterus, and fallopian tubes); and gonads consisting of ovarian stroma characterized by a few identifiable primordial follicles. The proband had one normal Y and one X characterized by duplication of a portion of the X short arm (Xp).

Analysis of other family members revealed that the identical abnormal X chromosome was present in the proband's mother, sister, and maternal grandmother. All had a second normal X, however, and all were phenotypically normal. Amniocentesis in a subsequent pregnancy by the proband's mother revealed a chromosomal complement identical to that of the proband, for which reason the pregnancy was terminated. The aborted fetus showed many of the features present in the proband: hypertelorism, micrognathia, low-set ears, cleft palate, ventricular septal defect, and prominent heels. External genitalia were again unequivocally female, Müllerian derivatives were present, and the ovary contained primary follicles (Figure 1). In the family of Bernstein et al.[30] H-Y antigen was present in neither

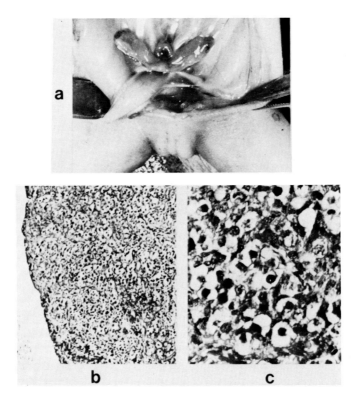

FIGURE 1. Female anatomy in having a Y chromosome and duplication of the X short arm fetus. (a) Gross anatomy of the external and internal genitalia; (b) histology of fetal ovary (magnification × 90); (c) histology of fetal ovary (magnification × 360) showing numerous primordial follicles. (From Bernstein R., Jenkins, T., Dawson, B., Wagner, J., Dewald, G., Koo, G., and Wachtel, S., *J. Med. Genet.*, 17, 291, 1980. With permission.)

the proband nor the fetus. Because the Y chromosome was intact, the authors suggested that Xp contained a locus capable of suppressing H-Y when present in duplicate. That is, sex reversal was postulated to be the result of a double dose of a suppressor locus on Xp.

Bernstein et al.[30] alluded to several other reports documenting similar clinical findings. Their allusion to these cases indicated that they considered the possibility of a genetic syndrome. However, data were inadequate to determine whether the clinical findings in their sibs could be explained on that basis.

Since the 1981 report of Bernstein et al.[30] additional publications have substantiated the likely existence of a syndrome characterized by genital, palatal, and cardiac anomalies (genito-palato-cardiac) (Table 2). Greenberg et al.[31] in particular, reported two affected sibs, and recommended that this condition be called the Gardner-Silengo-Wachtel syndrome, honoring three individuals who contributed greatly to its delineation. Greenberg et al.[31] considered the first such case to be that of Gardner et al.[32] who in 1971 briefly described a phenotypic female with cleft palate, micrognathia, kyphoscoliosis, and equinovarus. External genitalia were female, and Müllerian derivatives were present; however, the gonads consisted of testes. Consideration of this particular case was obfuscated by the claim of Gardner et al.[32] that the etiology involved ingestion of oral contraceptives. We now recognize this possibility as extremely unlikely.[33,34]

In 1974, Silengo et al.[35] reported a 46,XY phenotypic female with low-set ears, downward slanting palpebral fissures, anteverted nares, depressed nasal bridge, carp-shaped mouth,

Table 2
GENITO-PALATO-CARDIAC SYNDROME (GARDNER-SILENGO-WACHTEL)

Sex Differentiation in 46,XY Individuals

• Female external genitalia
• Müllerian derivatives
• Gonads: ovaries (sex reversal) or testes

Somatic Anomalies

• Facial dysmorphia
  (slanting palpebral fissures, micrognathia, low-set ears)
• Cleft palate
• Cardiac defects
  (Ventricular septal defect)
• Prominent heels

Inheritance

• X-linked or autosomal recessive inheritance

and cleft alveolar ridge. Uterus and fallopian tubes were present, and gonads most closely resembled dysgenetic testes with Sertoli cells. Wolman et al.[36] reported a 46,XY phenotypic female having cleft palate, micrognathia, club foot, and thoracic abnormalities. Uterus and fallopian tubes were present; two small "ovary like" gonads were said to contain seminferous tubules but no Leydig cells, no follicles, and no ova. H-Y antigen was present.

In addition, Lindenbaum had also communicated to Bernstein et al.[30] knowledge of three individuals with microcephaly, cleft palate, low-set ears, ventricular septal defect (one of three cases), and postaxial polydactyly.

In 1987, Greenberg et al.[31] reported two 46,XY sibs, who despite discordant gonads, appeared to have a similar disorder. The proband (fetus) was detected *in utero* after anomalies were observed in prenatal ultrasonography performed on a 28-year-old woman. At pregnancy termination, the fetus was shown to have micrognathia, intact palate, low-set ears, flexion deformities of the thumbs and great toes, double-outlet right ventricle, and ventricular septal defect. External genitalia were female, Müllerian derivatives were present, and gonads consisted of ovaries. Chromosomal analysis of multiple tissues (amniotic fluid, blood, skin fibroblasts) confirmed the 46,XY complement. In a subsequent pregnancy by the same woman, ultrasound at 16 weeks gestation revealed oligohydromnios, intrauterine growth retardation, hydrops fetalis, and bilateral renal hydronephosis. Another pregnancy termination was performed. The second fetus showed micrognathia, low-set ears with malformed right auricles, bilateral cleft lip and palate, bilateral flexion deformities of the thumbs, transposition of the great vessels, bilateral cystic kidneys, agenesis of the gall bladder, intestinal malrotation, and "dysgenesis" of the urinary bladder. External genitalia consisted of first degree penile hypospadias and an empty scrotal sac. Testes were located in the pelvic cavity. Neither a uterus nor fallopian tubes were present.

Also of relevance is the report of Beemer and Ertbruggen,[37] who described two dysmorphic 46,XY sibs of consanguineous parents. The first sib showed various facial abnormalities (bulbous nose, downward-slanting palpebral fissures, low-set ears, retrognathia) and tetrology of Fallot. Polydactyly was not present. External genitalia were *female* with enlarged clitoris, but gonads consisted of testes. The second 46,XY sib was a phenotypic *male* with hydrocephalus, small auricles, and double-outlet right ventricle; testes were present.

Whether the two sibs of Beemer and Ertbruggen[37] had the same disorder is uncertain. Moreover, whether the disorder(s) those sibs manifested corresponds to that described by Gardner et al.,[32] Silengo et al.,[35] Bernstein et al.,[30] and Greenberg et al.,[31] is also uncertain. If the family of Beemer and Ertbruggen[37] did represent the same condition as shown by others,[30-32,35] it would be another example of genital findings being discrepant among affected XY sibs. Unlike Beemer and Ertbruggen,[37] however, Greenberg et al.[31] showed both genital *and* gonadal discrepancies. If, furthermore, the cases described by Wolman et al.[36] and Bernstein et al.[30] have the same disorder, we have several other discrepancies to explain, namely differences in H-Y antigen status and gonadal histology.

In aggregate, however, at least one previously unrecognized distinct sex-reversal syndrome(s) would seem to exist. The syndrome might be said to be characterized by the following somatic abnormalities: micrognathia, facial asymmetry, cleft palate, cardiac defects, and prominent heels. Neither postaxial polydactyly nor campomelic dysplasia are characteristic. External genitalia are female, and Müllerian derivatives are present. Strangely, gonads vary from ovaries to testes. All affected individuals have been 46,XY, raising the possibility of X-linked recessive etiology. On the other hand, ascertainment biases (more ready detection of males because of their genital abnormalities) surely exist; thus, the disorder could still be inherited in autosomal recessive inheritance. (Indeed, the parents in Beemer and Ertbruggen[37] were consanguineous.) Inasmuch as the only cytogenetic abnormalities detected were those of Bernstein et al.,[30] such findings could well be coincidental. Alternatively, perturbations of Xp in that family could have produced hemizygosity, a postulate consistent with X-linked recessive inheritance.

What is the more general lesson, especially for scientists who are not physicians? Individuals with gonadal abnormalities and somatic abnormalities could be abnormal as a result of a mutant gene, irrespective of associated cytogenetic findings that sometimes render cause and effect postulations almost irresistible. Again, caution is necessary before assuming that unusual phenotypes are always explained by associated unusual findings, be they DNA hybridization patterns, H-Y antigen status, or cytogenetic abnormalities.

V. SMITH-LEMLI-OPITZ SYNDROME, TYPE II

The genito-palato-cardiac (Gardner-Silengo-Wachtel) syndrome cannot be discussed without at least brief allusion to the Smith-Lemli-Opitz syndrome. This syndrome is a well-recognized dysmorphic syndrome,[38] initially reported in 1964. The authors described four individuals with characteristic facial dysmorphia (low-set ears, eyelid ptosis, slanting palpebral fissures, epicanthal folds, esotropia, anteverted nares, broad nasal tip), syndactyly between the second and third toes, and severe mental retardation. Prenatal and postnatal growth retardation exists. Both males and females can be affected, indicating autosomal recessive inheritance. Females show no genital abnormalities, but males typically display mild to severe hypospadias and cryptorchidism. However, as typically defined, genital abnormalities in 46,XY individuals with Smith-Lemli-Opitz syndrome do not extend to female-like external genitalia (sex reversal). There are no examples of a uterus persisting in an affected male, and gonads have always consisted of testes.

In contrast to the above description, Curry et al.[39] recently proposed the existence of a related but distinct condition, which they termed Smith-Lemli-Opitz syndrome Type II. (Potential controversy in designating the individuals in this fashion has been discussed by Opitz and Lowry,[40] and need not be reiterated here.) Of relevance, however, are the somatic and genital findings in the proposed disorder, for they are highly reminiscent of those in the genito-palato-cardiac (Gardner-Silengo-Wachtel) syndrome.

In support of their proposal for existence of Smith-Lemli-Opitz syndrome Type II (Table 3), Curry and colleagues[39] collected 19 cases from several institutions. The 19 cases showed

Table 3
SMITH-LEMLI-OPITZ SYNDROME, TYPE II

Sex Differentiation
   46,XY individuals
      Hypospadias (male pseudohermaphroditism) or female external
        genital
      Lack of Müllerian derivatives
      Testes
   46,XX individuals
      Female external genitalia
      Müllerian derivatives
      Ovaries

Somatic Anomalies (46,XY and 46,XX)
   Mental retardation
   Facial dysmorphia (similar to Smith-Lemli-Opitz, Type I)
   Cleft palate
   Cataracts
   Cardiac defects
   Unilobate lungs
   Large adrenals
   Pancreatic islet hyperplasia
   Hirschprung disease
   Intestinal malrotation
   Renal anomalies

Inheritance
   Autosomal recessive

facial features reminiscent of those in Smith-Lemli-Opitz, Type I. However, other more severe somatic anomalies were present: cleft palate, small tongue, cataracts, cardiac defects, postaxial polydactyly, and one or more unusual internal anomalies: unilobate lungs, large adrenals, renal anomalies, pancreatic islet cell hyperplasia, absent parathyroids, hypoplastic gall bladder, and Hirschsprung disease. Sometimes limbs were short and flexion contractures present.

Genital changes were also more severe in the 46,XY individuals described by Curry et al.[39] than they were in males with Smith-Lemli-Opitz, Type I.[38] Genital abnormalities extended beyond the hypospadias characteristic of Type I to female external genitalia (sex reversal) (Figure 2). However, gonads always consisted of testes, unlike genito-palato-cardiac syndrome. Androgen receptor studies were normal, and no other endocrine abnormalities were present. Of the nine 46,XY individuals described by Curry et al.,[39] two showed Müllerian derivatives. One of these two individuals showed a uterus and a septate vagina with urogenital sinus, whereas the other had a rudimentary uterus. Interestingly, one of the ten affected 46,XX females had no right ovary and an abnormal right fallopian tube.

In addition to the 19 cases reported by Curry et al.,[39] Belmont et al.[41] described two unrelated individuals with the same apparent syndrome. Both were 46,XY phenotypic females, albeit characterized by clitoral hypertrophy. Both had postaxial polydactyly and many of the other somatic findings described by Curry et al.,[39] including unilobate lung. Again, gonads consisted of bilateral testes. One affected individual showed Wolffian derivatives, whereas the other showed uterus duplex bicornus.

Because it manifests a different spectrum of somatic anomalies (e.g., polydactyly, unilobate lung, cataracts, lack of gonadal sex reversal), it would appear that "Smith-Lemli-Opitz, Type II" is an entity distinct from the genito-palato-cardiac (Gardner-Silengo-Wachtel) syndrome. Of course, the argument about "lumping" or "splitting" in clinical delin-

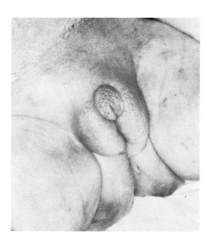

FIGURE 2.   Sex reversal in XY individual reported by Curry et al.[39] An example of Smith-Lemli-Opitz, Type II.

eation of rare syndromes is a familiar one to clinical geneticists, who have learned to accept such controversy as a heuristic exercise devoid of expectations for definitive answers. Such answers require measurable markers, hopefully the gene product itself. Yet the significance in the current context is that existence of genital (albeit not gonadal) sex reversal in Smith-Lemli-Opitz Syndrome, Type II could potentially confuse those seeking informative experiments of nature. An erroneous diagnosis of genito-palato-cardiac syndrome could easily be made.

VI. BROSNAN SYNDROME

Two apparently unique 46,XY sibs were reported by Brosnan et al.[42] in 1980. The two sibs showed streak gonads and a similar pattern of anomalies: unusual facial features, cardiac anomalies, renal defects, mental retardation, and ectodermal abnormalities such as scalp defects. The latter findings are especially unusual. Again, the underlying feature of note is the presence of somatic anomalies and gonadal dysgenesis in 46,XY individuals. Surely other cases exist, necessitating caution by investigators who might unwarily mix cases of Brosnan syndrome with cases deserving other diagnoses (e.g., typical XY gonadal dysgenesis).

VII. XY GONADAL DYSGENESIS IN INDIVIDUALS WHOSE RELATIVES HAVE GENITAL AMBIGUITY, BILATERAL TESTES, AND MÜLLERIAN DERIVATIVES

In three families,[43-45] an individual with XY gonadal dysgenesis had a sib with genital ambiguity, bilateral testes, and Müllerian duct derivatives (uterus). In one of these families[44] an affected male also had a sister whose son had hypospadias. Possibly another family was reported by Lowry et al.[46] Inasmuch as the phenotype suggests 45,X/46,XY mosaicism, one possible explanation is the presence of a gene causing mitotic nondisjunction. The monosomy X (45,X) cell line might be difficult to detect. Indeed, familial mosaicism is known to exist.[47] Irrespective of that, a distinct etiology can be assumed in these three families. It follows that sporadic cases of similar phenotype could be of similar etiology.

Table 4
DIFFICULTY IN
DETECTING 45,X/46, XY
MOSAICISM

	Number of cells	
	45,X	**46,XY**
Blood	0	25 (100%)
Skin	0	15 (100%)
Gonad	19 (76%)	6 (24%)

Note: Cytogenetic studies of a 45,X/46,XY phenotypic female studied at the University of Tennessee, Memphis (Laboratory of Dr. Avirachan Therapel).

VIII. DISTINGUISHING XY GONADAL DYSGENESIS FROM 45,X/46,XY MOSAICISM

Presence in a single individual of 45,X as well a 46,XY cells is associated with a spectrum of phenotypes. The phenotypic spectrum includes females completely indistinguishable from those with Turner syndrome as well as males with simple hypospadias. 45,X/46,XY mosaicism may also be associated with genital ambiguity; such individuals usually having unilateral streak gonad and a bilateral testis (mixed or asymmetric gonadal dysgenesis). Presumably the basis for the phenotypic spectrum is the relative tissue distributions of 45,X and 46,XY cells. However, this is not evident from analysis of lymphocytes alone.

Of various 45,X/46,XY individuals, those with female external genitalia concern us most at present. These individuals usually have bilateral streak gonads. If they show short stature or other features of the Turner stigmata, the presence of a 45,X cell line would not be surprising. A diligent search for a monosomic X cell line would be conducted in other tissues if lymphocyte studies revealed only 46,XY cells. However, a more confusing situation might arise if a sporadic 45,X/46,XY individual of relatively normal stature not only lacked somatic anomalies but showed only 46,XY cells in lymphocytes. Such an individual could easily be categorized as an example of XY gonadal dysgenesis.

That the above phenomenon is not rare is illustrated by our own experience at the University of Tennessee, Memphis. At our institution, nine 45,X/46,XY cases have been studied. In five of the nine, genital ambiguity was present; a male sex of rearing was chosen on the basis of penoscrotal hypospadias. The four other individuals were raised as females; in three, mild clitoromegaly was present, but in the fourth, completely normal female external genitalia were noted. One especially instructive case involved a 17-year-old female whose cytogenetic data are shown in Table 4. This 58-inch tall subject had mild clitoromegaly and normal pubic hair, but only a small amount of palpable breast tissue. Her lymphocytes were 46,XY. However, a monosomic line was suspected because the phenotype did not correlate with karyotype, i.e., the slightly diminished stature was unexplained. Cytogenetic analysis of skin also revealed 46,XY cells. However, fibroblasts from the left gonad revealed 45,X cells. The obvious message is caution before assuming that a sporadic XY gonadal dysgenesis case is truly nonmosaic.

IX. CONCLUSION

In a variety of disorders, sex reversal can exist in 46,XY individuals. Genetic and even causal heterogeneity exists. Arriving at the precise diagnosis is thus not so obvious as one might imagine. In addition to genetic heterogeneity, cytogenetic pitfalls (e.g., undetected mosaicism) may also exist. Although XY gonadal dysgenesis and related disorders remain useful for elucidating human sex determination and isolating *TDF*, caution remains appropriate in evaluating results of molecular studies.

REFERENCES

1. **Simpson, J. L.,** Phenotypic-karyotypic correlations of gonadal determinants: current status and relationship to molecular studies, *Proc. VII Int. Congress Human Genetics,* Springer-Verlag, Berlin, in press.
2. **Vergnaud, G., Page, D. C., Simmler, M. C., Brown, L., Rouyer, F., Noel, B., Botstein, D., de La Chapelle, A., and Weissenbach, J.,** A deletion map of the human Y chromosome based on DNA hybridization, *Am. J. Hum. Genet.,* 38, 330, 1986.
3. **Krauss, C. M., Turksoy, R. N., Atkins, L., McLaughlin, C., Brown, L. G. and Page, D. C.,** Familial premature ovarian failure due to an interstitial deletion of the long arm of the X chromosome, *N. Engl. J. Med.,* 317, 215, 1987.
4. **Swyer, G. I. M.,** Male pseudohermaphroditism: a hitherto undescribed form, *Br. Med. J.,* 2, 709, 1955.
5. **Simpson, J. L., Christakos, A. C., Horwith, M. and Silverman, F. S.,** Gonadal dysgenesis in individuals with apparently normal chromosomal complements: tabulation of cases and compilation of genetic data, *Birth Defects Orig. Artic. Ser.,* 7(6), 215, 1971.
6. **Simpson, J. L.,** *Disorders of Sexual Differentiation,* Academic Press, New York, 1976, 276.
7. **Chan, C. L. K., Cameron, I. T., Findlay, J. K., Healy, D., Leeton, J. F., Lutjen, P. J., Renou, P. M., Rogers, P. A., Trounson, A. O., and Wood, E. C.,** Oocyte donation and in vitro fertilization for hypergonadotropic hypogonadism: clinical state of art, *Obstet. Gynecol.,* 42, 350, 1987.
8. **Rosenwaks, Z.,** Donor eggs: their application in modern reproductive technologies, *Fertil. Steril.,* 47, 895, 1987.
9. **Cussen, L. J. and MacMahon, R. A.,** Germ cells and ova in dysgenetic gonads of a 46,XY female dizygotic twin, *Am. J. Dis. Child.,* 133, 373, 1979.
10. **Evans, E. P., Ford, C. E. and Lyon, M. F.,** Direct evidence of the capacity of the XY germ cell in the mouse to become an oocyte, *Nature (London),* 267, 430, 1977.
11. **Disteche, C. M., Casanova, M., Saal, H., Friedman, C., Sybert, V., Graham, J., Thuline, H., Page, D. C., and Fellous, M.,** Small deletions of the short arm of the Y chromosome in 46,XY females, *Proc. Natl. Acad. Sci., U.S.A.,* 83, 7841, 1986.
12. **Page, D. C.,** Sex reversal: deletion mapping the male-determining function of the human Y chromosome, *Cold Springs Harbor Symp. Quant. Biol.,* 51, 229, 1986.
13. **Sternberg, W. H., Barclay, D. L., and Kloepfer, M. W.,** Familial XY gonadal dysgenesis, *N. Engl. J. Med.,* 278, 695, 1968.
14. **Espiner, E. A., Veale, A. M. O., Sands, V. E., and Fitzgerald, P. H.,** Familial syndrome of streak gonads and normal male karyotype in five phenotypic females, *N. Engl. J. Med.,* 283, 6, 1970.
15. **Simpson, J. L., Summitt, R. L., German, J. and Merkatz, I. R.,** Etiology of XY gonadal dysgenesis, *Gynecol. Invest.,* 7, 37, 1976.
16. **German, J., Simpson, J. L., Chaganti, R. S. K., Summitt, R. L., Reid, L. B., and Merkatz, I. R.,** Genetically determined sex-reversal in 46,XY, *Science,* 202, 52, 1978.
17. **Mann, J. R., Corkery, J. J., Fisher, H. J. W., Cameron, A. H., Mayerova, A., Wolf, U., Kennaugh, A. A., and Woolley, V.,** The X linked recessive form of XY gonadal dysgenesis with a high incidence of gonadal cell tumours: clinical and genetic studies, *J. Med. Genet.,* 20, 264, 1983.
18. **Wachtel, S. S.,** *H-Y Antigen and the Biology of Sex Determination,* Grune & Stratton, New York, 1983.
19. **Simpson, J. L. and Photopulos, G.,** The relationship of neoplasia to disorders of abnormal sexual differentiation, *Birth Defects Orig. Artic. Ser.,* 12(1), 15, 1976.
20. **Verp, M. S. and Simpson, J. L.,** Abnormal sexual differentiation and neoplasia, *Cancer Genet. Cytogenet.,* 25, 191, 1987.
21. **Simpson, J. L., Blagowidow, N., and Martin, A. O.,** XY gonadal dysgenesis: genetic heterogenecity based upon clinical observations, H-Y antigen status, and segregation analysis, *Hum. Genet.,* 58, 19, 1981.
22. **Lusuka, T., Fryns, J. P., and Van den Berghe, H.,** Gonadoblastoma and Y-chromosome fluorescence, *Clin. Genet.,* 29, 311, 1986.
23. **Nazareth, H. R. S., Moreira-Filho., C. A., Cunha, A. J. B., Viera-Filho, J. P. B., Lengyei, A. M. J., and Lima, H. C.,** H-Y antigen in 46,XY pure testicular dysgenesis, *Am. J. Med. Genet.,* 3, 149, 1979.

24. **Allard, S., Codotte, M., and Boivin, Y.,** Dysgenesie gonadique pure familiale at gondoblastome, *Union Med. Can.,* 101, 448, 1972.
25. **Eicher, E. M., Washburn, L. L., Whitney, J. B., III, et al.,** *Mus poschiavinus* Y chromosome in the C57BL/6J murine genome causes sex reversal, *Science,* 217, 535, 1982.
26. **Washburn, L. L. and Eicher E. M.,** Sex reversal in XY mice, caused by dominant mutation on chromosome 17, *Nature (London),* 303, 338, 1983.
27. **Houston, C. S., Opitz, J. M., Spranger, J. W., et al.,** The Campomelic syndrome: review, report of 17 cases, and follow-up on the currently 17-year-old boy first reported by Maroteaux, et al. in 1971, *Am. J. Med. Genet.,* 15, 3, 1983.
28. **Bricarelli, F. D., Fraccaro, M., Lindsten, J., Müller, U., Baggio, P., Carbone, L. D., Hjerfe, A., Lindgren, F., Mayerova, A., Rinsertz, H., Ritzen, E. M., Rovetta, D. C., Siccchero, C., and Wolf, U.,** Sex-reversed XY females with campomelic dysplasia are H-Y negative, *Hum. Genet.,* 57, 15, 1981.
29. **Puck, S. M., Haseltine, F. P., and Francke, U.,** Absence of H-Y antigen in an XY female with campomelic dysplasia, *Hum. Genet.,* 57, 23, 1981.
30. **Bernstein, R., Jenkins, T., Dawson, B., Wagner, J., Dewald, G., Koo, G., and Wachtel, S.,** Female phenotype and multiple abnormalities in sibs with a Y chromosome and partial X chromosome duplication: H-Y antigen and Xg blood group findings, *J. Med. Genet.,* 17, 291, 1980.
31. **Greenberg, F., Gresik, M. W., Carpenter, R. J., Law, S. W., Hoffman, L. P., and Ledbetter, D. H.,** The Garnder-Silengo-Wachtel or Genito-Palato-Cardiac syndrome: male pseudohermaphroditism with micrognathia, cleft palate, and conotruncal cardiac defect, *Am. J. Med. Genet.,* 26, 59, 1987.
32. **Gardner, L. I., Assemany, S. R., and Neu, R. L.,** 46,XY female: anti-androgenic effect of oral contraceptive?, *Lancet,* 2, 667, 1970.
33. **Carson, S. A. and Simpson, J. L.,** Virilization of female fetuses following maternal ingestion of progestional and androgenic steroids, in *Hirsutism and Virilization,* Mahesh, V. B. and Greenblatt, R. L., Eds., PSG Publishing Co. Littleton, Mass., 1984, 177.
34. **Simpson, J. L.,** Relationship between congenital anomalies and contraception, *Adv. Contraceptives,* 1, 3, 1985.
35. **Silengo, M., Kaufmann, R. L., and Kissane, J.,** A 46,XY infant with uterus, dysgenetic gonads, and multiple anomalies, *Humangenetik,* 25, 65, 1974.
36. **Wolman, S. R., McMorrow, L. E., Roy, S., Koo, G., Wachtel, S., and David, R.,** Aberrant testicular differentiation in 46,XY gonadal dysgenesis: morphology, endocrinology, serology, *Hum. Genet.,* 55, 321, 1980.
37. **Beemer, F. A. and Ertbruggen, I. V.,** Pecular facial appearance, hydrocephalus, double-outlet right ventricle, genital anomalies, and dense bones with lethal outcome, *Am. J. Med. Genet.,* 9, 391, 1984.
38. **Smith, D. W.,** *Recognizable Patterns of Human Malformation: Genetic, Embryologic & Clinical Aspects,* 3rd ed., W. B. Saunders, Philadelphia, 1982, 98.
39. **Curry, C. J. R., Carey, J. C., Holland, J. S., Chopra, D., Fineman, R., Golabi, M., Sherman, S., Pagon, R. A., Allanson, J., Shulman, S., Barr, M., McGravey, V., Dabiri, C., Schimke, N., Ives, E., and Hall, B. D.,** Smith-Lemli-Opitz Syndrome-Type II: multiple congenital anomalies with male pseudohermaphroditism and frequent early lethality, *Am. J. Med. Genet.,* 26, 45, 1987.
40. **Opitz, J. M. and Lowry, R. B.,** Lincoln vs. Douglas again; comments on the papers by Curry et al., Greenberg et al., and Belmont el al., *Am. J. Med. Genet.,* 26, 69, 1987.
41. **Belmont, J. W., Hawkins, E., Hejtmancik, J. F., and Greenberg, F.,** Two cases of severe lethal Smith-Lemli-Opitz syndrome, *Am. J. Med. Genet.,* 26, 65, 1987.
42. **Brosnan, P. G., Lewandowski, R. C., Toguri, A. G., Payer, A. F., and Meye, W. J.,** A new familial syndrome of 46,XY gonadal dysgenesis with anomalies of ectodermal and mesodermal structures, *J. Pediatr.,* 97, 586, 1980.
43. **Baron, J., Rucki, T., and Simm, S.,** Familial gonadal malformations, *Gyneacologia,* 153, 290, 1962.
44. **Barr, M. L., Carr, D. H., Plunkett, E. R., Soltan, H. C., and Wiens, R. G.,** Male pseudohermaphroditism and pure gonadal dysgenesis in sisters, *Am. J. Obstet. Gynecol.,* 99, 1047, 1967.
45. **Chemki, J., Carmichael, R., Stewart, J. M., Geer, R. H., and Robinson, A.,** Familial XY gonadal dysgenesis, *J. Med. Genet.,* 7, 105, 1970.
46. **Lowry, R. B., Honore, L. H., Arnold, W. J. D., Johnson, H. W., Kilman, M. R., and Marshall, R. H.,** Familial true hermaphroditism, *Birth Defects Orig. Artic, Ser.,* 11(4), 105, 1975.
47. **Hsu, L. Y. F., Hirschhorn, K., Goldstein, A., and Barcinski, M. A.,** Familial chromosomal mosaicism, genetic aspects, *Ann. Hum. Genet.,* 33, 343, 1970.

Chapter 26

LIVESTOCK EMBRYO SEXING: PAST, PRESENT, AND FUTURE

Keith J. Betteridge

TABLE OF CONTENTS

I. INTRODUCTION

This paper has two purposes. First, it aims to remind colleagues investigating the fundamentals of sex-determining mechanisms that there are some very practical reasons for wishing to predetermine, or diagnose, the sex of unborn farm animals.[1] Second, it is intended as a background for other papers in this book which deal with more specific aspects of the subject.[2,3]

Embryo sexing is just one of the several topics in animal breeding that, as "biotechnologies", have received new impetus from the remarkable advances in developmental and molecular biology over the past few years. The wider implications of biotechnology in livestock production, which have been the subject of numerous recent symposia and reviews,[4-9] have become very apparent to the agricultural industry. This is reflected in increasingly widespread industrial sponsorship of interdisciplinary research in agriculture at universities and other research centers, illustrating how new technologies necessitate interactions between previously disparate scientific disciplines, as well as between industry and academia. It is, therefore, interesting to consider why embryo sexing is of commercial importance, what general approaches to embryo sexing are currently available, how these methods evolved, and which approaches augur well for the future.

II. THE NEED FOR SEXED EMBRYOS

Clear distinction should be made between the prediction or control of genetic and phenotypic sex for the purposes of animal breeding, as will be discussed here, and the alteration of phenotypic sex to change behavioral and production characteristics. The possibilities of exploiting the latter procedure will not be considered except to record an interesting suggestion made by Dr. Maria New. With reference to her own studies of errors in secondary sex differentiation in humans,[10] she raised the possibility that future genetic engineering might arrange the deletion of the gene controlling the enzyme 21-hydroxysteroid dehydrogenase, thereby causing the phenotypic masculinization of all fetuses *in utero,* irrespective of genetic sex.

It is the modern agricultural industry itself that has identified its need for an ability to predict, or control, the sex of offspring before birth. The validity of the need may be arguable; whereas sex control is often given high priority in cattle producers' "wish lists" for example, cost analyses reveal that many economic factors need to be taken into account in assessing the worth of any given procedure.[11] However, it is cost analyses that have also indicated how valuable it would be to exporters and importers of livestock, especially cattle, to know the sex of embryos being moved internationally. Several facts contribute to this situation.

The first fact to be appreciated is that there is already a thriving international trade in livestock genes and a definite place for embryos in that trade. Until the past decade, the commodities traded have necessarily been either entire animals (often pregnant females) or semen. However, the advent of routine embryo transfer during the 1970s opened up the possibility of moving embryos as well, and the logistics of doing so have been enormously simplified since the success rates achievable with cryopreserved ("frozen") embryos have edged toward those obtained with fresh ones. There are several advantages of using embryos for international trade. As with entire animals, the buyers can see the effects of their purchase immediately rather than waiting out a gradual breeding up process after buying semen. At the same time, it is obviously much less expensive to ship canisters of liquid nitrogen containing dozens of embryos than it is to ship just one adult cow or even calf. Even more importantly, the risks of transmitting disease with properly handled embryos are very much reduced, if not eliminated.[12-14] The dangers of transmitting disease internationally along with animals have increased in proportion to the ease of international travel, of course. However, the problem has been around for a long time; the first person documented to have imported

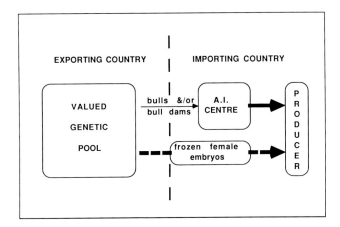

FIGURE 1. Movement of livestock of high genetic worth across international boundaries is often in the form of breeding stock (proven bulls and/or potential mothers of bulls, for example) which is multiplied in the importing country. The market for such stock is very limited in comparison to the market amongst the producers. To enter the latter market with embryos, an exporter needs to provide the required product at a competitive price.

disease across the Atlantic with his livestock was Christopher Columbus, and it is now believed that swine influenza, rather than small pox, was largely responsible for decimating the American Indians at the end of the 15th century.[15]

Another factor that explains why embryo sexing is important in the international livestock trade is that the market for the exporting country or company could become self-limiting unless it is economically possible to reach the final buyer with the product required — for example, a frozen, viable embryo of known sex, often female (Figure 1).

A further, indirect, reason for linking embryo sexing with international trade is that the ability to export frozen, sexed embryos requires technical and marketing expertise that can only be sustained by a buoyant domestic market in the exporting country. There is good evidence of the requisite buoyancy in North America, as can be inferred from the growth of embryo transfer in dairy cattle in Canada, for example (Figure 2). However, the continuation of this expansion will depend more and more on demonstration of the real genetic worth of embryo transfer as the novelty value, which helped marketability in the early years, wears off.

The contributions to genetic improvement that can be expected of the proper use of MOET (multiple ovulation and embryo transfer) have been analyzed in detail on several occasions in recent years.[16-18] In the present context, it should be noted that with current use of artificial insemination the genetic gain derived from selection on the male side of cattle breeding programs is much greater than gains made from selection on the female side. This is because a bull can readily produce 50,000 times more progeny in a year than can a cow. There are two important consequences of this physiological difference. First, the accuracy of genetic selection, which depends on the number of relatives available for testing, is almost 1.5 times greater on the sire's side than the dam's side. Second, the intensity of selection can also be much higher on the male side because far fewer males than females need to be retained to maintain a given population; only the best 5 to 10% of breeding stock are required to provide the next generation of bulls, whereas at least 70% of cows have to be used to maintain the female (milking) population. Current use of MOET is largely for producing dams of bulls, thereby allowing even more intense selection on the male side and improving the rate of genetic change by a factor of 1.11 (Table 1). If MOET can be made sufficiently cheap and effective to breed commercial herds so that only the top 25% of cows need be retained, the

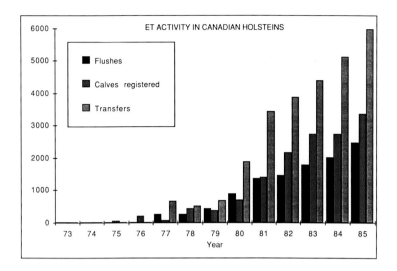

FIGURE 2.   The growth of embryo transfer activities in a single dairy breed, the Canadian Holstein, in Canada between 1973 and 1985.

Table 1
THE RELATIVE CONTRIBUTIONS TO GENETIC GAIN IN CATTLE THAT CAN BE ANTICIPATED THROUGH THE MALE AND FEMALE SIDES OF BREEDING PROGRAMS USING ARTIFICIAL INSEMINATION (A.I.) WITH OR WITHOUT MOET

	Current progeny testing[a]	With MOET of bull dams[a]	With MOET of bull dams and commercial cows[b]
Through A.I.			
Sires of bulls	40	40	40
Sires of cows	26	26	26
Dams of bulls	30	41	41
Through cow herd			
Dams of cows	4	4	26[c]
	100	111	133

[a]   After Van Vleck.[17]
[b]   C. Smith, University of Guelph, Ontario, Canada, personal communication, 1987.
[c]   Selecting 0.25 as dams.

genetic change could be increased further to 1.27 times current rates (Table 1). Thus, the need for sexed embryos in cattle breeding is but one component of an overall need for more efficient embryo production.

Interesting paradoxes can be encountered in examining the call for sexed embryos. In beef production, for example, one might have expected that faster growing male embryos would be the creatures of choice. Instead, it has been shown that a "single-sex bred heifer" (SSBH) system would increase food conversion efficiency by about 27% in the relatively intensive management systems practiced in the U.K. and would also provide an improved means of genetic selection.[19,20] In essence, this scheme depends upon heifers always producing heifer calves, thereby replacing themselves immediately as breeding stock and al-

FIGURE 3. Democritus of Abdera (470 to 402 B.C.), one of the first authors to discuss the mechanisms responsible for the determination of sex. (From Russell, B., *Wisdom of the West*, Foulkes, P., Ed., Crescent Books, New York, 1959. With permission.)

lowing all the members of each generation to be slaughtered for beef once weaned of their calves. The cost of maintaining adult breeding females in the herd is thus eliminated.

III. HISTORIC AND FOLKLORIC ASPECTS OF SEX CONTROL

Although the need for sexed embryos in livestock production is associated with modern technology, it would be quite wrong to presume that the desire to control the sex of one's animals' offspring, or of one's own children, is a modern phenomenon. The fascination of how sex is determined runs deeper and is probably as old as man himself. It is described in the earliest known medical writings and extends right up to the present day. Such a persistent interest surely justifies a brief consideration of the history of the subject and the folklore that it has engendered. In presenting such a summary, I make no pretense of being exhaustive and I shall rely heavily on Smith's history of veterinary literature,[21] as I have done before.[22-23]

About 2500 years ago, in the time of Socrates, there were really two schools of thought on how sex was determined. According to Smith,[21] the first documents on the subject come from Democritus of Abdera (470 to 402 B.C. Figure 3) who was probably the first person to write on anatomical subjects. He also wrote on animal husbandry and although his original works have been lost, they have come down to us indirectly through a Roman citizen named Columella who recorded them in his 12-book "Husbandry", completed in about 55 A.D. In Democritus' view, male or female offspring can be obtained quite readily, simply by binding up one or the other testicle, since females originate from the left testis and males from the right. This statement was repeated by various writers for many centuries and the concept of "rightness" being associated with "maleness" has persisted in various guises, which are often frankly discriminatory; Mittwoch[24] has drawn attention to the associations made between right (adroit, dexterous, righteous) and male, and left (gauche, sinister, "left-handed") and female.

The other Socratic school of thought on how sex is determined also involved leftness and rightness, but of the uterus rather than testicles. It was based on observations of the sow uterus which purported to show that male piglets are to be found predominantly in the right horn and females in the left! This led to medical recommendations for those human couples wishing to have boys: boys could be conceived if the woman lies on her right side during intercourse, thus allowing semen to flow to the right side of the uterus.[25,26] Smith[21] ascribes this alternative explanation of how "left" and "right" affect sex to two other contemporaries of Socrates, the philosophers Anaxogoras and Parmenides. However, Mittwoch[24] described Anaxagoras as considering leftness and rightness to act via the male's testes and also tells us that the plausibility of both theories was enhanced by the ideas of Empendocles, who felt the right side to be hotter than the left, and males to be hotter than females!

Leftness and rightness in relation to sex determination was also considered of veterinary importance in Roman times. One of the vehicles through which Greek philosophical writings passed into Roman civilization was Pliny's "Natural History" which asserts that if, after a service, a bull dismounts from the cow on the right side, a bull calf results; if on the left, a heifer.

Even today, the human equivalent of "binding up" an animals' testicle as recommended by Democritus is perhaps reflected in certain practices that are supposedly still used in various parts of the world for predetermining the sex of children. These include the custom of holding one testicle or the other during intercourse or, much less direct, biting one's wife's right ear at the appropriate moment when a boy-child is to be invoked.[25]

In tracing the historic association between left and right, femaleness and maleness, Mittwoch[24,27] points out that the Greeks had perhaps, by intuitive thinking,

arrived at an explanation of sex determination in man which contained a basic element of truth. In the embryo, the relationship of male to female bears a dominance relationship which is essentially similar to that manifested by right and left sides.[24]

Her conclusion is also of some interest in the context of this book:

. . . it may be possible to trace back any male chauvinism encountered in postnatal life to the early difficult days, when the embryo had to undergo its development in an alien hormonal environment. It could thus be regarded as a belated reaction to the dominant relationship of the mammalian mother vis-à-vis her embryo![24]

The Romans evidently felt that the wind direction at the time of service also influenced the sex of calves; bull calves resulting when the wind is from the north, heifers when from the south. This often-forgotten piece of information was first recorded in Book XVII of "Geoponica", one of a pair of works (the other being "Hippiatrika") compiled in the 10th century under Emperor Constantine VII which conserved for Europe whatever knowledge of veterinary science and agriculture had been derived from the Greeks and practiced in the Later Roman (Byzantine) Empire. Geoponica also has remedies for impotent bulls (application of burnt deer's antler mixed with urine) and declares that the double hump of the Bactrian Camel results from impregnation by a wild boar!

These, and other bizarre assertions and treatments, would appear to have got on the wrong side of Major General Sir Frederick Smith, K.C.M.G., C.B., for his veterinary history[21] conveys considerable indignation in his statement:

We can quite imagine how superstitions and filthy practices appealed to the degenerates of the Later Roman Empire, and how readily they would be adopted by equally superstitious Western Europe when the information reached them.

"The information" from Geoponica, Hippiatrika, and similar Byzantine works passed mostly into Arabian custodianship up until the 12th century. Printed versions of Geoponica and Hippiatrika did not appear in Europe until the 16th century, first in Latin and Greek and, much later (1805), in English. Smith holds the Church responsible for the prevalence of superstition and active discouragement of research during what he calls the "Ecclesiastical Period" in veterinary medicine.[21] Small wonder, then, that the next two references to sex control that we come across, just after the end of the Middle Ages, still contain liberal helpings of superstition, this time in relation to the moon. However, the treatments of the superstition by the two authors were diametrically and interestingly opposed.

It had been believed for centuries that conception before or after a full moon produced offspring of opposite sexes. Writing in *The Boke of Husbandries*, dedicated to Henry VIII in 1523, the agriculturist Fitzherbert reported what must be the first scientific experiment in sex control. He used 160 mares to show that there was no truth in the belief of lunar control of sex. On the other hand, more than 50 years after Fitzherbert's refutation, George Turberville, a sportsman, poet and lawyer, published (1576) a treatise on hunting in which he dealt with the breeding of hounds. According to him, mostly male pups could be obtained from matings taking place after the moon had passed the full and under the Zodiacal signs of Gemini (May and June) or Aquarius (January and February). A further advantage of breeding hounds at this time, apparently, was that the resulting puppies were less prone to madness.

But as we smile and think of the quaintness of these ancient beliefs, we should remember that the public's taste for mystery and superstition is by no means satiated. Our 20th century newspapers and magazines are full of how to choose the sex of babies with the aid of vinegar or baking soda douches and of how to test for their sex by treating the urine of the mother-to-be with proprietary drain-cleaner. In agriculture, we have never been short of weird, wonderful, but unproven methods for separating X- and Y-bearing sperm which, only now, may be giving way to more soundly based methods.[28] Geoponica can still be recognized when a cattle breeder interviewed in an English pub describes how he controls the sex of his calves by correct compass orientation of his cows at the time of insemination. Smith cogently summed up these predilections:

. . . the public mind, then as now, preferred the fanciful and unreal to prosaic truth. Signs, symbols, wonders and mystery appealed to it more than plain, unvarnished common sense.

The difference in the approach to sex control in this century has been the increased emphasis on scientifically founded alternatives to "signs, symbols, wonders and mystery".

One of the first scientists rumored to have achieved such control was Walter Heape, the man who initiated embryo transfer.[22] Like many others before and since, he hoped to define the environmental factors that selectively favored the survival of male, or of female, gametes. To this end, he studied species as diverse as dogs, canaries, and humans. Walter Heape did not succeed on this side of his work, of course, but the embryo transfer side of his legacy has been developed as a means of fulfilling his dream.

IV. EMBRYO SEXING DURING THE COURSE OF TRANSFER

The role of embryo transfer in fulfilling our desire for selecting the sex of farm animals before birth has been reviewed many times recently,[29-34] and here, I shall only make three general points about the procedures and how they might develop in the future.

First and foremost, it should be acknowledged that separation of X- and Y-bearing spermatozoa would have enormously greater impact on sex selection in farm animals than is ever likely to be possible through sexing embryos. However, that separation has yet to be achieved in any authenticated, reproducible field trials, despite persistent commercial claims to the contrary.

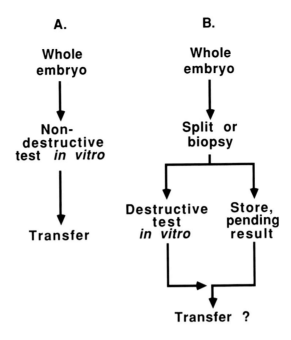

FIGURE 4.   The two general approaches to embryo sexing (and viability testing): (A) by nondestructive means; (B) by sacrificing a part of the embryo to the test.

The second point is to illustrate (Figure 4) the two general approaches to embryo sexing and their dependence on parallel, and often unforeseen, technological developments. The first approach (Figure 4A) has usually been described as the ideal to strive after because it is nondestructive. Immunological tests for male-specific antigens[32-34] and biochemical tests that aim to detect sex-dependent differences in the activities of X-linked enzymes[35-36] fall into this category (see also reviews cited above), and certainly it is exciting to see the 80 to 90% degree of accuracy that can be achieved in several species by experienced workers with the immunological methods.

The second approach (Figure 4B) is the one that produced the first sexed rabbits, calves, and sheep when the test used was cytological. It is interesting to trace the way in which technological advances have affected this approach since it was used to produce the first sexed calf in 1975.[37] At that time, "embryo splitting" was unheard of and so the biopsy for chromosome analysis was taken from a hatched, elongated blastodermic vesicle. These have yet to be frozen and thawed successfully, but the possibility that they could be should not be dismissed considering that the very first frozen calf came from a hatched blastocyst.[38] This limitation led to attempts to biopsy embryos at the morula stage, before it becomes too difficult to separate off just a few cells and at a stage when it is possible to freeze the biopsied embryo. The limitation then becomes one of providing sufficient cells to obtain analyzable metaphase spreads,[30] a factor that would not affect the applicability of today's Y-specific DNA probes, for example.[2,39]

The advent of routine "embryo splitting" at the blastocyst stage[40,41] increased the number of cells, still at a freezable stage of development, that could be analyzed by sacrificing one half-embryo to the examination. This approach has produced the first sexed calves from frozen embryos,[42] but three important limitations remain to be overcome. First, half-embryos do not tolerate cryopreservation as well as whole embryos do. This limitation should be overcome by improved methods of culture during a recovery period.[43,44] Second, despite the increased number of cells available for analysis, only about 60% of half-embryos can be sexed by chromosome analysis. Again, this limitation should not apply when Y-specific

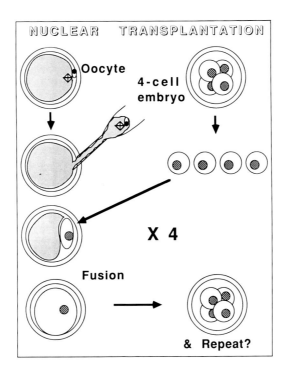

FIGURE 5. The general principles of nuclear transplantation for producing clones of a single embryo. Although a 4-cell embryo is depicted, 8- or 16-cell embryos have been used to produce identical lambs.[45]

DNA probes, or perhaps antibodies to male-specific antigens, are used. Third, a potential calf is lost for the sake of knowing the sex of just one other calf. However, that extravagant 1:1 ratio could probably be reduced if fewer cells could be used for accurate diagnosis with DNA probes, for example. Furthermore, other micromanipulative methods are capable of producing larger numbers of identical multiplets,[40,41] and the birth of lambs[45] and calves, following transplantation of nuclei from 8- and 16-cell blastomeres into enucleated oocytes (Figure 5), portends real cloning of early cleavage-stage embryos through repeated generations. Such a development, coupled with improved cryopreservation, would reduce the disadvantage of having to use a destructive test on one multiplet to determine the sex of the others.

This leads to the third and final point and I wish to make from this general description of embryo sexing, a point expressed diagramatically in Figure 6. The ultimate criterion in the assessment of the usefulness of embryo transfer, including embryo sexing, is its contribution to genetic improvement and increased efficiency of animal production systems. However, 12 years ago, who would have forecast that sex diagnosis in cattle embryos would depend on progress in the field of molecular biology? Similarly, which of us today can predict the techniques that will be routine at the turn of the century? What is certain is that progress will continue to depend on technical advances arising unpredictably from basic research in a wide variety of disciplines. That, I submit, is a point of considerable importance to all who plan and finance research strategies in animal breeding, and also a point that adds the zest of practical anticipation to the value of this book.

ACKNOWLEDGMENTS

I thank Drs. Charles Smith, John Gibson, and Alan Wildeman for helpful discussions

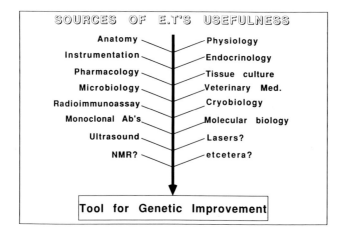

FIGURE 6. A dendritic representation of how various sciences and technologies have contributed, and will continue to contribute, to the usefulness of embryo transfer for bringing about the genetic improvement of livestock. NMR = Nuclear Magnetic Resonance.

and critical review; the Natural Sciences and Engineering Research Council of Canada, Semex Canada, and the Ontario Ministry of Agriculture and Food for financial support; Mrs. P. Fowle for manuscript preparation.

REFERENCES

1. **Seidel, G. E., Jr.,** Superovulation and embryo transfer in cattle, *Science,* 211, 351, 1981.
2. **Bradley, M. P. and Heslop, B. F.,** Recent developments in the serology and biochemistry of testicular sex-specific (H-Y) antigens, in *Evolutionary Mechanisms of Sex Determination,* Wachtel, S. S. , Ed., CRC Press, Boca Raton, Fla, 1988, chap. 13.
3. **Casanova, M., Cotinot, C., Kirszenbaum, M., Abbas, N., Bishop, C., Vaimon, M., and Fellous, M.,** Characterization of human XX males and sexing of cattle embryos by use of Y-specific DNA sequences, in *Evolutionary Mechanisms of Sex Determination,* Wachtel, S., Ed., CRC Press, Boca Raton, Fla., 1988, chap. 27.
4. **Evans, J. S. and Hollaender, A., Eds.,** *Genetic Engineering of Animals — An Agricultural Perspective,* Plenum Press, New York, 1986.
5. **Smith, C., King, J. W. B., and McKay, J. C., Eds.,** *Exploiting New Technologies in Animal Breeding: Genetic Developments,* Oxford Science Publications, New York, 1986.
6. **Womack, J.,** Genetic engineering in agriculture: animal genetics and development, *Trends Genet.,* 3, 65, 1987.
7. **Land, R. B. and Wilmut, I.,** Gene transfer and animal breeding, *Theriogenology,* 27, 169, 1987.
8. **Renard, J.-P. and Babinet, C.,** Genetic engineering in farm animals: the lessons from the genetic mouse model, *Theriogenology,* 27, 181, 1987.
9. **Church, R. B.,** Embryo manipulation and gene transfer in domestic animals, *Trends Biotechnol.,* 5, 13, 1987.
10. **New, M., Drucker, S., and Speiser, P.,** Nonclassical steroid 21-hydrolase deficiency as a cause of reproductive dysfunction, in *Evolutionary Mechanisms of Sex Determination,* Wachtel, S. S., Ed., CRC Press, Boca Raton, FL, 1988, chap. 24.
11. **Seidel, G. E., Jr. and Elsden, R. P.,** What will it cost for sexing embryos?, *Hoard's Dairyman,* August 25, 1985, 905.
12. **Singh, E. L.,** The disease control potential of embryos, *Theriogenology,* 27, 9, 1987.
13. **Atwell, J. K.,** The international movement of embryos and disease control: a regulatory perspective, *Theriogenology,* 27, 5, 1987.
14. **Gilbert, R. O., Coubrough, R. I., and Weiss, K. E.,** The transmission of bluetongue virus by embryo transfer in sheep, *Theriogenology,* 27, 527, 1987.

15. **Guerra, F.,** Cause of death of the American Indians, *Nature (London),* 326, 449, 1987.
16. **Nicholas, F. W. and Smith, C.,** Increased rates of genetic change in dairy cattle by embryo transfer and splitting, *Anim. Prod.,* 36, 341, 1983.
17. **Van Vleck, L. D.,** Potential genetic impact of artificial insemination, sex selection, embryo transfer, cloning and selfing in dairy cattle, in *New Technologies in Animal Breeding,* Brackett, B. G., Seidel, G. E., Jr., and Seidel, S. M., Eds., Academic Press, New York, 1981, 221.
18. **Smith, C.,** Use of embryo transfer in genetic improvement of sheep, *Anim. Prod.,* 42, 81, 1986.
19. **Taylor, St. C. S., Thiessen, R. B., and Moore, A. J.,** Single sex beef cattle systems, in *Exploiting New Technologies in Animal Breeding: Genetic Developments,* Smith, C., King, J. W. B., and McKay, J. C., Eds., Oxford Science Publications, New York, 1986, 183.
20. **Taylor, St. C. S., Moore, A. J., Thiessen, R. B., and Bailey, C. M.,** Efficiency of food utilization in traditional and sex controlled systems of beef production, *Anim. Prod.,* 40, 401, 1985.
21. **Smith, F.,** *The Early History of Veterinary Literature and Its British Development,* Vol. 1, Bailliére, Tindall and Cox, London, 1919, and J. A. Allen, London, 1976.
22. **Betteridge, K. J.,** An historical look at embryo transfer, *J. Reprod., Fertil.,* 62, 1, 1981.
23. **Betteridge, K. J.,** The folklore of sexing, *Theriogenology,* 21, 2, 1984.
24. **Mittwoch, U.,** To be right is to be born male, *New Scientist,* January 13, 1977, 74.
25. **Moreau, D.,** O bring forth men children only, *Vogue,* U.K., Winter, 56, 1971.
26. **Glaser, H.,** Geschicte der Wilkurlichen Geschlechtbestimung, *Wien. Med. Wochenschr.,* 113, 779, 1963.
27. **Mittwoch, U.,** Whistling maids and crowing hens — hermaphroditism in folklore and biology, *Perspect. Biol. Med.,* Summer, 595, 1981.
28. **Bradley, M., and Heslop, B. F.,** Recent developments in the serology and biochemistry of testicular sex-specific (H-Y) antigens, in *Evolutionary Mechanisms in Sex Determination,* Wachtel, S. S., Ed., CRC Press, Boca Raton, Fla., 1988, chap. 13.
29. **Hare, W. C. D. and Betteridge, K. J.,** Relationship of embryo sexing to other methods of sex determination in farm animals: a review, *Theriogenology,* 9, 27, 1978.
30. **Betteridge, K. J., Hare, W. C. D., and Singh, E. L.,** Approaches to sex selection in farm animals, in *New Technologies in Animal Breeding,* Brackett, B. G., Seidel, G. E., Jr., and Seidel, S. M., Eds., Academic Press, New York, 1981, 109.
31. **King, W. A.,** Sexing embryos by cytological means, *Theriogenology,* 21, 7, 1984.
32. **Wachtel, S., Nakamura, D., Wachtel, G., Felton, W., Kent, M., and Jaswaney, V.,** Sex selection with monoclonal H-Y antibody, *Fertil. Steril.,* 50, in press, 1988.
33. **Anderson, G. B.,** Identification of sex in mammalian embryos, in *Genetic Engineering of Animals — An Agricultural Perspective,* Evans, J. S. and Hollaender, A., Eds., Plenum Press, New York, 1986, 243.
34. **Anderson, G. B.,** Identification of embryonic sex by detection of H-Y antigen, *Theriogenology,* 27, 81, 1987.
35. **Rieger, D.,** The measurement of metabolic activity as an approach to evaluating viability and diagnosing sex in early embryos, *Theriogenology,* 21, 138, 1984.
36. **Williams, T. J.,** A technique for sexing mouse embryos by a visual colorimetric assay of the X-linked enzyme, glucose-6-phosphate dehydrogenase, *Theriogenology,* 25, 733, 1986.
37. **Hare, W. C. D., Mitchell, D., Betteridge, K. J., Eaglesome, M. D., and Randall, G. C. B.,** Sexing 2-week old bovine embryos by chromosomal analysis prior to surgical transfer: preliminary methods and results, *Theriogenology,* 5, 243, 1976.
38. **Wilmut, I. and Rowson, L. E. A.,** Experiments on the low-temperature preservation of cow embryos, *Vet. Rec.,* 92, 686, 1973.
39. **Leonard, M., Kirszenbaum, M., Cotinot, C., Chesné, P., Heyman, Y., Stinnakre, M. G., Bishop, C., Delouis, C. Vaiman, M., and Fellous, M.,** Sexing bovine embryos using Y-chromosome specific DNA probe, *Theriogenology,* 27, 248, 1987.
40. **Willadsen, S. M.,** Micromanipulation of embryos of the large domestic species, in *Mammalian Egg Transfer,* Adams, C. E., Ed., CRC Press, Boca Raton, Fla., 185, 1982.
41. **Fehilly, C. B. and Willadsen, S. M.,** Embryo manipulation in farm animals, in *Oxford Reviews of Reproductive Biology,* Vol. 8, Clarke, J. R., Ed., Clarendon Press, Oxford, 1986, 379.
42. **Picard, L., King, W. A., and Betteridge, K. J.,** Production of sexed calves from frozen embryos, *Vet. Rec.* 117, 608, 1985.
43. **Picard, L., King, W. A., and Betteridge, K. J.,** unpublished data, 1987.
44. **Chesné, P., Heyman, Y., Chupin, D., Procureur, R., and Ménézo, Y.,** Freezing cattle demi-embryos: influence of a period of culture between splitting and freezing on survival, *Theriogenology,* 27, 218, 1987.
45. **Willadsen, S. M.,** Nuclear transplantation in sheep embryos, *Nature (London),* 320, 63, 1986.
46. **Russell, B.,** *Wisdom of the West,* Foulkes, P., Ed., Crescent Books, New York, 1959.

Chapter 27

CHARACTERIZATION OF HUMAN XX MALES AND SEXING OF CATTLE EMBRYOS USING Y-SPECIFIC DNA SEQUENCES

**Myriam Casanova, Corinne Cotinot, Marek Kirszenbaum,
Nacer Abbas, Colin Bishop, and Marc Fellous**

TABLE OF CONTENTS

I. INTRODUCTION

In the human, XX individuals normally develop as phenotypic females, whereas XY and even XXY individuals develop as males under the dominant influence of the Y chromosome. A number of pathological conditions exist, however, in which phenotypic sex does not correlate with the apparent karyotype. Perhaps the best known example is XX male syndrome, originally described in 1964,[1,2] in which affected individuals develop as males despite an apparently normal XX karyotype. 46,XX men exhibit characteristic phenotypic features such as small testes devoid of spermatogonia and invariable sterility. Hypospadias may occur and there is a one in three incidence of gynecomastia.[3] The worldwide frequency of XX maleness is approximately 1/20,000 newborn boys.

Several hypotheses have been proposed to explain the occurrence of XX males. These include undetected low-level mosaicism involving XXY or XY cells, mutation or deletion of autosomal of X-linked genes involved in primary sex determination, and presence of cryptic Y-chromosomal material in the otherwise XX genome.[4,5]

In the human, the existence of some familial cases could argue in favor of the autosomal mutation hypothesis.[6] However, Xg and 12E7 allelic inheritance studies[7] and direct detection of Y-specific DNA sequences in XX males[8-10] demonstrate that one of the causes of XX maleness is the transfer of the primary sex-determining gene(s) from the Y chromosome into the otherwise female genome. It has been postulated that this transfer arises through an X-Y interchange during male meiosis.[4-5]

Improved cytogenetic analyses of patients with altered Y chromosomes has allowed the identification of strategic regions of the human Y involved in testis differentiation and maturation, as well as in spermatogenesis and fertility.[11] In several cases, long-arm iso-chromosomes (lacking the short arm of the Y) can be correlated with failure of testicular development.[11,12] In addition, simple deletion of the short arm of the Y chromosome leads to a female phenotype with streak gonads characteristic of the 45,X karyotype.[13] On the basis of these and similar observations, the testis-determining gene has been assigned to the Y short arm (Yp) although an additional locus on the proximal Y long arm (Yq) has been suggested.[11] Thus, it could be predicted that at least some of the Y-specific DNA sequences detected in 46,XX males originate in the short arm of the Y chromosome.

Using molecular techniques, numerous chromosome-enriched libraries have been cloned, including those enriched for the Y,[16,17] and it is now possible to isolate Y-specific sequences that can be used to determine presence of absence of the Y chromosome in cells.

In cattle, the identification of male and female embryos presents considerable economic advantage for the embryo-transfer industry especially with the development of techniques for cryopreservation. To be compatible with embryo transfer and freezing, sex determination must be performed on 7 to 8-day-old blastocysts. Until now, prenatal diagnosis of sex has been performed with some success by cytogenetic analysis of embryonic biopsies[14] and by immunological detection of the H-Y antigen on whole embryos.[15]

In this chapter, we describe the precise localization of probes on the human Y chromosome, and identify which segment of the Y, carrying the primary sex-determining gene(s), is present in the genome of XX males. We also describe the potential use of bovine Y DNA sequences for sexing embryos in bovine embryo transfer.

II. MATERIALS AND METHODS

A. DNA Isolation

Genomic DNA was extracted as described[18] from peripheral blood or lymphoblastoid cell lines of various origins: thirty-four 46,XX males provided mostly by Dr. C. Boucekkine and Dr. G. André,[19] two 46,XYp$^-$ females with features of Turner syndrome,[12-20] one 46,X psu dic (Y) (q11.1, q11.22) male,[12] and three 46,XYq$^-$ male patients[11-21] with nonfluorescent

Table 2
MALE-SPECIFIC DNA SEQUENCES IN THE GENOME OF HUMAN XX MALES

Loci	Y DNA probes	Normal male	XX Males				
			a (3)	b (6)	c (6)	d (7)	e (12)
DXYS5	p47b + c	2 (T)	2	2	2	2	–
DYS7	p50f2	4 (T)	2	2	2	–	–
DYS3	p52D	3 (T)	2	2	2	–	–
DYS8	p118	8 (T)	4	4	4	–	–
DYS6	p48	1 (H)	1	1	–	–	–
DYS5	p27	1 (T)	1	–	–	–	–
	p64a7	1 (E)	–	–	–	–	–
	p37c	1 (E)	–	–	–	–	–
DYS1	p49f	3 (T)	–	–	–	–	–
DYS11	p12f3	1 (T)	–	–	–	–	–

Notes: The results refer to Taq 1 (T), Hind III (H), or EcoR 1 (E) genomic blots following hybridization with different human Y genomic probes and low stringent washing conditions except for probes p64a7, p37c, and p49f. Under XX males, in each vertical column, the numbers indicate the male-specific bands detected with each probe. Numbers in parentheses correspond to the number of patients belonging to each group.

B. Characteristics and Origin of the Y-Specific DNA Sequences Found in Human XX Males

The detection of Y-specific DNA in the genome of four sporadic XX males was extended to 34 unrelated 46,XX males. Lymphocytes or lymphoblastoid cell line DNAs originating from these patients were digested with EcoR 1, Taq 1, or Hind III, fractionated on agarose gels, and blotted on Zetapore membranes as described.[17] The filters were successively hybridized to ten different genomic probes detecting Y-specific DNA sequences and washed after each hybridization under conditions of low stringency, allowing identification of DNA fragments not strictly homologous to the probes.

Of the 34 XX males tested, DNA from 22 (66%) reacted with some of the probes used, but to a different extent (Table 2). Among the positive patients, four groups were distinguishable depending on the number of probes giving a positive signal. Probes detecting multiple Y-specific bands in normal male DNAs revealed only some of these bands when hybridized to DNA from 46,XX males. However, all of the putative Y-specific bands detected with a given probe remained the same in all positive patients and appeared to be the same size as comparable bands from normal males (within the limits of three restriction enzymes tested).

Hybridization patterns of Y-derived probes to EcoR 1 digested genomic DNA from XX males and from Yp⁻ or Yq⁻ patients were compared in order to localize the Y-specific restriction fragments found in XX males. Figure 2 shows as an example the hybridization patterns of probe p50f2 with EcoR 1 genomic blots from members of a family of one XX male and from various Yp⁻ or Yq⁻ patients. This probe detects 5Y specific bands in 46,XY normal male DNA. Of these bands only the 10 kb and 8 kb bands were detected in the XX male patient (lane 3). The same two bands were missing in DNA from a 46,XYp⁻ female (lane 10) indicating that they can be localized to Yp.

The 6 kb and 1.7 kb Y-specific bands were still present in DNA from two 46,XYq⁻ males: lane 11 (46, X del (Y) (Yq11.1 – Yq11.23) and lane 12 (46,X del (Y) (Yq11.1 – Yq11.22). These bands could not be detected in the DNA from a 46,X,psu dic (Y) (q11.1 – Yq11.22): lane 13, or a 46,X,del (Y) (Yq11.1 – Yq11.21): lane 14. These bands could

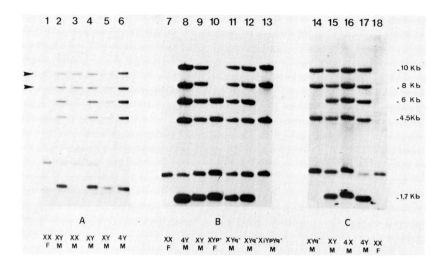

FIGURE 2.   Hybridization patterns of EcoR 1 genomic blots with probe p50f2. 1: 46,XX female (mother of patient BS); 2.4: 46,XY males (brothers of patient BS); 3: 46,XX male BS; 5: 46,XY male; 6: 49,XYYYY male; 7: 46,XX female; 8: 49,XYYYY male; 9: 46,XY male (father of patient AMc); 10: 46,XYp⁻ female AMc; 11: 46,XYq⁻ male BU8; 12: 46,XYq⁻ male BB; 13: 46,X psu dic Y male CC; 14: 46,XYq⁻ male LL226; 15: 46,XY male; 16: 49,XXXXY male; 17: 49,XYYYY male; and 18: 46,XX female.

therefore be localized to the Yq11.1 – Yq11.21 region. The 4.5 kb Y-specific band, present in all the DNAs from both Yp and Yq deletions, mapped to the region between Yp11 and Yq11.23.

These data demonstrate that the Y-specific sequences detected by p50f2 in the genome of XX males originated from the distal short arm of the Y; the same sequences are missing from the genome of the 46,XYp⁻ female.

The origin of the Y material detected in XX males by probes p47b + c, p52D, and p118 was similarly investigated. As indicated in Table 3, with the exception of two bands that could not be definitively mapped, all the Y-specific DNA sequences detectable in XX males originated from the short arm of the Y chromosome.

C. Bovine Y DNA Sequences in the Sexing of Cattle Embryos

For this purpose we used a DNA-cloning strategy based on the deletion enrichment method.[22] The resulting clones were analyzed by hybridization to Southern blots of male and female bovine genomic DNA. Of 200 clones tested, two, named B.C.1.2 and B.C.1.34, were entirely male specific, six gave a male-female differential hybridization pattern, and the others reacted similarly with male and female DNA. Interspecies (hamster/bovine) somatic cell hybrids studies and chromosomal *in situ* hybridization confirmed that the B.C.1.2 male-specific probe was derived from the Y chromosome. When we tested the specificity on DNA from different mammals, this Y probe remained *Bos* specific. The densitometric analysis of Southern blots containing increasing amounts of B.C.1.2 plasmid and genomic male DNA indicated that B.C.1.2-related sequences were repeated about 2500 times in the male genome. This estimation agrees with the frequency of B.C.1.2 positive clones in the male bovine genomic library. No transcription has been observed on RNA blots of different bovine tissues tested including testis, ovary, kidney, and liver.

The male specificity and repeated nature of the B.C.1.2 sequence has enabled us to use it as a molecular probe for sex determination on small numbers of embryonic cells. We have developed a simple and rapid screening method for prenatal sex diagnosis using *in situ* hybridization with a biotinylated B.C.1.2 DNA as probe. The accuracy of the tests on

Table 3
LOCATION OF Y-SPECIFIC DNA FRAGMENTS
DETECTABLE IN EcoR 1 DIGESTED DNA FROM
46,XX MALES

Probes	EcoR 1 Y-specific fragments (kb)	Location on the Y chromosome	XX males a, b, c, or d
p47b	7.8	Yp	+
p50f2	10	Yp	+
	8	Yp	+
	6	Yq	−
	4.5	Yq	−
	1.7	Yq	−
p52d	10	Yq	−
	1.8	Yp	+
	0.8	Yc	+
p118	7.8	Yq	−
	7	Yp	+
	6	Yc	+
	3.2	Yp	+
	2.6	Yp	+
p12f2	5.6	Yq	−
	5.2	Yq	−
	or		
	3.2	Yq	−
	1.5	Yq	−

Notes: The bands scored on XX males correspond to sequences localized on the Y chromosome. Their location is indicated as Yp (short arm), Yq (long arm), or Yc (precentromeric region). 46,XX Males a, b, c, or d refer to different groups defined in Table 2.

embryonic cells obtained by microdissection has been confirmed by cytogenetic analysis in the remaining cells. This method offer several advantages over conventional diagnosis:

1. As few as ten cells are sufficient for analysis
2. DNA extraction cell culture and metaphase spreads are not necessary
3. Diagnosis can be obtained within 12 hr

Furthermore, this nonradioactive detection system should allow widespread and routine application of the method.

IV. DISCUSSION

The results obtained here represent direct molecular evidence that the presumably sex-determining Y chromatin detected in approximately 66% of human XX males tested originated from the distal short arm of the Y chromosome. The Y sequences found in the 22 positive XX males appeared as totally inclusive and overlapping chromosomal sequences allowing a division of the patients into four positive groups and one negative group depending on the amount of DNA detected (Table 2). This genetic heterogeneity could not be related to the detailed clinical and pathological manifestations observed. As the sequences detected by probe 47b + c were present in all positive patients, it could be concluded that this locus is close to the Y located sex-determining gene(s).

It has recently been demonstrated that in humans, an obligatory reciprocal X-Y interchange involving the distal tips of Xp and Yp occurs during normal male meiosis.[19] Hence, XX maleness may arise from a rare abnormal recombination event transferring a large piece of the Y carrying sex-determining factors to the X, as originally hypothesized by Ferguson-Smith.[4] This would imply that the overlapping sequences in XX males could be orientated with respect to the telomere. Hence, *TDF (testis-determining factor)* might be located between p47b + c and the boundary of the X-Y pseudoautosomal region. Alternatively, accidental transpositions or exchanges of different amounts of Y short arm material between the X (or autosomes) could explain the genetic heterogeneity found in XX males. We feel this to be unlikely in view of the ordered inclusive overlapping arrangement of the DNA found.

X-Y interchange mechanisms predict that the Y-specific DNA detected in XX males should be found on the paternal X chromosome. Such a hypothesis is supported by *Xg* and *MIC2* allele inheritance studies.[7] In addition, although parental controls were lacking, Evans et al.[5] reported finding heteromorphic X chromosomes with additional material on the X in a high percentage of human XX male karyotypes. Preliminary results with direct *in situ* hybridization to XX male metaphase preparations also show the presence of Y-derived sequences on the distal short arm of one X chromosome.[12,24,25]

Finally, we have segregated the X chromosome from one XX male by constructing somatic cell hybrids with Chinese hamster cells (CHO). The maternal and paternal X chromosomes were then distinguished using an X-specific polymorphic DNA probe (Factor IX). Probing the hybrids with p47b + c showed that the Y sequences were located only on the paternal X.[26]

We have been able to detect Yp-chromosomal material in the genome of 22 of 34 human XX males tested. We favor the hypothesis of abnormal X-Y recombination as a possible mechanism. Our data also indicate a distal Yp location of the testis determining gene(s). With the isolation of probes close to the Y-located testis-determining locus, Y-specific material will undoubtedly be found in many of the presently negative patients. However, this category of XX males cannot be regarded as examples of true "sex-reversal" because they carry a partial Y chromosome. It is tempting to speculate that some XX males do not contain Y-chromosomal material and that they represent cases of mutation or deletion in autosomal, or X, or even X-Y common testis-determining loci. It should be pointed out that the existence of two non-Y testis-determining loci has been convincingly demonstrated in the mouse.[27] In this respect, the occurrence of cases of familial XX males is intriguing.

Our experiments in cattle show that the Y chromosome can be recognized by *in situ* hybridization in embryo biopsies containing at least ten cells. The overall efficiency of this method for the production of calves from frozen sexed embryos could be increased with improvement of the cryopreservation techniques. However, our sex determination diagnosis with biotinylated probes may be performed within 12 hr; and in that case the freezing of embryos is not necessary.

ACKNOWLEDGMENTS

We acknowledge the cooperation of Dr. J. Weissenbach and the skilled technical assistance of Mrs. A. Abadie. We thank Dr. M. C. Hors-Cayla for providing the hamster/bovine somatic cell hybrid.

This work was supported by grants from INSERM U-276, by grant HD-07997 from the National Institute of Health and Maternal and Child Health Service Mental Retardation Training grant MCT-000 920. The research on sexing of bovine embryos was supported by grant 3517A from ANVAR.

REFERENCES

1. **de la Chapelle, A., Horting, H., Neimi, J., and Wennstrom, J.,** XX sex chromosomes in a human male. First case, *Acta Med. Scand. Suppl.,* 412, 25, 1964.
2. **Thekerlsen, A. J.,** Sterile male with the chromosome constitution 46,XX, *Cytogenetics,* 3, 207, 1964.
3. **de la Chapelle, A.,** Analytical review: nature and origin of males with XX sex chromosomes, *Am. J. Hum. Genet.,* 24, 71, 1972.
4. **Ferguson-Smith, M. A.,** X-Y chromosomal interchange in the aetiology of true hermaphroditism and of XX Klinefelter's syndrome, *Lancet,* 1, 475, 1966.
5. **Evans, H. J., Buckton, K. E., Spowart, G., and Carothers, A. D.,** Heteromorphic X chromosomes in 46,XX males, evidence for the involvement of X-Y interchange, *Hum. Genet.,* 49, 11, 1979.
6. **de la Chapelle, A., Schröder, J., Murros, J., and Tallqvist, G.,** Two XX males in one family and additional observations bearing on the aetiology of XX males, *Clin. Genet.,* 11, 91, 1977.
7. **de la Chapelle, A., Tippelt, P. A., Wetterstrand, G., and Page, D.,** Genetic evidence of X-Y interchange in a human XX male, *Nature (London),* 307, 170, 1984.
8. **Guellaen, G., Casanova, M., Bishop, C. E., Geldwerth, D., André, G., Fellous, M., and Weissenbach, J.,** Human XX males with Y single copy DNA fragments, *Nature (London),* 307, 172, 1984.
9. **Vergnaud, G., Page, D., Simmler, M. C., Rouyer, F., Noel, B., Botstein, D., de la Chapelle, A., and Weissenbach, J.,** The male determining region of the human Y chromosome: deletion mapping by DNA hybridization, *Cytogenet. Cell Genet.,* 40, 593, 1985.
10. **Müller, U., Lalande, M., Donlon, T., and Latt, S. A.,** Moderately repeated DNA sequences specific for the short arm of the human Y chromosome are present in XX males and reduced in copy number in an XY female, *Nucleic. Acids Res.,* 14, 1325, 1986.
11. **Bühler, E. M.,** A synopsis of the human Y chromosome, *Hum. Genet.,* 55, 145, 1980.
12. **Magenis, E., Brown, M. G., Donlon, T., Olson, S. B., and Sheebry, R.,** Structural aberrations of the Y chromosome including the nonfluorescent Y: cytologic origin and consequences, in *Progress and Topics in Cytogenetics: the Y Chromosome Part A: Basic Characteristics of the Y Chromosome,* Sandberg, A. A., Ed., Alan R. Liss, New York, 1984, 537.
13. **Rosenfeld, R. G., Luzzati, L., Hintz, R. L., Miller, O. J., Koo, G. C., and Wachtel, S. S.,** Sexual and somatic determinants of the human Y chromosome: studies in a 46,XYp− phenotype female, *Am. J. Hum. Genet.,* 31, 458, 1979.
14. **Picard, L., King, W. A., and Betteridge, K. J.,** Production of sexed calves from frozen-thawed embryos, *Vet. Rec.,* 117, 603, 1985.
15. **Wachtel, S. S.,** H-Y antigen in the study of sex determination and control of sex ratio, *Theriogenology,* 21, 18, 1984.
16. **Müller, C. R., Davies, K. E., Cremer, C., Rappold, G., Gray, J. W., and Ropers, H. H.,** Cloning of genomic sequences from the human Y chromosome after purification by dual beam flow sorting, *Hum. Genet.,* 64, 110, 1983.
17. **Bishop, C. E., Guellaen, G., Geldwerth, D., Voss, R., Fellous, M., and Weissenbach, J.,** Single copy DNA sequence specific for the human Y chromosome, *Nature (London),* 303, 831, 1983.
18. **Casanova, M., Leroy, P., Boucekkine, C., Purello, M., Siniscalco, M., Weissenbach, J., Bishop, C., and Fellous, M.,** A human Y linked DNA polymorphism and its potential for estimating genetic and evolutionary distance, *Science.,* 230, 1403, 1985.
19. **Seboun, E., Leroy, P., Casanova, M., André, G., and Fellous, M.,** Le syndrome des hommes XX: un modéle d'étude de la détermination génétique du sexe masculin, *Presse Med.,* 29, 1355, 1986.
20. **Distéche, C. M., Casanova, M., Saal, H., Freidman, C., Sybert, V., Graham, J., Thuline, H., Page, D. C., and Fellous, M.,** Small deletions of the short arm of the Y chromosome in 46,XY female, *Proc. Natl. Acad. Sci. U.S.A.,* 83, 7841, 1986.
21. **Pierson, M., Gilgenkrantz, S., and Olive, D.,** Anomalies de taille et de structure du chromosome Y, *Pédiatrie,* 23, 543, 1968.
22. **Lamar, E. E. and Palmer, E.,** Y-encoded species-specific DNA in mice: evidence that the Y chromosome exists in two polymorphic forms in inbred strains, *Cell,* 37, 171, 1984.
23. **Rouyer, F., Simmler, M. C., Johnsson, C., Vergnaud, G., Cooke, H. J., and Weissenbach, J.,** A gradient of sex linkage in the pseudoautosomal region of the human sex chromosomes, *Nature (London),* 319, 291, 1986.
24. **Bishop, C. E., Weissenbach, J., Casanova, M., Bernheim, A., and Fellous, M.,** DNA sequences and analysis of the human Y chromosome, *Progress and Topics in Cytogenetics: The Y Chromosome Part A: Basic Characteristics of the Y Chromosome,* Sandberg, A. A., Ed., Alan R. Liss, New York, 1985, 141.
25. **Buckle, V., Boyd, Y., Craig, I. W., Fraser, N., Goodfellow, P. N., and Wolfe, J.,** Localization of Y chromosomal sequences in normal and "XX" males, *Cytogenet. Cell Genet.,* 40, 593, 1985.
26. **Casanova, M., Cotinot, C., Kirszenbaum, M., Abbas, N., Bishop, C., Vaiman, M., and Fellous, M.,** A molecular approach to the study of the human Y chromosome and anomalies of sex determination in man, *Cold Spring Harbor Symposia on Quantitative Biology,* Vol. 1, 1986, 237.

27. **Washburn, L. L. and Eicher, E. M.,** Sex reversal in XY mice caused by dominant mutation in chromosome 17, *Nature (London),* 303, 338, 1983.

Index

INDEX

A

AA cells, hijacking gene in *Drosophila* spp. and, 77
Acetone, 6
Acne, 254
Aconitase-1, 52—54
Acrocentricity, 43, 120, 195
ACTH, see Adrenocorticotropic hormone
Activin, 76, 212
Adenine, 11, 15—21, 38, 171—177
Adenosine triphosphate (ATP), 132
Adrenal(s), 212, 235, 254, 256—259, 273
Adrenarche, precocious, 254
Adrenocorticotropic hormone (ACTH), 256—257
Agarose, 154, 293
Agglutinin, wheat germ, 210
Aggression, 193
Aging, postnatal oocyte, 207
AKR mice, 87, 183—185
Alanine, 246
Alcohol dehydrogenase-2, 53
Aldosterone, 254
Allele(s)
   aconitase-1, 52—53
   differentiation, 106
   genetic control of previously indifferent sex, 71
   sex-linked peptidase gene, 52
   -spotted side (Sp) sex-linked pigment marker, 8
   Sxl, 27, 31—33
   tra-2, 31
   X chromosome, 78
Alligator mississipiensis, 9, 10
Alligators, 63, 64, see also individual species
Allocyclic sex-linked behavior, 73
Alveolar ridge, cleft, 271
Ambyostoma mexicanum, 39, 42
Ambyostoma tigrinum, 39, 42
Ambyostomidae, sex-determining mechanisms and
      sex-specific chromosomes found in, 42
Amenorrhea, 191
AMH, see Anti-Müllerian hormone
Amino acid(s), 237, 249
   human rhodopsin gene sequence, 243
   H-Y antigen sequence, 153
   S-antigen sequence , 246
   sequence homology, 32
Ammonium sulfate, 146
Amniocentesis, 269
Amniote vertebrates, evolution and variety of sex-
      determining mechanisms in, 57—65
Amniotic fluid, 271
Amphibia, see Amphibian(s) and individual species
Amphibian(s)
   absence of dosage compensation in, 52
   fossils, 38
   gene mapping on sex chromosomes and, 53
   genetic linkage in, 52

heterogametic gonads in, 7
heteromorphic sex chromosomes in, 48
homomorphic sex chromosomes and, 40
karyotypes, 39
morphological differentiation within, 43
origin and evolution of sex chromosomes in, 37—
      54
phylogeny and classification of, 38
sex-determining mechanisms in, 39—50
sex-linked genes in, 52—54
sex-specific H-Y antigen and, 8—9
W chromosomes, 9
Y chromosomes, 9
Amphipathy, 115
Amplification, 173
AN3 mice, *tda-1* XY sex reversal and, 184—185
Anaphase I, 51
Anastomosis, vascular, 17, 21
Anatomy of eye, 242
Androgen(s), 7, 126, 222—227, 273
Androstenedione, 257
Aneides ferreus, 42, 48—49
Aneides spp., evolution of sex chromosomes in, 50
Aneuploidy, 78, 202
Angiotensin, 212
Anlage(s), 4, 9
Antibodies, 116, see also Heteroantibodies
   12E7, 101
   anti-antigen A IgM, 20
   anti-epitope testicular, 211
   anti-H-Y, 8, 21, 132—133, 136, 138
   anti-sheep red blood cell, 133
   B cell, 20—21
   fluorescent, 137
   gw 16, 133—134
   lymphocyte subset-specific 132
   monoclonal, 136, 137, 210
Anti-epitope technology, 211
Antigen(s)
   A, 20—21
   B1, 135, 137
   blood group, 147
   cell-surface, 95, 103
   cross reactive, 40
   HLA, 259
   HLA-A, 115
   HLA-B, 115
   H-Wt, 114
   H-Y, 83, 90, 132, 155, 191—192, 196—197, 220,
      267—268, 270, 272, 292
      biochemistry of, 139—148
      cell membrane-associated, 40
      dogmas and enigmas concerning, 111—119
      evolutionary conservation of sex-specific, 8—9
      female cell, 135
      flow cytometry and, 133—135
      function of, 152—153

U

Y

Z

RANDALL LIBRARY-UNCW

3 0490 0092579 5